AF292778

Monographs in Visual Communication

Stephan von Bechtolsheim

TEX in Practice

Volume IV:
Output Routines, Tables

Springer-Verlag
New York Berlin Heidelberg London
Paris Tokyo Hong Kong Barcelona Budapest

Stephan von Bechtolsheim
c/o Springer-Verlag New York
175 Fifth Avenue
New York, NY 10010
USA

Library of Congress Cataloging-in-Publication Data
Bechtolsheim, Stephan von
    TEX in practice / Stephan von Bechtolsheim
        p.   cm. (Monographs in visual communication)
    On t.p. E is subscript
    Includes bibliographical references and indices.
    Contents: v. 1. Basics v. 2. Math and fonts v. 3. Tokens, macros v. 4. Output
    routines, tables.
    ISBN-13: 978-1-4613-9144-9   e-ISBN-13: 978-1-4613-9144-9
    DOI: 10.1007/978-1-4613-9144-9
    1. TeX (computer system) 2. Computerized typesetting 3. Mathematics printing.
    I. Title. II. Series.
    Z253.4.T47B4 1992
    686.2"2544   dc20   90-10034

Printed on acid-free paper.

Production supervised by Kenneth Dreyhaupt, manufacturing coordinated by Vincent
Scelta.
Photocomposed copy prepared from the author's TEX files.

9 8 7 6 5 4 3 2

To My Brothers Sebastian and Florian

The present letter is a very long one, simply
because I had no leisure to make it shorter.

Blaise Pascal

Preface

Although you only have one volume in front of you, writing four volumes and 1600 pages on a single subject needs some form of justification. And then on the other hand, why write even more?! Can't, at least, the preface of something that long be short?!

Very well, so let's keep it short. It is my sincere hope that the series "TeX in Practice" will be useful for your own TeX work. But *please*, before you get started, read the "Notes on 'TeX in Practice'," because it instructs you how to use this series. You will find these notes on pages xxvii–xxxvi.

The fourth and last volume deals with two different subject areas. First of all, there are the so-called output routines which are responsible for putting together the pages as generated by TeX. You will be amazed at how many different things can be done with TeX's output routines. The second subject area we are dealing with in this volume are tables. About a hundred different tables you can choose from should provide you with a starting point in the selection of tables.

Stephan von Bechtolsheim

Arlington Heights, Illinois
May 1993

Table of Contents

List of Figures

Acknowledgements

A long list of people is what a long book deserves. So let's get started.

My editors, Kim Miller (who did the bulk of the work) as well as Kris Schlenker and Judith Lewis, tried patiently to make this series sound like "real English." If it doesn't sound right always, blame me (a quote from one of my students: "by now you are probably *bilingually illiterate*"). Last minute corrections on my part may also have introduced errors.

At Purdue University Chris Hoffman and Bob Lynch (Computer Sciences Department), Mark Senn (Statistics Department; by the way he is the person who brought TeX to Purdue), and Steve Samuels (also Statistics Department) helped; there was also Ed Ramsey (National Pesticide Center of Purdue University) who let me use his SPARCStation to process this series. Clients of mine such as Siemens-Nixdorf Corporation (Jörg Heinemann, Joachim Haenel, Bernd Stümke), Jörg Steffenhagen, of Munich, West Germany, and Technical Typesetting, Inc. (Bill Taylor and Harry Kirk) located in Philadelphia allowed me to publish macros I had developed for them. Gerhard Rossbach, Mark Hall, and Rüdiger Gebauer of Springer-Verlag all waited patiently for me to finish my series. Dave Rogers (United States Naval Academy, Anapolis) gave some valuable hints. Chris Thompson (Cambridge University Computing Service) and Thomas Reid (Computing Services Center, Texas A&M University) provided some input too. A.C. Conrad of the Menil Foundation in Houston hired me as a consultant. Some of the TeX problems I solved for him found their way into this series. I used Oliver Schoett's idea (Institut für Informatik, TU München, München (or should I have written "Munich"), Federal Republic of Germany), designing macro \PrintHyphens. One of the tables in the table chapters was submitted to me by Cynthia Rodriguez, Computer Center, University of Illinois at Chicago. Michael Doob, University of Manitoba in Canada, helped with a problem in the output routine chapters. Kabelschacht's article (Kabelschacht (1987)) was used in a chapter in this series.

To thank Donald E. Knuth, the author of TeX and METAFONT, warrants a separate paragraph, in particular, because he replied to my mail instantaneously even after he had announced that he would not be involved in the further development of TeX. Not only that, *he* provided TeX to all of us! Thank you *very much*.

Leslie Lamport also deserves special mentioning. When I started to write this series, there were plenty of things I did not understand. For instance, verbatim mode was one of the things I initially simply ignored. What I did was rather straightforward: I started to write this series using LaTeX. As time proceeded I replaced one part of LaTeX after another by my own macros. You will probably also recognize some macro names in this series which remind you of LaTeX macro names such as \Section (written \section in LaTeX) and \normalsize (called the same in LaTeX). Even now some of the diagrams are done with LaTeX's picture environment. Other things I stole / borrowed / adopted include the definition of macro \frac, the idea of a font selection scheme which separates size and type face, etc.

Victor Eijkhout, Center for Supercomputing Research and Development, University of Illinois at Urbana-Champaign, also deserves his own paragraph: he actually went through the whole series. Maybe he did not read every page, but he certainly read *almost* every page. There were a lot of little things he pointed out to me.

Of course, what would be a series on TeX without mentioning Barbara Beeton of the American Mathematical Society in Providence, Rhode Island, who helped me with a lot of small details. Ray Goucher of the TeX Users Group, who was a good boss while I taught classes for the TeX Users Group, deserves acknowledgement too.

Addison-Wesley gave permission to reprint and distribute (in modified form) some macros of the TeXbook (see 16.8, p. II-297). Penguin Books gave permission to use Shakespeare (1605) for examples in Chapter 10 (this material is in the public domain, but I thought you might want to know where I got it from).

The Free Software Foundation must be mentioned too, because it provides the GNU Emacs to which I am addicted by now. GNU Emacs is the only editor you want to use after you have used it a couple of times. It certainly has done wonders for me. I designed my own "TeX-mode" and that made the tedious task of editing hundreds of pages much more tolerable.

Finally my current employer (Datalogics Incorporated in Chicago, a subsidiary of Frame Technology of San Jose, California) allowed me to use one of their printers for printing the last n revisions of this series ($n < 10$).

Trademarks

The following trademarks are used in this series.

$\TeX$ is a trademark of the American Mathematical Society; METAFONT is a trademark of Addison-Wesley Publishing Company; PostScript is a trademark of Adobe Systems Incorporated; UNIX is a registered trademark of AT&T; DEC, VAX, and VMS are registered trademarks of Digital Equipment Cooperation; Apple LaserWriter, Macintosh are trademarks of Apple, Inc.; IBM is a registered trademark of International Business Machines Corporation; HP LaserJet is a trademark of Hewlett Packard; SUN is a trademark of Sun Micro Systems, Inc.; Mercedes is a registered trademark of Mercedes-Benz AG, Stuttgart, Federal Republic of Germany; SPARC is a registered trademark of SPARC International, Inc., licensed exclusively to Sun Microsystems, Inc.; SPARCStation is a registered trademark of Sun Microsystems, Inc.; Porsche is a registered trademark of Dr. Ing. h.c. F. Porsche AG.

Other brand or product names are the trademarks or registered trademarks of their respective holders.

A Note About the Dedication

I dedicate this volume to my two brothers (I am the oldest of three sons). I don't think my brothers thought always very highly of my way of treating them when we were all young (what an excuse...). Now we appreciate each other very much and I hope that this will remain so for as long as possible.

Stephan von Bechtolsheim

Arlington Heights, Illinois
May 1993

General Notes on "TEX in Practice"

This part of the book contains some general comments about the series "TEX in Practice." It is reprinted in the beginning of every volume of this series and provides kind of a "general introduction and series-wide table of contents." I urge you to read the following material carefully.

1 Some Friendly Advice Up Front Or What To Do If You Hate to Read Lots of Material

If you hate to read lots of material you are in trouble, obviously. But here are some hints anyway (who knows whether you even read this paragraph ...):

- Continue reading this part, "General Notes on 'TEX in Practice'," please, please.
- Quickly look through the tables of contents of all four volumes. I suggest limiting yourself to reading the title of all chapters and sections, but to ignore subsections and further subdivisions.
- Read the overview of all chapters on p. xxix.

  Thank you.

2 Terminology, Conventions, Other General Remarks

Let me make some general comments about the terminology used in "TEX in Practice" and some conventions I followed.

1. The series "TEX in Practice" is subdivided into *four volumes*. Each volume is identified by a *capital Roman numeral*. Each volume is subdivided into *chapters*. Chapters are identified by *arabic numerals* and numbered across volumes. Each chapter is further subdivided into *sections* (the second section of chapter 12 would be numbered 12.2), and sections, in turn, are divided into *subsections* (the fourth subsection of section 12.2 would be numbered 12.2.4). There are also *subsubsections*, which are numbered accordingly.

    Note that in cross-references to sections, subsections, and subsubsections the words section, subsection or subsubsection respectively are dropped. So a reference to Section 12.3 might read "see 12.3, p. 23, for further details." References to other entities like chapters, figures or tables are "spelled out" that is a reference to a figure might read "Fig. 3.4."

2. Arabic page numbers are used with the exception of the preliminaries of each volume in which Roman page numbers are used. The page numbering starts at page 1 in every volume.

    In references to page numbers, the volume number precedes the page number; in references to the *current* volume the volume number is dropped. For instance, "p. 23" refers to page 23 in the current volume, and "p. I-12" refers to page 12 of the first volume (in any volume other than the first).

3. The table of contents of each volume lists all chapters, sections and subsections of the volume, but not the subsubsections.

4. "TEX in Practice" tries to give a comprehensive and detailed view of TEX, but it does *not* claim to be complete. For very tricky questions I suggest that you first consult this series, but then also look into the TEXbook. I hope that reading this series will make understanding the TEXbook much easier, but I skip over some of the more bizarre details (which the TEXbook does not).

5. In this series I need to occasionally refer to the series itself. I do so using the expression "this series."

6. In general in this series I refer to TEX 3.0. The extensions implemented in TEX 3.0 are described in Knuth (1990). Note that TEXbooks published prior to 1990 do not contain a description of these extensions. The 3.0 extensions are minor, but nevertheless important. In some instances I will identify those differences by a short phrase such as "a TEX 3.0 feature" to identify a TEX 3.0 extension (*not all* such extensions are identified as such).

3   The Intended Audience of "TEX in Practice"

Let me now discuss whom this series was written for. The more of the following statements apply to you, the better:

*   You want to learn TEX thoroughly because you have to typeset complicated documents. Your text might include mathematical formulas and tables; both areas are very well covered by TEX.

- You are a curious person: when you use a tool (and TEX, of course, is a typesetting tool), then you wish to *understand* the tool. You do *not* only want to apply it.
- You believe that having lots of examples will allow you to understand something better. This series contains plenty of examples.
- You think that designing your own macros is something you may want to do, now or in future, because none of the existing macro packages fulfill your needs precisely. This series should come in handy, because it already contains the basics of a fairly sophisticated macro package plus it contains many basic utility type of macro definitions.
- You recognize that WYSIWYG (<u>W</u>hat-<u>Y</u>ou-<u>S</u>ee-<u>I</u>s-<u>W</u>hat-<u>Y</u>ou-get) systems, where your typeset output is immediately displayed on the screen, are limited in their power in that they cannot be easily adopted to specialized needs.

  Also note, that WYSIWYG systems cannot solve the real problem of writing, that is to present thoughts in an orderly and organized fashion, either. Owning a Porsche doesn't make you necessarily a good driver (though probably a happy one), if you know what I mean.

4   A Brief Overview of "TEX in Practice"

The series "TEX in Practice" is divided into four volumes. An overview of each volume follows.

1. Volume I:

    - Chapter 1 explains the existing relationship in the complexities of problems and tools to solve these problems. It explains that the complexity of TEX is a direct consequence of its power.
    - Chapter 2 provides a general introduction into TEX. It explains the basics of entering data into TEX and processing a document.
    - Chapter 3 begins the discusses of registers concentrating on counter registers primarily. Generic counter macros are discussed and an overview of all TEX's parameters is given.
    - Chapter 4 continues the register discussion looking at dimension registers and box registers. Also glue registers are discussed
    - Chapter 5 discusses glue (a very important concept of TEX), rules and leaders.
    - Chapter 6 discusses horizontal boxes. This is important in itself and also forms the basis of the discussion of the following two chapters. We also discuss how to do tables using hboxes.
    - Chapter 7 discusses vertical boxes, their dimensions, and different types line spacing and struts.
    - Chapter 8 discusses more advanced concepts relating to vertical boxes such as \vsplit and the shifting of reference points of vboxes.

- Chapter 9 discusses boxes and rules and introduces macros to draw rules around boxes. Boxes with rules are a very useful debugging feature of TeX.

2. Volume II:

- Chapter 10 discusses how paragraphs are built. You find the very basics of typesetting text in this chapter.
- Chapter 11 continues the discussion of the previous chapter and shows how more complex paragraph shapes are typeset.
- Chapter 12 concludes the discussion of the typesetting of paragraphs.
- Chapter 13 discusses the basics of typesetting mathematical formulas.
- Chapter 14 continues the discussion of typesetting mathematical formulas.
- Chapter 15 discusses fonts. This discussion includes an introduction to the Computer Modern fonts and how special fonts can be used in TeX.
- Chapter 16 continues the discussion of fonts. A major point in this chapter is the proper organization of fonts.
- Chapter 17 deals with the "environment" of TeX. This includes the discussion of METAFONT and the WEB system.

3. Volume III:

- Chapter 18 discusses tokens, grouping and category codes. You will also find an explanation of verbatim modes.
- Chapter 19 continues the discussion of the previous chapter explaining cross-referencing, list macros, and the use of token registers.
- Chapter 20 discusses token lists and arrays.
- Chapter 21 explains the basics of macro definitions, based on \def with delimited and undelimited parameters.
- Chapter 22 continues the discussion with macro definitions based on \edef, nested macro definitions and goes into more details about macro arguments.
- Chapter 23 discusses \let, \futurelet and \afterassignment.
- Chapter 24 discusses \expandafter and its applications as well as \the.
- Chapter 25 discusses conditionals.
- Chapter 26 discusses date and calendar related macros.
- Chapter 27 discusses more additional macro related issues such as recursion, environments and error handling.
- Chapter 28 discusses input and output. This includes a discussion of delayed and immediate writes, writing of table of contents files, and literal writing.
- Chapter 29 continues the preceding chapter discussing index file and verbatim file macros. It also discusses table of contents and endnotes generation.
- Chapter 30 discusses how the partial processing of a document can be administered (one document's text is stored in separate files which also

should be processed separately).

- Chapter 31 continues the discussion of the preceding macro. It in particular offers some rather sophisticated and interesting cross-reference macros.

4. Volume IV:

- Chapter 32 begins the discussion of tables typeset with TEX's \halign instruction.
- Chapter 33 also discusses tables, including vertical spacing, struts and rules in tables.
- Chapter 34 discusses the centering of tables, tables and paragraphs and preamble related macros.
- Chapter 35 discusses numerical computations in tables, splitting tables, \valign and double tables.
- Chapter 36 discusses the determination of page breaks by TEX.
- Chapter 37 discusses the basics of output routines, and introduces concepts like logical and physical pages.
- Chapter 38 presents some simple output routines.
- Chapter 39 discusses the output routine of the plain format and variations of it.
- Chapter 40 discusses output routines with insertions.
- Chapter 41 discusses double column output routines.

Each volume furthermore contains two prefaces, the first one is common to all volumes, the second one is a preface particular to the specific volume. The bibliography and the acknowledgements are identical in all volumes as is the "General Notes on 'TEX in Practice,'" the part you are currently reading.

One other brief point: I found it *extremely* difficult to present TEX in a linear order, without forward references. There is of course the possibility of making two or three passes over a subject such as TEX, each time with increasing difficulty. If I had done that, this series would have become even longer. At least in the approach I took I found it unavoidable and regrettable to have a fair number of forward references.

5   Using the Macros Presented in "TEX in Practice"

I hope that the macros presented in this series are useful, and that you will actually apply these macros. Here are some general comments about these macros, how they should be used, and so forth.

1. All macro source files presented in this series have the file extension .tip for "TEX in Practice." The file extension .tex is *not* used by any of the macro

source files. Some of the example source files use file extension `.tex`.

I *ask* everyone to *not* use the file extension `.tip` for any files other than the files of this series. I also ask you to *not* change any of the source code files of this series. If you need to change any of the published macro source code files (which you are encouraged to do, because one of the main purposes of this series is to use all the presented macros), then please *rename* the file you are using, and please also *change the file extension* at that point.

2. There are three classes of TEX source files presented in this series:

- *Published* files. Those are those files, which are available to you in machine readable form. The published macro source files are identified by a $\mathcal{P}$ in the line preceding the beginning of the source code listing. The file names of these files are at most 8 characters long, a restriction which is probably useful for the PC environment.

    Here is an example of a published source code file.

$\mathcal{P}$ • publish.tip •

```
19  \def\SampleMacro{%
20      ... Sample of a macro which is published on a diskette.
21  }
```

• End of publish.tip •

- Most of the published files are pre-loaded as part of the `texip` format which is defined in 31.3, p. III-612. Those macros *need not* be loaded, *if* the format defined at the given reference is used. Those macros are marked by a $\mathcal{P}'$.

    See item 13, p. xxxv, for some additional information.
- *Unpublished* files. In most cases unpublished files are example files, which apply one or more of the published files. These files are obviously *not* marked by $\mathcal{P}$.

3. The file names of those macro source files, which were designed particularly for the publication of this series, start with "`ts-`," which stands for "TEX in Practice Series."

4. To get your own copy of the published macro source files contact one of the following places:

    (a) The TEX Users Group (see further below for the address).
    (b) Your favorite TEX supplier.
    (c) The standard places for TEX source code on the networks, if you are hooked up to one of the national or international networks.

    If you acquire those macros commercially, you must *not* be charged any fee other than a nominal fee for media and shipping (the same way you do not have to pay extra for the plain format or LATEX macro source files). One more request: if you modify a macro, you are required to rename it, even if the modification is minor. With the exception of the conditions laid down in this paragraph you can do anything reasonable with these macros you want.

5. The following additional information applies, in particular, if you retrieve

the macro source files using a network (I try to establish a convention here and I hope everyone will follow it).

(a)   The macros will be stored in a separate subdirectory called `texip` (all lower case).

(b)   With the macros comes a file `readme.tip` which you should read first. This file contains the following information:

   i.   A release date of all macros.
   ii. A version number for all the macros together. This version number is a "combined" version number of all individual macro source files: it is the maximum version number of any individual source file's version number.
   iii. A short history of the changes made to any of the macro source files.
   iv. A complete list of *all macro source files* and their version numbers listed in *alphabetical order*.

6. Every macro source file starts out with the same standardized text, which reads as follows (the last four lines are, of course, different from file to file, also note that the file name such as `reg2.TEX` as well as the line numbers refer to my own source code files and are therefore of secondary interest for the user of the macros in this series):

```
 1   % This macro source file is from the four volume series
 2   % "TeX in Practice" by Stephan von Bechtolsheim, published
 3   % 1993 by Springer-Verlag, New York.
 4   % Copyright 1993 Stephan von Bechtolsheim.
 5   % No warranty or liability is assumed.
 6   % This macro may be copied freely if no fees other than
 7   % media cost or shipping charges are charged and as long
 8   % as this copyright and the following source code itself
 9   % is not changed. Please see the series for further information.
10   %
11   % Version 1.0
12   % Date: May 1, 1993
13   %
14   %
15   % This source code is documented in 4.1.11, p. I-91.
16   % Original source in file "reg2.TEX", starting line 587
17   \wlog{L: "absdimen.tip" ["reg2.TEX," l. 587, p. I-91]}%
18   % This file DOES belong to format "texip."
```

The previous text is 18 lines long. The text is *not* reprinted in the series itself anywhere else, in particular not with every source code listing. Therefore, to synchronize the line numbers of the source code as listed in this series and the source code of the `.tip` files, all *listings* of published source code files start with line 19, rather than line 1.

7. To *use* any of these macros, *always* use the following approach:

(a)   First load file `inputd.tip` using `\input` (only only):

```
1   \input inputd.tip
```

(b) Next load all macro files you want to load by calling the macro \InputD (do *not* use \input to load any .tip macro source files with the exception of inputd.tip). The \InputD macro has one parameter, #1, the name of a source code file to load.

For instance, to load file box-mac.tip, enter \InputD{box-mac.tip}. Do *not* omit the file extension tip. The beginning of your TeX source file would now look as follows:

```
1   \input inputd.tip
2   \InputD{box-mac.tex}
```

More \InputD calls could follow at this point.

8. Note that this series defines a format which has most of the relevant macros defined in this series pre-loaded. See 31.3, p. III-612, for details.

9. The series contains the reprint of numerous log files. The version of the TeX program I was running for the processing of this series was changed to print shortened log file lines to accommodate the width of the pages in this series.

10. In case you use LaTeX with any of my macros, you should not have any problems. Enter the instruction to load inputd.tip and the subsequent calls of the input macro \InputD *after* the \documentstyle command and *before* \begin{document}.

Here is a brief example of what your LaTeX source code file might look like:

```
1   \documentstyle{article}
2   \input inputd.tip
3   \InputD{box-mac.tip}
4   \InputD{...}
5   ...                  % Your macros and initializations go here.
6   \begin{document}
7      ...
8   \end{document}
```

Two more remarks about using these macros with LaTeX:

(a) Be aware that LaTeX uses \everypar. Therefore, if you use any of the macros from this series which change \everypar, you must be aware of potential conflicts.

(b) The plain format and LaTeX both have a \line macro with two totally different and unrelated meanings.

11. Remember, this series teaches you how to use TeX. The macros do *not* form a complete format including style files. In a typical application you will load macros and then write yourself additional macros. This series is like a cookbook: you find lots of recipes, but you still need to get out the utensils yourself, decide on the menu and clean up later. If you need something ready to use I suggest using LaTeX.

12. I cannot assume any *liability* for the macros. But I did my best to provide error free code. Please contact me in case you find any errors (in the source

code of the presented macros or otherwise in this series). The next Section contains my address.

13. If loading individual macros, etc., seems to be a little cumbersome to you, we provide to you (as part of the macros) a special file called `texip-ex.tip` (the file name stands for `texip.tip`, _expanded_). Here the word _expanded_ means that this is _one_ macro source file, which (with the exception of `plain.tex`) contains _all_ macro source files marked with $\mathcal{P}'$.

    Whether you run `initex` on `texip.tip` (which loads another 200 or so files) or run `initex` directly on `texip-ex.tip` does not make any difference as far as the end result is concerned. It's just easier in that if you just pick up `texip-ex.tip` from somewhere then you don't need to pick up other files, at least initially.

    I suggest that if you use `texip-ex.tip` that you rename the resulting `.fmt` file from `texip-ex.fmt` to `texip.fmt`.

6   Contacting the Author

If you wish to get in touch with me, please write to the following address. I am interested in your feedback, suggestions, clarifications, corrections and so forth. After all, one of _your_ suggestions may make it into the second edition (if there is such)! Please, _write_ or _send email_. Do _not_ call, and please _do not_ use of Springer's FAX numbers. Please do not contact me at my current employer.

Here is the address at which I can be contacted:

Stephan von Bechtolsheim

Springer-Verlag

175 Fifth Avenue

New York, NY 10010

Email: `texip@cs.purdue.edu`

The above email address will either forward to me or to someone at Springer-Verlag who will subsequently get in touch with me.

7   TEX Users Group

TEX is a great product and it's free! I do encourage you to show your support by joining the TEX Users Group, which can be contacted at the following address:

TEX Users Group

P.O. Box 869

Santa Barbara, CA 93102

(805) 899 4673

Internet: `tug@math.ams.com`

Corrections to this series will be distributed by the TeX Users Group. This is yet another reason to join this organization.

8   Rewriting Code for Improved Readability

Let me point out that I took the liberty of reformatting plain format macro code for improved readability. For instance, to set a register (counter or dimension register) to zero, the plain format declares a dimension register `\@z` which is initialized to zero:

```
1  \newdimen\z@
2  \z@ = 0pt
```

This register can be used to set another register, let's say `\dimen0`, to zero, as in `\dimen0 = \z@`. This, in turn, can be abbreviated to `\dimen0 \z@`. The *dimension* register `\z@` can also be used to set a *counter* registers to zero, as in `\count0 = \z@`. Finally, `\z@` can be used in place of a number 0, that is you can write `\count3 = \z@` or `\count3 \z@`, for short. This method of using `\z@` has also the advantage that no space needs to be written after the constant.

Now back to the changes to the plain format macros I applied in this series, when I reprint any plain format macro or macros. For instance, `\z@` is used in the definition of `\allowbreak` as follows:

```
1  \def\allowbreak{\penalty\z@}
```

In this series though, I state that this macro is defined as follows:

```
1  \def\allowbreak{\penalty 0 }
```

I think the second form of presenting this macro is more *readable*, and this is really all that this is about. I prefer readability and clarity in computer code over efficiency.

9   Summary

I hope you have a good time reading this series.

32
The Page Breaking Algorithm

The *page breaking algorithm* is that part of the TeX program that determines page breaks. Once a page break is determined by TeX, an output routine is called to write the current page to the `dvi` file. This chapter discusses the page breaking algorithm and the following five chapters discuss output routines.

32.1 Definitions and Other Basics

Out discussion of the very basics of the computation of page breaks begins with some definitions. Then we will explore the "main vertical list" in greater detail and discuss *where* TeX can insert page breaks.

The `\vsplit` primitive represents a simplified page breaking algorithm and is discussed in 8.2, p. I-267; I suggest reading about it there before proceeding with this chapter.

32.1.1 Definitions Relating to the Page Breaking Algorithm

The following are definitions and other basic points:

1. If we use the word "print" in this chapter (or in the following chapters on output routines) then this word is not to be taken literally, but in the sense of writing a page to the `dvi` file using the `\shipout` instruction.
2. The *main vertical list* is where TeX collects the material that has not yet been printed (written to the `dvi` file). The material collected on the main vertical list is logically divided into two parts:
    (a) The *current page*. This is the material TeX intends to print on the current page.
    (b) The *recent contributions*. This is the material, that will be committed to the current page part later.

3. TEX's *page breaking algorithm* is the part of the TEX program that determines the placements of page breaks.

4. The *output routine* is the part of the currently executing TEX code that deals with writing a page to a `dvi` file. The output routine is normally defined by the format being used.

    The activity of the page breaking algorithm is coordinated with that of the output routine. *After* the page breaking algorithm has determined *where* a page break should occur, the material of the current page to be printed is moved from the "current page" part of the main vertical list to box register 255. Then the output routine is called. The output routine takes the contents of this register and writes it to the `dvi` file (typically running headers and foot are added and the page number is incremented at this point).

    Since output routines are dealt with in much more detail in the five chapters following this one, we will present the simplest (and still useful) output routine here. This output routine takes the current page, stored in box register 255, and writes it to the `dvi` file. This example also shows how an output routine is set up: enclose the code of the output routine inside curly braces and assign it to \output (a token parameter that is expanded whenever the page breaking algorithm invokes the output routine):

```
1   \output = {\shipout \box255}
```

5. *Insertions* are the pieces of material that are not to be printed at the "current location," that is, where the regular text is printed. Instead, TEX saves this material and prints it at some other location as was explained above. For details see 32.7, p. 28.

32.1.2   The Main Vertical List

As mentioned before, the main vertical list can be subdivided into the *current page* and the *most recent contribution* part. The following items can be found in each of those parts.

1. The main vertical list can contain *boxes*. The potential sources of such boxes are:

    (a) A line from a paragraph as computed by the line breaking algorithm.
    (b) A formula in display math mode which is made into a box by TEX.
    (c) A row of a table produced by \halign, which becomes a box when processed by TEX.
    (d) An hbox generated, for instance, by a call to \line, \centerline, etc.
    (e) A vbox, for instance, generated by \vbox or \vtop.
    (f) A rule generated by an \hrule command. Rules are not really boxes, but in the discussion about the page breaking algorithm we look at them as a special form of a box.

2. The main vertical list can contain *whatsits*. There are two different types of whatsits:

    (a) *Delayed* (not immediate) *writes* generated by \write. The token list of a delayed \write is *not* written-out until the surrounding material of a \write makes it to the output routine where a \shipout is executed. Therefore, the \write token list has to be stored on the main vertical list.

    Delayed \openout and \closeout are also whatsits, but they are not discussed in further detail here, because the same rules apply to them as apply to \write.

    (b) \special. The token list of a \special command is stored with the main vertical list because the token list needs to be written to the dvi file. This happens, as in the case of \write, at the time of the \shipout.

3. *Marks*, which are used to generate running heads; see 34.7, p. 81, for details.

4. *Vertical glues.* Vertical glue can be either *implicit vertical glue* (such as \baselineskip or \parskip) or *explicit vertical glue* inserted by the user, for instance, by calling a macro like \bigskip.

    Understanding vertical glue in the context of the computation of page breaks is *very* important, because page breaks usually occur at some glue (TEX deletes this glue as side effect of performing a page break); see 32.4, p. 10.

    Leaders, of course, fall into the same category; see 5.6, p. I-145, for details.

5. *Kerns* ("vertical kern" of course). The relationship between kern and glue in the case of the determination of "vertical break points" (page breaks) and the relationship between horizontal kerns and line breaks is similar. The main difference between a kern and glue is that TEX does not generate a page break at a kern, unless that kern is followed by glue. Vertical kerns are not discussed in any further detail in this series. See 5.4.8, p. I-140, for an explanation of horizontal kerns.

6. Vertical *penalties.* A penalty can be an *explicit vertical penalty.* Here is an example: \penalty -200. Or it can be an *implicit vertical penalty*, being added by TEX. Penalties in general were discussed in 5.8, p. I-157. Penalties in the context of page breaks are discussed in 32.5, p. 15.

The preceding list naturally leads to a discussion of where TEX inserts page breaks, which is the subject of the next Subsection.

32.1.3   Where Page Breaks Can Occur

TEX *can* (not necessarily will) break a page at the following points. The term "at" means that when the break occurs at such an item, this item is subsequently removed. Everything preceding the item goes on to the current page, everything following it on the next page.

1. At a *glue*, if this glue is *preceded* by a non-discardable item. The qualifying clause "preceded by a non-discardable item" means that if a glue A is preceded by another glue B, no break at glue A will occur because the preceding item, glue B, is discardable. The break, if at all, occurs at glue B. Glue A is discarded too, because it follows a glue (glue B) that disappeared due to a break.
2. At a (vertical) *kern*, but only if this kern is immediately followed by glue.
3. At a (vertical) *penalty*.

32.1.4   Tracing TeX's Page Break Computations, \showlists, \tracingpages

There are two ways to trace TeX's page break computations in detail and you should know about these methods in case you want to explore the page breaking algorithm thoroughly. (None of the following two methods are used in this chapter because the page breaking algorithm is not discussed in that much detail.)

1. Call \showlists. Invoking this primitive causes TeX to display the main vertical list. The main vertical list will be display in two parts, the current page and the recent contributions part. See the TeXbook for details. See 14.1.1, p. II-189, for an application of \showlists in the case of processing mathematical equations.
2. Call \tracingpages. This primitive causes TeX to log page-break-related computations continuously in the log file. See the TeXbook for details.

32.2   The Page Breaking Algorithm in General

32.2.1   Accounting for the Page Dimensions, \pagetotal and \pagegoal

TeX has two dimension parameters, \pagetotal and \pagegoal, which are crucial for the understanding of the page breaking algorithm. Later we will discuss other factors considered in the computation of page breaks. In particular, I currently *ignore insertions* for the purpose of simplifying the discussion.

1. \pagetotal is the sum of the height and depth of the material accumulated of the "current page" on the main vertical list. This value therefore reflects the vertical size of the material collected so far for the current page (insertions are ignored).

2.   \pagegoal is the length of the vertical material of the current page that the page breaking algorithm tries to fit on the current page before a page break is generated.

   Normally TeX tries to fill the page completely, that is, up to \vsize. Therefore, at this point of the discussion it is *not* clear why TeX does not refer to \vsize *directly*, but to \pagegoal (of which the "default" value is \vsize). An explanation will follow shortly.

The important conclusion from the preceding statements is that *the difference between* \pagegoal *and* \pagetotal *is the amount of free space still available on the current page*. This statement is also true once insertions enter the picture as you will see later.

32.2.2   Updating \pagegoal and \pagetotal

Let us now discuss when \pagegoal and \pagetotal are updated by TeX as material is contributed to the current page part of the main vertical list.

   First, when there is *no* material on the current page part of the main vertical list, \pagetotal is zero and \pagegoal is \maxdimen (\pagegoal is *not* \vsize as you might assume).

   Immediately before the first item is added to a previously empty current page part, \pagegoal is set to \vsize because, in general, one tries to fill the current page completely. \pagetotal is set to zero because nothing has been contributed to the main vertical list yet.

   Each time material is added to the main vertical list, \pagetotal is increased by the vertical length (the sum of height and depth) of this material.

   A page break occurs when \pagetotal reaches \pagegoal, or in other words, when the current page is filled. This statement ignores penalties and insertions, which will be discussed later.

32.2.3   Printing \pagetotal and \pagegoal, \LogPageTG

The macro \LogPageTG writes the current values of \pagetotal and \pagegoal into the log file. This macro is very useful for tracing TeX computations of page breaks. The next Subsection contains an application of the macro.

   The macro \LogPageTG also writes an additional message, given as its only parameter #1 to the log file.

$\mathcal{P}'$ • lpagetg.tip •

```
15   \def\LogPageTG #1{%
16       \wlog{\string\LogPageTG [#1]:}%
17       \wlog{\string\pagetotal: \the\pagetotal,
18          \string\pagegoal: \the\pagegoal}%
```

¹⁹   `}`

• End of `lpagetg.tip` •

32.2.4   Computing the Available Vertical Space on a Page, \ComputeFreeSpaceOnPage, \FreeSpaceConditional

The macro \ComputeFreeSpaceOnPage computes the amount of free space on the current page by computing the difference between \pagegoal and \pagetotal. This macro is useful for applications where one has to decide whether there is sufficient room to print an item at the current point.

This macro (more precisely, macro \FreeSpaceConditional which is based on \ComputeFreeSpaceOnPage) was previously used in 12.3.2, p. II-115. The example at this reference reflects a typical application of this macro. In the example, a figure needs to be placed in the middle of a paragraph with part of the paragraph's text before the figure and part after. The placement of the figure in the middle of a paragraph only makes sense if there is sufficient space on the page to print the part of the paragraph that appears before the figure, and the figure itself (the second part of the paragraph may appear, partially or completely, on the next page). If there is *insufficient* room, the requirement is to print that figure on top of the next page.

To determine whether there is sufficient space, call macro \ComputeFree-SpaceOnPage (no parameters). The macro takes into account that \pagetotal and \pagegoal are zero and \maxdimen respectively, if the current page part of the main vertical list is empty. This macro's result, the amount of free space available, can be found in dimension register \ComputeFreeSpaceOnPage after the call to this macro is completed.

Here is the source code of the macro \ComputeFreeSpaceOnPage:

$\mathcal{P}'$ • `freespac.tip` •

¹⁵   `\InputD{lpagetg.tip}`   `% 32.2.3, p. 5.`

The \ComputeFreeSpaceOnPage macro stores the amount of free space available on the current page in dimension register \FreePageSpace which is now declared.

¹⁶   `\newdimen\FreePageSpace`

The definition of macro \ComputeFreeSpaceOnPage begins here.

¹⁷   `\def\ComputeFreeSpaceOnPage{%`

Reading the values of \pagegoal and \pagetotal is only correct if TeX is in vertical mode.

¹⁸   `\par`

Print the current values of \pagetotal and \pagegoal.

¹⁹   `\LogPageTG{\string\ComputeFreeSpaceOnPage}%`

Compute the available space.

```
20        \ifdim\pagetotal = 0pt
```

The *current page part* of the main vertical list is *empty*.

```
21            \FreePageSpace = \vsize
22        \else
```

The *current page part* of the main vertical list is *not empty*.

```
23            \FreePageSpace = \pagegoal
24            \advance\FreePageSpace by -\pagetotal
25        \fi
26  }
```

Now define a conforming conditional type of macro so you can use the preceding macro directly in a comparison. The following macro should be preceded by \ifdim and then be followed by one of the relational operators =, > and < in turn followed by a dimension or dimension register. For instance, a call to this macro could look as follows: \ifdim\FreeSpaceConditional > 1in ...\else ...\fi.

```
27  \def\FreeSpaceConditional{%
```

Terminate preceding \ifdim, similar to what is done in the case of a conforming conditional. Note though that the conditional preceding \FreeSpaceConditional is expected to be \ifdim. The result of the following conditional (true in our case) does not really matter.

```
28        0pt = 0pt \fi
```

Next compute the amount of free space.

```
29        \ComputeFreeSpaceOnPage
```

Now use the result from the preceding macro call stored in dimension register \FreePageSpace.

```
30        \ifdim\FreePageSpace
31  }
```

• End of <code>freespac.tip</code> •

The \ComputeFreePageOnSpace macro *cannot* be called *within* a paragraph to determine whether there is enough space to print the part of the paragraph on the current page. The reason is that TeX treats paragraphs as single units. The user cannot intervene while a paragraph is built and while the lines of a paragraph are copied to the main vertical list. Note the \par at the very beginning of this macro.

32.2.5 \pagegoal and Insertions

Insertions have no influence on the value of \pagetotal. I stated before that the initial value of \pagegoal is \vsize, reflecting the fact that the full page is available for the regular text of the current page. This Subsection discusses the changes of \pagegoal in the presence of insertions.

When TeX encounters an insertion, typesetting takes place as usual. The resulting insertion material is first written to the most recent contribution part of the main vertical list (not to the current page part). Note that the typesetting of the insertion provides TeX with the length of the insertion (the space this material will occupy on a page when it is printed later).

The insertion material of each insertion class is more or less treated separately. The easiest model of imagening the workings of the page breaking algorithm is to assume separate queues for each insertion class, and a separate queue for the material of the "current page." The term queue is used in computer science to describe exactly that: a list of elements where elements can be added on one end and taken off from the other end, and the order of the elements of the queue are always preserved.

The output routine combines the material from the current page (stored in box register 255) and the insertions (stored in other box registers) into one page, which is then written to the dvi file.

After the insertion has been typeset, TeX decides whether the current insertion's material should be printed on the current page. How this decision is made is explained later. Obviously one criterion is whether there is sufficient room on the current page in the first place. This can be computed from the difference between \pagegoal and \pagetotal.

Let us now assume that TeX decided to place the current insertion material on the current page; the material is therefore moved from the recent contributions to the current page part. TeX has to account for the vertical space required when this insertion is printed, and therefore the vertical size of the insertion material is subtracted from \pagegoal.

From this we conclude two important facts:

- \vsize − \pagetotal is the amount of vertical space used by insertions on the current page.
- In 32.2.1, p. 5, we stated that the difference between \pagegoal and \pagetotal is the amount of free space on the current page. This statement is obviously still true, even after insertions were added to the discussion.

Therefore, macro \ComputeFreeSpaceOnPage of p. 6 also works in the presence of insertions.

32.3   Page Layouts "Bottom Flush" and "Ragged Bottom"

I now discuss page layouts. The values of the parameters of the page breaking algorithm and the design of the output routine must be synchronized with each other as we will see shortly.

32.3.1   The Page Layouts in General

There are two choices for page layout:

1. *Bottom flush.* TeX tries to push pages together or stretch pages out vertically, with the goal of achieving a page length of \vsize. This corresponds to right and left flush text as far as determining line breaks in a paragraph is concerned.

    This stretching out or pushing together is done by the output routine, which might contain code along the following lines (the \vbox to \vsize is what generates a bottom flush page; the \unvbox unwraps the current page so the vertical glues of the page can stretch or shrink to adjust the page length).

```
 1   \output = {
 2       \shipout
 3           \vbox{
 4                \hbox{... running head ...}
 5
 6                \vbox to \vsize{
 7                     \unvbox 255
 8                }
 9           }
10   }
```

    In the discussion of printing right and left flush text, it was stated that the interword glue is the main source for adjusting the space between words. In the case of bottom flush text, the following vertical glues are the main sources for adjusting the page length and therefore should be set to glue values with non-zero stretchabilities and shrinkabilities:

    (a)   \parskip, the glue between paragraphs.
    (b)   \abovedisplayskip, \belowdisplayskip, etc., giving the glues above and below displayed equations.
    (c)   Vertical glue around headings.
    (d)   Vertical glue around figures and tables.

    A bottom flush page layout is the plain format's default.
2. *Ragged bottom.* In this case, TeX fills the current page with material as far as possible, but it will not at the end stretch out the page to \vsize. The amount by which a page is shorter than \vsize will be different from page to page, although TeX will try to minimize this amount.

32.3.2   Page Layouts in the Plain Format

Both types of page layout, as discussed in the preceding Subsection, are available in the plain format. For a ragged bottom page layout the user needs to call macro \raggedbottom, while for a bottom flush page layout the user needs to call \normalbottom. The latter is also the default.

The following conditional is true if and only if ragged bottom is the current page layout.

```
1    \newif\ifr@ggedbottom
```

The following macro \raggedbottom, when called, initiates ragged bottom page layout.

```
2    \def\raggedbottom{%
3        \topskip = 10pt plus 60pt
4        \r@ggedbottomtrue
5    }
```

The following macro \normalbottom, when called, initiates flush bottom page layout.

```
6    \def\normalbottom{%
7        \topskip = 10pt
8        \r@ggedbottomfalse
9    }
```

Flush bottom is also the default.

```
10    \normalbottom
```

The purpose and settings of \topskip are explained in 32.4.2, p. 12.

32.4   Vertical Glues of the Main Vertical List

Vertical glues of the current page part of the main vertical list play an important role in adjusting the page length and determining a page break. Before we study those glues, we will discuss a method of "shedding more light" on the main vertical list.

32.4.1   Writing the Current Page Part of the Main Vertical List to a Log File

Writing the current page part of the main vertical list to the log file at the time the output routine is called, is a very useful thing to do when developing a new output routine. Look at the definition of \output below: the \Show-BoxDepthOne{255} macro call can be added easily to any output routine.

• ex-mainvl.tip •

```
1    \input inputd.tip
2    \InputD{shboxes.tip}                        % 4.5.15, p. I-111.
```

The output routine below first shows the main vertical list by calling macro
\ShowBoxDepthOne. See 35.8, p. 120, for a version of the plain output routine,
which incorporates the setup below.

```
3   \output = {
4       \ShowBoxDepthOne{255}
```

Next the page is written out to the dvi file.

```
5       \shipout\box 255
6   }
```

The following settings are identical to the defaults of the plain format. They
are repeated here for improved documentation of the following example.

```
7   \baselineskip = 12pt
8   \parskip = 0pt plus 1pt
9   \topskip = 10pt
```

Here is a macro which generates a short paragraph.

```
10  \def\APar{%
11      This is just some short paragraph. This is just some short
12      paragraph. This is just some short paragraph. This is just
13      some short paragraph. This is just some short paragraph.
14      This is just some short paragraph.
15      \par
16  }
```

Now \APar is called three times.

```
17  \hbox{\vrule height 30pt depth 30pt width 2pt}
18  \APar
19  \APar
20  \APar
```

A second page is generated with a first line that was made artificially too
high. In that case a negative \topskip is at first computed, but the value inserted
is actually zero.

```
21  \vrule height 30pt depth 0pt width 2pt
22  \bye
```

• End of `ex-mainvl.tip` •

When the above TeX source code is executed, the following log file is generated. This particular log file is not very interesting, because all that is shown
in the preceding example is how an output routine is designed to call \Show-
BoxDepthOne of box register 255. In the log file you see one line for every glue
or box on the main vertical list.

• `ex-mainvl.log` •

```
1   This is TeX, C Version 3.14 (...)
2   **&/usr/local/tex/lib/fmt/plain ex-mainvl.tip
3   (ex-mainvl.tip (inputd.tip
4   (namedef.tip
5   ) (inputdl.tip
```

```
 6  \@InputDStream=\write0
 7  ))
 8  (shboxes.tip
 9  )
10  > \box255=
11  \vbox(643.20255+0.0)x469.75499, glue set 482.31366fill
12  .\glue(\topskip) 0.0
13  .\hbox(30.0+30.0)x2.0 []
14  .\glue(\parskip) 0.0 plus 1.0
15  .\glue(\lineskip) 1.0
16  .\hbox(6.94444+1.94444)x469.75499, glue set - 0.63487 []
17  .\penalty 300
18  .\glue(\baselineskip) 3.11111
19  .\hbox(6.94444+1.94444)x469.75499, glue set 8.94862fil []
20  .\glue(\parskip) 0.0 plus 1.0
21  .\glue(\baselineskip) 3.11111
22  .\hbox(6.94444+1.94444)x469.75499, glue set - 0.63487 []
23  .\penalty 300
24  .\glue(\baselineskip) 3.11111
25  .\hbox(6.94444+1.94444)x469.75499, glue set 8.94862fil []
26  .\glue(\parskip) 0.0 plus 1.0
27  .\glue(\baselineskip) 3.11111
28  .\hbox(6.94444+1.94444)x469.75499, glue set - 0.63487 []
29  .\penalty 300
30  .\glue(\baselineskip) 3.11111
31  .\hbox(6.94444+1.94444)x469.75499, glue set 8.94862fil []
32  .\glue(\parskip) 0.0 plus 1.0
33  .\glue(\lineskip) 1.0
34  .\hbox(30.0+0.0)x469.75499, glue set 447.75499fil []
35  .\glue 0.0 plus 1.0fill
36
37  <to be read again>
38  }
39  <output> ...wBoxDepthOne {255}
40  \shipout \box 255 }
41  \supereject ...\penalty -\@MM
42
43  \bye ...ar \vfill \supereject
44  \end
45  l.22 \bye
46
47  [1] )
48  Output written on ex-mainvl.dvi (1 page, 904 bytes).
```

32.4.2   Glue Inserted at the Top of Every Page, \topskip

If you read the log file of the previous example carefully, you discovered that
\topskip glue was inserted at the top of the page. The amount of glue inserted

(first page) is 3.05556 pt, *not* the initial value of \topskip which is 10 pt. The value of 3.05556 pt is computed by subtracting the height of the first box of the current page on the main vertical list (the height of the first line of text in the preceding example, 6.94444 pt) from the value to which \topskip is initialized (10 pt).

In order to understand what \topskip is about, we must go back to the discussion of \baselineskip (see 7.3.4, p. I-220). There we said that although the log file shows \baselineskip as the glue that is inserted, the amount of glue inserted is really \baselineskip minus the depth of the previous line minus the height of the next line. This computation is done to ensure constant line spacing of the output.

The \topskip glue can be described as the "every page's first line's \baselineskip." The amount of \topskip glue inserted is, therefore, \topskip minus the height of the first line. Regardless of the height of the first box (within "reasonable" limits of course, so that the height of the first line is less than or equal to \topskip), the baseline of this first line (or box) of a page will, therefore, always appear at the same vertical position on each page.

Note that \topskip is a *glue* parameter and its default value in the plain format is 10 pt, a glue *without* any stretchability and shrinkability. Also note that this coincides with the fact that a bottom flush page layout is used. In the case of a ragged bottom page layout, \topskip is set to a glue which is allowed to stretch (10 pt plus 60 pt). This tricks the page breaking algorithm into thinking that a relatively large amount of vertical stretchability is available on the current page. It is the output routine that actually places the material in such a way that it is printed ragged bottom; see 32.3.2, p. 9, for additional details (in particular for that part of the plain format output routine that handles the page layout). Note that in the case of the plain format (and most other output routines), the \topskip glue is never stretched or pushed together by the output routine.

As a final point I would like to draw your attention to a special output routine (34.6, p. 76), that sets \topskip to zero. See the discussion at the given reference for why this is done.

32.4.3   Implicit Vertical Glues and Page Breaks

Implicit vertical glues are very important for the discussion of page breaks and are listed here so that their proper settings can be discussed:

1. \baselineskip (\lineskip). The purpose of the \baselineskip glue is to ensure constant line spacing, mainly of the lines of a paragraph. This glue was discussed in 7.3.4, p. I-220. Occasionally, \lineskip glue is inserted between lines instead of \baselineskip glue. Again, see the above reference for details.

    \baselineskip (and \lineskip) is typically set to glue values *without* stretchability and shrinkability. If one allowed \baselineskip to stretch and shrink then pages would lose their uniform appearance in a bottom flush page layout. Line spacing would vary from page to page, which is obviously undesirable. Therefore, this glue is of little help when it comes to adjusting the page length in bottom flush page layouts because it cannot be stretched or pushed together.

2. `\parskip`. The `\parskip` glue is one of the most important sources for adjusting the page length in a bottom flush page layout, especially for text without displayed equations or tables and figures. See 10.8, p. II-25, for a more detailed discussion of this glue.

    For a bottom flush page layout, the `\parskip` glue should *always* be set to a glue value with a non-zero stretchability. The default of this glue in the plain format is `\parskip = 0pt plus 1pt`. Note that it is *not* set to a zero value *without* any stretchability.

    If you want to set up TeX in such a way that one empty line is left between paragraphs, you should specify `\parskip = 12pt plus 2pt minus 1pt`. You should *not* specify `\parskip = 12pt`, because then this glue could not be used to adjust the page length.

3. `\abovedisplayskip`, `\abovedisplayshortskip`, `\belowdisplayskip`, and `\belowdisplayshortskip`. These glues are inserted before and after displayed equations. They are another very important source for adjusting the page length in a bottom flush page layout. Observe that these glues are also inserted before and after centered tables, if the centering is done by enclosing the table in a vbox, which is centered using display math mode. See 14.9.1, p. II-216, for a more detailed discussion of these glues.

    Again, when these glues are set up, is should be done with non-zero stretchability and shrinkability.

4. `\topskip`. This glue is normally set to a value which does not allow for stretching and shrinking unless a ragged bottom page layout is chosen. See 32.4.2, p. 12, for further details.

32.4.4   Explicit Vertical Glues and Page Breaks

Besides implicit vertical glues inserted by TeX, the user can generate vertical glue with `\vskip`, also by referring to macros `\smallskip`, `\medskip`, and `\bigskip` (5.5.2, p. I-142).

    Another frequent source of explicit vertical glues are the vertical glues inserted by heading macros. Therefore, vertical glues before and after headings must be allowed to stretch or shrink for a bottom flush page layout.

32.4.5   Closing Remark

It is important to realize why we discussed vertical glue on the main vertical list: in the case of a bottom flush page layout (the most common page layout chosen in documents) the glue surrounding headings, between paragraphs, around equations, and around tables should always have some stretchability and shrinkability so that it is possible to adjust the page length.

32.5 Penalties Controlling Page Breaks

Note that besides the size of vertical material on the current page part of the main vertical list and the vertical glues on the same list, penalties are *very* important when determining a page break. In general, the further away a potential page break point is from the bottom of the current page, the larger negative the penalty must be, to "convince" the page breaking algorithm to insert a page break at the current location.

We are only interested in vertical penalties here because we are discussing page breaks and not line breaks. Before you insert an explicit penalty (using \penalty) which is intended to be an explicit *vertical* penalty, you must make sure that TeX is in vertical mode. There are also implicit vertical penalties that are automatically inserted by TeX.

Following are some macros of the plain format which deal with vertical spacing and penalties.

32.5.1 Macros Combining Vertical Glue and Vertical Penalties, \smallbreak, \medbreak, and \bigbreak

The plain format defines three macros \smallbreak, \medbreak, and \bigbreak, each of which combines the insertion of a vertical glue and a vertical penalty before this glue. The macros also take into account which vertical glues have already been inserted on the current vertical list.

• pl-brk.tip •

These macros do not use explicit glue values, but retrieve their values from the following glue registers. Those glue registers are defined and initialized first.

```
1    \newskip\smallskipamount \smallskipamount=  3pt plus 1pt minus 1pt
2    \newskip\medskipamount    \medskipamount  =  6pt plus 2pt minus 2pt
3    \newskip\bigskipamount    \bigskipamount  = 12pt plus 4pt minus 4pt
```

Next we will repeat the definitions of macros \smallskip, \medskip, and \bigskip here. Note that we define macro \smallskip and *not* \smallbreak. The same remark applies to the other two macros.

```
4    \def\smallskip{\vskip\smallskipamount}
5    \def\medskip{\vskip\medskipamount}
6    \def\bigskip{\vskip\bigskipamount}
```

Here is the definition of macro \smallbreak.

```
7   \def\smallbreak{%
```

Force vertical mode first (if TeX is already in vertical mode, then \par has no effect).

```
8         \par
```

Now check what amount of vertical glue can be found on the current vertical list. There are two possibilities:

1. Less than \smallskipamount (which includes none at all). In this case remove the glue (more precisely, neutralize it by calling macro \removelastskip) and then insert a penalty and vertical glue in the amount of \smallskipamount.
2. More than (or equal to) \smallskipamount. Do nothing in that case.

Here is the source code implementing \smallbreak:

```
9         \ifdim\lastskip < \smallskipamount
```

Remove the vertical glue inserted last and insert a small negative penalty followed by a small amount of vertical glue. This encourages a page break although (because the penalty is quite small not by very much).

```
10            \removelastskip
11            \penalty -50
12            \smallskip
13        \fi
14  }
```

The macro \medbreak is defined in a similar way, only the threshold glue value and the inserted penalty are different.

```
15  \def\medbreak{%
16        \par
17        \ifdim\lastskip < \medskipamount
18            \removelastskip
19            \penalty -100
20            \medskip
21        \fi
22  }
```

The macro \bigbreak is also defined along the same lines, although again the threshold glue value and the inserted penalty are different.

```
23  \def\bigbreak{%
24        \par
25        \ifdim\lastskip < \bigskipamount
26            \removelastskip
27            \penalty -200
28            \bigskip
29        \fi
30  }
```

The macro \removelastskip removes the last vertical glue inserted on the main vertical list by inserting a compensating glue (the same amount but in the opposite direction). This macro is defined as follows (see 8.5.2, p. I-306, for an explanation of \lastskip). The last vertical glue is not really removed, but an "undoing" vertical glue is inserted.

```
31  \def\removelastskip{%
32      \ifdim\lastskip = 0pt
33      \else
34          \vskip -\lastskip
35      \fi
36  }
```

• End of pl-brk.tip •

The \...skip and \...break macros can be compared in terms of the vertical glue they generate. Assume that the current vertical list does *not* contain any vertical glue so far. If you write

```
1  \smallskip
2  \bigskip
3  \smallskip
```

then you can make the following two observations:

1. The effective length of the vertical glues inserted is the sum of the three vertical glues, one of which is inserted by each of the macros.
2. There are no penalties inserted by the preceding \...skip macros.

Now look at a similar example using the \...break macros instead:

```
1  \smallbreak
2  \bigbreak
3  \smallbreak
```

Next make the following observations:

1. The preceding code is equivalent to writing \bigbreak only.
2. The effective vertical glue is the *largest* amount of vertical glue generated by any of the three macros called, which in this case is the vertical glue inserted by \bigbreak.
3. There are penalties inserted by all three macros, but similar to the point just made it is the largest penalty, the one inserted by \bigbreak, that really matters.

32.5.2   Explicit Vertical Penalties, \filbreak, \goodbreak, \eject

Let us first repeat the definitions of some general penalty-related macros. The three macros \break, \nobreak, and \allowbreak can be used to control line breaks as well as page breaks (depending on whether TeX is in horizontal or vertical mode):

```
1    \def\break{\penalty -10000 }
2    \def\nobreak{\penalty 10000 }
3    \def\allowbreak{\penalty 0 }
```

The following macros involve penalties and apply to page breaks only. Observe that the first instruction in all of the following four macros is \par, which forces TeX into vertical mode if it is not already in vertical mode (if TeX is already in vertical mode, \par has no effect). This is to ensure the later following inserted penalty is a vertical penalty.

Here are the four macro definitions:

1. \def\filbreak{\par\vfil\penalty-200\vfilneg}:
   Let us first describe the effect of this macro before we discuss how it works. If a call of this macro is inserted at some point A in the text, a page break occurs at point A, unless *all* of the following material up to a following \filbreak at point B also fits on the current page. Then the page break occurs at point B, unless there is more material and another \filbreak C and all the material from A to C would fit onto one page, and so forth.

   Thus text enclosed by two \filbreaks is printed on one page, unless that text occupies more than one page in the first place.

   The macro \filbreak works by pushing up all the material on the current main vertical list by adding \vfil glue. Then it indicates that the current point is a good page break point (\penalty -200) and it finally neutralizes the effect of \vfil by inserting \vfilneg.

   There are two possibilities. If There is a page break at the \filbreak inserted penalty, then the current page's material is pushed upwards by the \vfil inserted by \filbreak. The \vfilneg inserted by \filbreak is subsequently ignored, because the \vfilneg follows a break immediately If, on the other hand, there is *no* page break at the \filbreak inserted penalty, then \vfil and \vfilneg neutralize each other.

   In a letter format it may desirable "to keep paragraphs together." This can be achieved by inserting \filbreak at the end of every paragraph. This can be done automatically as follows:

```
1    \InputD{aevpar.tip}                    % 25.1.23.2, p. III-349.
2    \AfterEveryPar = {\filbreak}
```

2. \def\goodbreak{\par\penalty-500 }:
   A call of this macro indicates a good potential page break point because a significant negative penalty of -500 is inserted.

3. \def\eject{\par\break}:
   This macro forces a page break by inserting a penalty of -10000. Usually this macro is called immediately after a \vfill (\vfill\eject). This way the current part of the current text page is moved-up on the current page.

   If one uses \eject by itself (that is without a preceding \vfill) and a page break is forced, the current page will be vertically spread-out, if the page layout is bottom flush. This is identical to the effect of a \break in the middle of a line of a paragraph, which leads to a spaced-out line (see 10.10.5, p. II-42), that is usually undesirable.

Despite the fact that \vfill\eject is quite short and easy to write I found it useful to have a macro \NewPage which expands to \vfill\eject. Here is the source code of this macro definition:

$\mathcal{P}'$ • newpage.tip •

```
15    \def\NewPage{%
16        \vfill
17        \eject
18    }
```

• End of newpage.tip •

4. \def\supereject{\par\penalty-20000 }:
    There is no real difference between \eject and \supereject as far page breaks are concerned. A penalty value of less than -10000 has the same effect on the computation of a page break as a penalty of -10000.

    Because this penalty value is "reported" to the output routine in the counter parameter \outputpenalty, the output routine is able to distinguish between \eject and \supereject; see 33.6.1, p. 52.

32.5.3   Applying Explicit Vertical Penalties

Explicit vertical penalties have several applications:

1. To prevent a page break from occurring at some point, enter \par\nobreak (\par can be omitted if TEX is already in vertical mode).
2. To control page breaks in the middle of a paragraph, a different route must be taken. Note that there is no way in TEX to intervene while a paragraph is being processed, particularly, while the lines of the current paragraph are written-out to the main vertical list. However, \vadjust (see 12.3, p. II-112) can be used to add a vertical penalty to the main vertical list from within a paragraph. Here are two applications:

    (a) \vadjust{\nobreak} *prevents* a page break between the line where the \vadjust occurred and the next line.
    (b) \vadjust{\break}, on the other hand, *forces* a page break between the line where \vadjust occurred and the next line.

32.5.4   Implicit Vertical Penalties

Implicit vertical penalties are automatically inserted by TEX. If you design a format, it is important that you understand those penalties so that you program

TEX in such a way that page break computations are done automatically as much as possible.

32.5.4.1 Club and Widow Lines

Let me define two terms here:

1. A *club line* occurs when a page break takes place in the middle of a paragraph, such that the *first* line of the paragraph appears on the *bottom* of a page with the remainder of the paragraph on top of the following page. This first line on the bottom of a page is called a club line.
2. A *widow line* is generated when a page break occurs such that the *last* line, the widow line, of a paragraph appears on the *top* of a page (the remaining lines of this paragraph are obviously printed on the bottom of the preceding page).

Page breaks resulting in club and widow lines are undesirable in good typesetting as are page breaks where the *last* line on the bottom of the current page is hyphenated. Trying to avoid these three types of page breaks is why TEX defines certain implicit vertical penalties.

In addition to paragraph-related implicit vertical penalties, there is another set of penalties relating to displayed equations.

32.5.4.2 A List of Implicit Vertical Penalties

Here is the complete list of TEX's implicit vertical penalties. A discussion of what happens if more than one penalty applies to a particular case follows later. The listed default values apply to the plain format.

1. *Paragraph-related implicit vertical penalties.*

    (a) \brokenpenalty. This penalty is inserted after a hyphenated line and the next line of a paragraph. It therefore influences (that is prevents if set to 10000) page breaks after a hyphenated line (such a page break would lead to a hyphenated line on the bottom of a page).

    The default of this penalty in the plain format is 100. To process this series a value of 1000 is used.

    (b) \clubpenalty. This penalty is inserted between the first and the second line of a paragraph, therefore influencing page breaks resulting in a club line.

    The default of this penalty in the plain format is 150. The value for printing this series is 1000.

    (c) \widowpenalty. This penalty is inserted between the last two lines of a paragraph, influencing page breaks resulting in a widow line.

The default of this penalty in the plain format is 150. The value in this series is 1000.

(d) \interlinepenalty. This penalty is inserted between every two lines of a paragraph.

This penalty can be used, for instance, "to keep paragraphs together." If you set this penalty to 10000, page breaks inside a paragraph are completely prevented (this should be combined with a ragged bottom type of page layout). A different solution to the same problem using \filbreak was discussed in 32.5.2, item 1, p. 18.

The default of this penalty in the plain format is 0. The value used in this series is also 0.

(e) There is *no* penalty associated with a page break right before or after a paragraph, so there is nothing like a \preparpenalty or \postparpenalty. Of course, nothing prevents the user from inserting penalties manually.

2. *Displayed equation-related penalties.* Some implicit vertical penalties are associated with displayed equations:

(a) \displaywidowpenalty. This is the penalty for creating a widow line before a displayed equation. It is inserted between the last two lines of a paragraph if this paragraph is followed by a display (that preceding paragraph must have at least two lines, otherwise the penalty is not inserted). The default of this penalty is 50. The same value is used for the printing of this series.

(b) \predisplaypenalty. This penalty is inserted before a displayed equation and determines the penalty associated with a page break before a display. The default for this penalty is 10000. This prevents page breaks before displayed equations which is also the value used in printing this series.

(c) \postdisplaypenalty. This is a penalty for a page break after a display. The default of this penalty is 0 and the same value is used for the printing of this series.

In the case where *more than one* of the above penalties applies, the *sum* of all applicable penalties is inserted by TEX. For example, the penalty inserted between the two lines of a two line paragraph is the sum of \clubpenalty, \widowpenalty and \interlinepenalty.

32.5.5 Illustrating \interline-, \club-, \widow- and \brokenpenalty

The following example illustrates the various paragraph-related penalties. The content of a vertical list generated by typesetting various paragraphs is printed.

I chose some "clever" values for the four penalties \interlinepenalty, \club-penalty, \widowpenalty, and \brokenpenalty so that you can easily see which penalty or penalties were inserted; the values chosen for these penalties have nothing to do with penalties you would use in a real TeX application. Remember that sums of penalties are inserted when more than one kind of penalty applies.

Here is the source of the example:

• vpenalty.tip •

```
1   \input inputd.tip
2   \InputD{shboxes.tip}                    % 4.5.15, p. I-111.
```

The following penalty values were chosen so you can easily see which penalty or penalties were inserted.

```
3   \interlinepenalty =    1
4   \clubpenalty =         10
5   \widowpenalty =        100
6   \brokenpenalty =    1000
```

Generate four paragraphs now.

```
7   \setbox0 = \vbox{%
```

First, generate a paragraph representing paragraphs with four or more lines (the \hfil\breaks force line breaks and have no influence on the implicit vertical penalties discussed here). The \specials inserted below have no other purpose than to label the examples (those \specials appear in the log file):

```
8       \special{Example 1}
9       This is fun,\hfil\break
10      and really,\hfil\break
11      and more,\hfil\break
12      That's it.\par
```

Second, generate a three-line paragraph.

```
13      \special{Example 2}
14      A line,\hfil\break
15      next one,\hfil\break
16      Third one.\par
```

Third, generate a two-line paragraph.

```
17      \special{Example 3}
18      A line,\hfil\break
19      A second one.\par
```

Fourth, generate a paragraph with a hyphenated line.

```
20      \special{Example 4}
21      We need a hyphenated line. Some more short words here.
22      Mathematicsmathematicsmathematicsmathematics%
23              mathematicsmathematics, which is an artificially
24      long and non-existent word.
25      Now let us write a little more text. Just plain text.
26      Now let us write a little more text. Just plain text.
27      Now let us write a little more text. Just plain text.
```

```
28       Now let us write a little more text. Just plain text.
29       Now let us write a little more text. Just plain text.
30       Now let us write a little more text. Just plain text.
31       \par
```

Finally, show that if a zero penalty value is computed, then no penalty (rather than a zero penalty) is inserted.

```
32       \interlinepenalty = 0
33       \clubpenalty =       0
34       \widowpenalty =      0
35       \brokenpenalty =     0
36       \special{Example 5}
37       Let us write a short example here. Note the
38       example needs to be very short only. Not that
39       short, but still pretty short.
40   }
```

Now show the contents of the vertical list of the vbox stored in box register 0.

```
41   \ShowBoxDepthOne{0}
42   \bye
```

• End of `vpenalty.tip` •

The above source code generated the following log file:

• `vpenalty.log` •

```
1    This is TeX, C Version 3.14 (...)
2    **&/usr/local/tex/lib/fmt/plain vpenalty.tip
3    (vpenalty.tip (inputd.tip
4    (namedef.tip
5    ) (inputdl.tip
6    \@InputDStream=\write0
7    ))
8    (shboxes.tip
9    )
10   > \box0=
11   \vbox(186.94444+1.94444)x469.75499
12   .\special{Example 1}
13   .\glue(\parskip) 0.0 plus 1.0
14   .\hbox(6.94444+1.94444)x469.75499, glue set 399.92157fil []
15   .\penalty 11
16   .\glue(\baselineskip) 3.11111
17   .\hbox(6.94444+1.94444)x469.75499, glue set 424.17155fil []
18   .\penalty 1
19   .\glue(\baselineskip) 3.11111
20   .\hbox(6.94444+1.94444)x469.75499, glue set 425.83823fil []
21   .\penalty 101
22   .\glue(\baselineskip) 3.11111
23   .\hbox(6.94444+0.0)x469.75499, glue set 428.58824fil []
24   .\special{Example 2}
25   .\glue(\parskip) 0.0 plus 1.0
26   .\glue(\baselineskip) 5.05556
27   .\hbox(6.94444+1.94444)x469.75499, glue set 420.58827fil []
```

```
28    .\penalty 11
29    .\glue(\baselineskip) 3.90477
30    .\hbox(6.15079+1.94444)x469.75499, glue set 429.47713fil []
31    .\penalty 101
32    .\glue(\baselineskip) 3.11111
33    .\hbox(6.94444+0.0)x469.75499, glue set 423.61601fil []
34    .\special{Example 3}
35    .\glue(\parskip) 0.0 plus 1.0
36    .\glue(\baselineskip) 5.05556
37    .\hbox(6.94444+1.94444)x469.75499, glue set 420.58827fil []
38    .\penalty 111
39    .\glue(\baselineskip) 3.11111
40    .\hbox(6.94444+0.0)x469.75499, glue set 408.86601fil []
41    .\special{Example 4}
42    .\glue(\parskip) 0.0 plus 1.0
43    .\glue(\baselineskip) 5.05556
44    .\hbox(6.94444+1.94444)x469.75499, glue set - 0.12361 []
45    .\penalty 1011
46    .\glue(\baselineskip) 3.11111
47    .\hbox(6.94444+1.94444)x469.75499, glue set 0.4876 []
48    .\penalty 1
49    .\glue(\baselineskip) 3.11111
50    .\hbox(6.94444+1.94444)x469.75499, glue set 0.30411 []
51    .\penalty 1
52    .\glue(\baselineskip) 3.11111
53    .\hbox(6.94444+1.94444)x469.75499, glue set 0.30411 []
54    .\penalty 101
55    .\glue(\baselineskip) 3.11111
56    .\hbox(6.94444+1.94444)x469.75499, glue set 99.78212fil []
57    .\special{Example 5}
58    .\glue(\parskip) 0.0 plus 1.0
59    .\glue(\baselineskip) 3.11111
60    .\hbox(6.94444+1.94444)x469.75499, glue set 0.0933 []
61    .\glue(\baselineskip) 3.11111
62    .\hbox(6.94444+1.94444)x469.75499, glue set 395.14374fil []
63
64    <to be read again>
65    }
66    1.41 \ShowBoxDepthOne{0}
67
68    )
69    No pages of output.
```

Let us briefly analyze the above log file and specify the rules for the computation of implicit vertical penalties.

1. The penalty between any two lines of a paragraph is \interlinepenalty.
2. The penalty between the first and the second line of a paragraph is the sum of \interlinepenalty and \clubpenalty.
3. The penalty between the last two lines of a paragraph is the sum of \interlinepenalty and \widowpenalty.

4. In case the paragraph has only two lines, the penalty between those two lines is the sum of all three penalties \interlinepenalty, \clubpenalty, and \widowpenalty.

5. In case there is a hyphenated line in a paragraph, the value of \brokenpenalty is added on top of the penalty value computed according to rules 1–4.

6. Note that if a penalty computation results in a zero penalty, *no* penalty is inserted (rather than a *zero* penalty).

32.5.6   Page Breaks Around Headings

Rules for page breaks around headers usually read as follows:

1. There should be no page break within the header (this applies only to the case where the header is more than one line long).

2. There should be no page break *after* the header and *before* the first line of text that follows the header.

3. There should be at least two lines of text after the header. Sometimes the specification requires even three lines.

The first requirement can be implemented fairly easily by inserting a \nobreak after the header's text. This does prevent a page break after the header and the following text, although certain other problems must still be considered; see 11.5, p. II-90, for details.

The second requirement of having at least two lines of text after the heading can be implemented fairly easily by setting \clubpenalty to 10000.

Here is an example for a typical setup of heading macros.

• <code>ex-heading1.tip</code> •

```
1   \input inputd.tip
2   \InputD{shboxes.tip}                    % 4.5.15, p. I-111.
```

Set the club penalty to 10001. This value is used so it stands out in the log file reprinted below (the page breaking algorithm interprets it as if it had a value of 10000).

```
3   \clubpenalty = 10001
4   \widowpenalty = 0
5   \interlinepenalty = 0
```

Here is a header macro with one argument, #1, which is the text of the header.

```
6   \def\HeaderOne #1{%
```

Terminate the preceding paragraph if it was not terminated before. Then print the header.

```
7        \par
8        \leftline{\bf #1}
```

Now insert a penalty to prevent a page break following the header.

```
 9        \penalty 10000
10   }
```

Build a vertical list and then show it. This also demonstrates the penalties generated by the preceding source code.

```
11   \setbox 0 = \vbox{
12       This is fun.
13       \HeaderOne{Let's Go for It.}
14       And here is the text of the paragraph following this header.
15       And also I hope you have a lot of luck. So let's go for it.
16       And more of this.
17   }
18   \ShowBoxDepthOne{0}
19   \bye
```

• End of <code>ex-heading1.tip</code> •

Here is the log file generated by the preceding source code. Note that the penalty of 10000 (line 11 of the log file) will protect the two following vertical glues (lines 12 and 13 of the log file) from being valid break points for page breaks.

• <code>ex-heading1.log</code> •

```
 1   This is TeX, C Version 3.14 (...)
 2   **&/usr/local/tex/lib/fmt/plain ex-heading1.tip
 3   (ex-heading1.tip (inputd.tip
 4   (namedef.tip
 5   ) (inputdl.tip
 6   \@InputDStream=\write0
 7   ))
 8   (shboxes.tip
 9   )
10   > \box0=
11   \vbox(42.94444+1.94444)x469.75499
12   .\hbox(6.94444+0.0)x469.75499, glue set 399.92157fil []
13   .\glue(\baselineskip) 5.05556
14   .\hbox(6.94444+0.0)x469.75499, glue set 393.04568fil []
15   .\penalty 10000
16   .\glue(\parskip) 0.0 plus 1.0
17   .\glue(\baselineskip) 5.05556
18   .\hbox(6.94444+1.94444)x469.75499, glue set - 0.0748 []
19   .\penalty 10001
20   .\glue(\baselineskip) 3.11111
21   .\hbox(6.94444+1.94444)x469.75499, glue set 328.75473fil []
22
23   <to be read again>
24   }
25   l.18 \ShowBoxDepthOne{0}
26
27   )
28   No pages of output.
```

A closing note at this point: there is no direct way in TEX to enforce *three* lines of text after the heading without a page break, because no penalty such as "\afterclubpenalty" exists.

32.5.7   Other Approaches to Header Macros

There are two other approaches to control the vertical space after a header. Assume, for example, that you determined that you would like to have at least 30 pt of vertical space available on the current page. If there is less than 30 pt, a page break *before* the header is printed should be generated. There are two approaches to solving this problem:

1. Use the macro \ComputeFreeSpaceOnPage of 32.2.4, p. 6, to check on the availability of sufficient space. Generate a page break, if necessary, with \eject (bottom flush text layout) or \vfill\eject (ragged bottom page layout). Then print the header and the following text.
2. Insert the following sequence of instructions before the header line is generated: \vskip 30pt \vskip -30pt. This causes TEX to skip down and then immediately back by the same amount. One of the following two things will happen:

    (a) If there is *sufficient space* on the current page, the two \vskip instructions neutralize each other.
    (b) If there is *insufficient space* on the current page, the first \vskip triggers a page break. This \vskip is therefore removed, because it is the glue associated with a page break point. Now observe that all glues immediately following a glue which was removed by a page break are also removed. Therefore TEX also removes the \vskip -30pt glue. Therefore, this second glue is eliminated and it does *not* cause the text on the following page to be moved up by 30 pt.

    The preceding two alternatives correspond to the two alternatives discussed in the application of \filbreak (see 32.5.2, item 1, p 18).

32.6   Other Page Break Algorithm Controlling Parameters

32.6.1   Some Basic Parameters

In addition to the previously mentioned, the following parameters are also relevant when it comes to the computation of page breaks.

1. \vsize, the vertical size. The dimension stored in this dimension parameter determines the length of a page. The value of \vsize is saved internally by TeX as soon as something has been added to the current page part of the main vertical list. Therefore, a change to \vsize has no effect on the current page (but on the next page only), if there is already material on the current page part of the main vertical list.

2. \maxdepth. This value determines the maximum depth of a page (if a page exceeds this value, the page is "moved up" in a process that is very similar to the adjustment of a reference point in a vbox when \boxmaxdepth is taken into account; see 7.5.4, p. I-246). This parameter is saved when TeX starts a new page, the same way \vsize is saved.

3. \outputpenalty. This counter parameter register is set by the page breaking algorithm after it determines the penalty associated with a page break. The value stored in this register can be analyzed by the output routine. See 33.3.1, p. 52, for a discussion of when this happens.

32.6.2   Other \page... Parameters

Besides \pagetotal, which keeps track of the total length of the accumulated material, TeX maintains the following parameters:

- \pagefilstretch
- \pagefillstretch
- \pagefilllstretch
- \pageshrink
- \pagestretch

These registers present the sum of all stretchabilities and shrinkabilities of any accumulated glue for the current page. For instance, \pagefillstretch contains the sum of all the fill type of stretchabilities of the vertical glue of the current page part of the main vertical list[1]. These values are taken into account by the page breaking algorithm when deciding on a page break.

32.7   Insertions

Insertions, as already explained, allow the user to instruct TeX to place material at some specific place on the page. The insertion text will *not* be printed following the text previously processed. Where it is actually printed depends on

---

[1] I know: too many "of"s; still better than translating this into German where all the parts would become one long word ....

the *insertion class*. For example, in the case of the *"insertion class footnote"* the insertion text is a group of footnotes and therefore the insertion text is placed at the bottom of a page.

Let me discuss two classes of insertions as they are commonly found in documents. This should make the concept of insertions easier to understand.

1. *Footnote insertions* or *footnotes*. Treating footnotes as insertions allows the user to insert the footnote information together with the text the footnote refers to. TeX moves the footnote text into an insertion, so that the footnote text can be printed on the bottom of the current page.
2. *Floating body insertions*. Floating bodies are tables or figures which must not be split up across pages and which frequently must be placed at specific locations (such as the top of pages). Again one would like to *enter* the table's or figure's text near where this table or figure occurs in the text, but leave the detail of placing the entity to TeX. In the case of a change to the text, a floating body might then be printed on a different page.

   There is a somewhat more complicated version of the same problem where one would like to instruct TeX to place the floating body right where it is defined, if there is *sufficient* room left on the current page. Only when there is *insufficient* space left, should the floating body be placed on top of the next page. This is discussed in 35.4.5, p. 104.

32.7.1   Insertion Classes Are Identified by Numbers

TeX identifies each insertion class by an *insertion class index*, a number n. This number is the index of a dimension, a counter, a glue, and a box register simultaneously; the purpose of these four registers will be discussed shortly. The registers allocated to an insertion class should not be used for any other purpose. Obviously, the insertion number of an insertion index must be in the range of $10 \ldots 254$. Indices 0 through 9 cannot be used because of the special meanings of the counter registers with those indices. Index 255 cannot be used, because box register 255 has a special purpose; when the output routine is called, it contains the current page part of the main vertical list.

32.7.2   \newinsert To Allocate a New Insertion Class

The macro \newinsert allocates a new insertion index. This macro is very similar to the other \new... type of instructions, such as \newcount or \newdimen. For instance, to establish insertion class \TestInsert write

```
1  \newinsert\TestInsert
```

The actual insertion index is assigned by \newinsert and written to the log file as is the case with most of the \new... macros.

Observe that TeX assigns registers with numbers starting at 10 when the instructions like \newcount, \newdimen, and so forth are used. The allocated register index is increased with each call to \newcount, \newdimen, and so forth. Because an insertion requires the same register index to be allocated for four different registers, it is natural to assign register indices for insertions starting at the *upper* end of the spectrum of available register indices. The assignment of register indices by \newinsert starts at 254 and proceeds in descending order. As explained in the preceding Subsection, indices 0–9 and 255 are excluded.

Therefore, the first call of \newinsert in TeX will assign insertion class index 254, the next call 253 and so forth. Normally you are not really concerned with an insertion class index because you refer to an insertion class by its name such as \TestInsert rather than by the index itself.

32.7.3   Registers in Insertions

We already mentioned that for insertion class n the following registers have a special meaning: box register n, counter register n, dimension register n and glue register n. The purpose of these registers is:

1. *Box register n* is loaded by the page breaking algorithm with whatever material from insertion class n is to be printed on the current page when the output routine is called.

   This corresponds directly to the purpose of box register 255. Box register 255 is loaded with all the material of the current page of the main vertical list, which is supposed to written to the dvi file when the output routine is called. For instance, the material of insertion class \TestInsert, which needs to be printed on the current page, can be extracted by writing \box\TestInsert inside the output routine.

   The conditional \ifvoid (25.1, item 5, p. III-322) can be used in the output routine to determine whether there is material from insertion class n to be printed on the current page.

2. *Dimension register n* (\dimen n) defines the *maximum vertical space* the material from insertion class n may occupy on one page. This allows one to *limit* the *vertical size* of insertion material to be printed on one page.

   This register should be loaded by the user after the \newinsert call to allocate an insertion class number. For instance, the instruction \dimen\TestInsert = 3in limits the maximum size of insertions of this insertion class to 3 in per page. If more than 3 in of insertion material has been produced and could be placed on the current page, material is held back and printed on later pages.

3. *Glue register n* (\skip n) defines the amount of glue to be inserted on the current page if *some* insertion material of class n printed on a page. If more than one insertion of insertion class n happens to be printed on the current page, the specified amount of glue will be reserved only once on the current page. The output routine inserts the specified amount of glue; the page

breaking algorithm takes the necessary amount into account. Therefore, this glue is subtracted from \pagegoal for every page where there is insertion material of a specific insertion class printed.

For example, assume an insertion class \FootIns for footnotes. Then \skip\FootInsert = 0.3in reserves 0.3 in of additional space for each page with at least one footnote. This additional space may be used to offset the text page and the footnotes vertically and to insert a horizontal rule between the text and the footnote or footnotes[2].

4. *Counter register n* (\count n) contains the "magnification factor" for the length of insertions of this insertion class. After the actual length of an insertion was determined, the length is *multiplied* with the magnification factor of this insertion (and divided by 1000, of course, the way magnification works in TEX) and then subtracted from \pagegoal, if the insertion is placed on the current page. (Note that previously I have ignored the magnification factor in saying that the length of the insertion is subtracted from \pagegoal, but this magnification factor is usually 1000 anyway).

Again, this is a value that must be set by the user after a \newinsert call.

See 36.1, p. 133, for an example of an insertion and a related output routine where the magnification factor is zero.

32.7.4   Generating Insertion Material (\insert)

To generate an insertion, the user invokes the \insert instruction, which is followed by the insertion index of the insertion class for which material is being generated. The text which is contributed to the insertion class follows next, enclosed in parentheses. For example, assume in insertion class \TestInsert, we would like to reserve 1 in of white space (to glue in a diagram into a document later). Then this would be done as follows:

```
1    \insert\TestInsert{%
2        \vbox to 1.0in{}
3    }
```

The material to be "inserted" can also be surrounded by \bgroup and \egroup instead of curly braces. So the above example can be rewritten as follows:

```
1    \insert\TestInsert\bgroup
2        \vbox to 1.0in{}
3    \egroup
```

The solution using \bgroup and \egroup instead of curly braces is important because it allows splitting the beginning of the contribution of some vertical material and the ending instruction into two separate macros, as the following example shows. Two macros \BeginTestInsert and \EndTestInsert are

---

[2] Here you have an example of what is being discussed.

defined next, with obvious meanings.

```
1    \def\BeginTestInsert{%
2        \insert\TestInsert\bgroup
3    }
4    \def\EndTestInsert{%
5        \egroup
6    }
```

The preceding macros can be used as follows:

```
1    \BeginTestInsert
2        \vbox to 1in{}
3    \EndTestInsert
```

Note that the material of an `\insert` is enclosed in an implicit group. See 19.4.10, p. III-106, for details on implicit groups.

32.7.5 Updating `\pagegoal` in Insertion Processing

The next example shows how `\pagegoal` is updated as *insertion material* is generated. Assume we use the insertion class `\TestInsert` for the placement of figures at the top of the current page. What follows is very similar to the plain format's `\topinsert` insertion class (35.4.3, p. 103).

Here is the example:

• `testinsert1.tip` •

Define the new insertion class `\TestInsert`.

```
1    \newinsert\TestInsert
```

Initialize all registers of this insertion class. The maximum length of insertion material printed on one page is 290 pt, and no additional space needs to be reserved if there is at least one insertion on one page, and the magnification factor of this insertion class is 1.0.

```
2    \dimen\TestInsert = 290pt
3    \skip\TestInsert = 0.0in
4    \count\TestInsert = 1000
```

• End of `testinsert1.tip` •

An output routine goes with this insertion class which needs to be defined next. Output routines will be discussed later in more detail; observe that the contents of the box register containing the saved-up insertion material is extracted by the output routine.

• `testinsert2.tip` •

```
1    \output = {%
2        \wlog{\string\output: height / depth of
3            box \string\TestInsert: \the\ht\TestInsert\space /
4            \the\dp\TestInsert
```

```
 5        }%
 6        \shipout\vbox{%
```

Get the contents from box register `\TestInsert` (if not empty) and box register 255 containing the current page. The following `\ifvoid` can be omitted. I chose to leave it in there, because it more clearly represents the intentions of the code below.

```
 7             \ifvoid\TestInsert
 8             \else
 9                 \box\TestInsert
10             \fi
11             \box255
12        }
```

Increment the page number.

```
13        \global\advance\pageno by 1
14    }
```

• End of `testinsert2.tip` •

The source code of the example itself starts here.

• `testinsert3.tip` •

```
 1    \input inputd.tip
 2    \InputD{emptybox.tip}                    % 4.5.12, p. I-103.
 3    \InputD{lpagetg.tip}                     % 32.2.3, p. 5.
```

Read-in the setup of the requested insertion class as well as the associated output routine.

```
 4    \InputD{testinsert1.tip}                  % 32.7.5, p. 32.
 5    \InputD{testinsert2.tip}                  % 32.7.5, p. 32.
```

Change some parameters to make TEX's log file a little easier to interpret.

```
 6    \vsize = 500pt
 7    \parskip = 0pt
 8    \baselineskip = 20pt
```

Some sample text is generated by the following macro.

```
 9    \def\Text{%
10        \par
11        This is now some text. This is now some text, this is text.
12        This is now some text. This is now some text, this is text.
13        This is now some text. This is now some text, this is text.
14        \par
15    }
```

Now regular text as well as insertion material is generated.

```
16    \LogPageTG{1} \Text
17    \LogPageTG{2} \Text
18    \insert\TestInsert{\EmptyBox{250pt}{0pt}{\hsize}}    \LogPageTG{3}
19    \insert\TestInsert{\EmptyBox{50pt}{0pt}{\hsize}}     \LogPageTG{4}
20    \insert\TestInsert{\EmptyBox{50pt}{0pt}{\hsize}}     \LogPageTG{5}
```

```
21   \insert\TestInsert{\EmptyBox{50pt}{0pt}{\hsize}}      \LogPageTG{6}
22   \bye
```

• End of <code>testinsert3.tip</code> •

This source code generates the following log file:

• <code>testinsert3.log</code> •

```
1   This is TeX, C Version 3.14 (...)
2   **&/usr/local/tex/lib/fmt/plain testinsert3.tip
3   (testinsert3.tip (inputd.tip
4   (namedef.tip
5   ) (inputdl.tip
6   \@InputDStream=\write0
7   ))
8   (emptybox.tip
9   ) (lpagetg.tip
10  ) (testinsert1.tip
11  \TestInsert=\insert252
12  )
13  (testinsert2.tip)
14  \LogPageTG[1]:
15  \pagetotal: 0.0pt, \pagegoal: 16383.99998pt
16  \LogPageTG[2]:
17  \pagetotal: 30.0pt, \pagegoal: 500.0pt
18  \LogPageTG[3]:
19  \pagetotal: 70.0pt, \pagegoal: 250.0pt
20  \LogPageTG[4]:
21  \pagetotal: 70.0pt, \pagegoal: 200.0pt
22  \LogPageTG[5]:
23  \pagetotal: 70.0pt, \pagegoal: 200.0pt
24  \LogPageTG[6]:
25  \pagetotal: 70.0pt, \pagegoal: 200.0pt
26  \output: height / depth of box \TestInsert: 300.0pt / 0.0pt
27  [1]
28  \output: height / depth of box \TestInsert: 100.0pt / 0.0pt
29  [2] )
30  Output written on testinsert3.dvi (2 pages, 708 bytes).
```

In the log file observe that we limited the maximum amount of insertion material on one page to 290 pt. TeX actually exceeded this value when it placed two pieces of insertion material on the first page (source code lines 18 and 19), but the following three pieces of insertion material (source code lines 20 and 21) were moved to the next page (note that \vsize is 500 pt).

32.7.6 \insertpenalties

The counter parameter \insertpenalties plays a role in the computations of page breaks in TeX. We will not go into details, but here is a brief discussion.

When page breaks are being computed this parameter contains the *sum of all penalties for all insertions on the page.* On the other hand, when the output routine is called the value in this register is the *number* of held-over insertions. Held-over insertions are insertions that are not being printed on the current page but that will be printed on later pages. See 35.4.7, p. 108, where you find an explanation of the \dosupereject macro.

32.7.7 \floatingpenalty

The counter parameter \floatingpenalty is the amount of penalty TeX will use in case it has to *split* material of one insertion. In the case of the plain format the value of this penalty is set in two instances:

1. It is set to zero in the case of \topinsert indicating that we don't care whether an insertion is split or not.
2. It is set to 20000 (a very large value, obviously) in the case of footnotes, to indicate that TeX, if at all possible, should avoid splitting the text of footnotes across different pages.

32.7.8 Counter Parameter \holdinginserts

TeX 3.0 has added one additional feature, the counter parameter \holdinginserts. If this register is at its default value of zero, TeX works as described so far. If this register is positive, *no* insertion material is copied to its respective box register. Thus only material for the current page (from box register 255) can be printed on the next page. As the name of this parameter suggests it allows "holding back" insertion material.

32.7.9 Concluding Remarks About Insertions

The discussion of insertions cannot be complete without a more detailed understanding of output routines. In this series I do not go into *all* the details of insertions. If you are interested in fully exploring the issue, here is my list of recommended readings:

1. 35.3, p. 96, discusses the output routine of the plain format and variations of it. Insertions play an important role there.
2. Chapter 36, p. 133, contains more output routines with insertions. See in particular 36.4, p. 157, for a discussion of the limitations of the whole insertion concept.

3. Go to the TeXbook and read the following pages:

    (a) Page 110, top half.
    (b) Page 111, bottom half.
    (c) Page 112, top paragraph.
    (d) Page 123, bottom half.
    (e) Page 124, top half.
    (f) Page 125, last paragraph.

4. David Solomon has written a series of articles on output routines. See Solomon (1990a) and Solomon (1990b).

32.8   Comparing the Line Breaking and the Page Breaking Algorithm

To conclude this chapter we will compare the line breaking and the page breaking algorithm. Both algorithms, in many respects, are very similar, but there are also quite a number of significant differences.

1. The line breaking algorithm tries to find the optimum line breaks by looking at *all* lines of a paragraph "simultaneously." For example, the algorithm takes into account the influence a change in the line break of lines 2 and 3 has on the line break of lines 10 and 11.

    The page breaking algorithm only looks at one page at a time. It does *not* take into account the influence the page break of page 2 has on the page break of page 10.

2. The line breaking algorithm uses demerits *and* penalties to make its line break decisions. The page breaking algorithm works with penalties only.

3. Usually, both algorithms break at break points formed by glue, and in both algorithms, glue following a break point is removed.

4. There is nothing comparable to hyphenation when it comes to page breaks.

5. The line breaking algorithm is really two algorithms in one. Line breaks without hyphenation are attempted first; then hyphenated line breaks are tried, if the first pass did not succeed.

6. Interword glue (also the extended glued after punctuation) is the usual resource for making line length adjustments (left and right flush text) by the line breaking algorithm.

    For the page breaking algorithm (bottom flush page layout), the main resources for adjusting the page length are the glue between paragraphs (\parskip) and the glue before and after equations (\beforedisplayskip, etc.). Other resources include the glue around headings. Also, the space before and after insertion material that is printed on a page is usually allowed to stretch and shrink, making yet another resource for adjusting the page length.

7. The paragraph layout of left and right flushed text corresponds to a bottom flush page layout. A ragged right paragraph layout corresponds to a ragged bottom page layout.

In ragged right text, the `\rightskip` glue is made stretchable (for left and right flush text it is rigid) and in ragged bottom text (page breaking algorithm) `\topskip` (see 32.4.2, p. 12) is made stretchable.

32.9   Summary

In this chapter we learned:

- The page breaking algorithm is that part of the TEX program that determines page breaks. It is, in many respects, an algorithm that is similar to the line breaking algorithm.
- The main vertical list is where TEX collects material of the current page and insertion material (recent contributions), that may be printed on following pages.
- Before calling the output routine, the page breaking algorithm loads box register 255 with the material from the current page and other box registers that are related to insertions. The output routine writes the current page to to the **dvi** file.
- `\pagetotal` contains the amount of vertical space accumulated on the current page (excluding recent contributions). `\pagetotal` increases as material is added to the current page. `\pagegoal`, on the other hand, contains the "goal value" up to which the current page will be filled before a page break is generated by the page breaking algorithm. Note that `\pagegoal`'s initial value is `\vsize` but it is decremented by the size of any recently contributed material printed on the current page.

  The difference of the two dimensions `\pagegoal` − `\pagetotal` reflects the amount of remaining vertical space on the current page.
- Two different page layouts were discussed: bottom flush (which is the default) and ragged bottom. The setup of the page breaking algorithm and the design of the output routine must be done together to accommodate them.
- Vertical glue (implicit or explicit) is the central source for adjusting the page length. It is therefore very important to understand its influence, particularly when a bottom flush type of page layout is chosen (which is usually the case).
- Penalties are an important ingredient in determining page breaks.
- There are various paragraph-related vertical penalties that can be used, for instance, to avoid widow and club lines, as well as pages where the last line on the page is hyphenated. There are also some penalties related to displayed equations.

- Some approaches to header macros using vertical glues prevent a page break between a header and the following text.
- Insertions in TEX allow the specification of material that will not be printed at the current location by TEX, but the material will be contributed to the "recent contributions" part of the main vertical list and printed later.
- Each insertion is identified by an insertion index, which is also the register index of four registers. The associated box register is used by the page breaking algorithm to transmit that part of the recent contributions that is printed on the current page to the output routine. The associated counter register is used to hold the magnification factor of the insertion. The associated dimension register holds the maximum amount of material that can be printed for an insertion on a page-per-page basis. Finally, the associated dimension register specifies additional glue, which is contributed to a page if there is some material printed on a page of the specified insertion.
- Insertion material is generated using \insert. This instruction is followed by the insertion index and the insertion material. The latter can be enclosed in curly braces or in a pair of \bgroup and \egroup.

33
Output Routines, Basics

In this and the following four chapters chapters we discuss output routines. For purposes of simplification insertions will initially be ignored.

Output routines are a complicated matter. In particular you will probably note that none of the output routines presented here will fit your needs precisely. Nothing can be done about that: there are simply too many possible variations. Hopefully you find at least a starting point in one of the presented output routines. Also note that there is really no way around presenting a lot of TeX code. You need to study that code more carefully than in other chapters of this series.

33.1   An Overview of the Presented Output Routines

If are already familiar with output routines then a brief overview of output routines discussed throughout the following chapters should enable you to locate quickly an output routine you might want to find.

The output routines presented in this and the following chapters can be subdivided into four parts:

1. *Output routines without insertions.*

   (a) A trivial output routine, which simply ships the content of box register 255 to the dvi file, is presented in 34.1, p. 57. This output routine is presented to illustrate the interaction of the page breaking algorithm and the output routine.

   (b) The same section also shows an output routine that increments the page number, something every output routine does normally.

   (c) 34.2, p. 58, shows an output routine that allows the user to define page number limits (a lowest and highest page number). These limits specify those pages which are actually written out to the dvi file. Pages outside these limits are discarded.

   (d) An output routine where each page is surrounded by double rules is presented in 34.3, p. 59.

(e)   An output routine with simple running head and footer lines is shown in 34.4, p. 60. This output routine also makes a distinction between left-hand side pages (even page numbers) and right-hand side pages (odd page numbers) shifting pages horizontally. This output routine also shows how crop marks can be added to a page.

(f)   34.5, p. 66, discusses an output routine to print library cards. This is the first output routine that shows that there is not necessarily one physical page (= one printed page) to one logical page (= page as provided and stored in box register 255 by the page breaking algorithm). The ideas presented in this output routine are basic for double-column output routines.

2.   *Output routines based on the plain format.* Note that there is really only *one* output routine of the plain format, but Chapter 35, p. 89, contains different variations of the plain format output routine. You also should study this chapter if you are interested in insertions. Here is a brief overview of the output routines presented in that chapter:

(a)   The plain format output routine in a simplified version with no insertions can be found in 35.2, p. 91.

(b)   The regular plain format output routine can be found in 35.3, p. 96.

(c)   A modification of \midinsert is presented in 35.7, p. 116. Under certain circumstances discussed in that Section, material generated with \midinsert is printed in reversed order as compared to the order of the input. The presented modification avoids this problem.

(d)   A version of the plain format output routine is presented which will write the current part of the main vertical list and also all lists of the insertions into the log file. This is a very useful feature for tracing the page break computations of TeX. See 35.8, p. 120, for details.

3.   *Output routines with insertions.* Chapter 36, p. 133, deals with more output routines which include insertions.

(a)   An output routine printing index terms into the text margins during the development cycle of an index is presented in 36.1, p. 133.

(b)   A fancy output routine with insertions can be found in 36.2, p. 137. In this output routine figures with captions being printed on either the left or the right side of the figure are presented (it is the output routine that places the figure caption on the proper side).

(c)   Footnote-related problems in output routines are discussed in 36.3, p. 156.

(d)   Also the general limitations of the insertion concept in TeX is discussed in 36.4, p. 157.

4.   *Double-column output routines.* All double-column output routines are presented in Chapter 37, p. 159. There are essentially two different approaches to double column output routines, which are represented by two different output routines in that chapter:

(a)   A double-column output routine, which treats the left and right column as two separate logical pages, can be found in 37.1, p. 159.

(b)   A double-column output routine, which facilitates switching back and
      forth between single and double column output, is presented in 37.2,
      p. 166.

      This output routine is also very interesting because it shows an
      example where output routines are switched *during* the processing of
      a document. In other words, more than one output routine is defined
      (at any point in time only one can be active of course).

33.2   The Page-Breaking Algorithm and Output Routine

Let us quickly review how the page-breaking algorithm and the output routine
interact (most of this was discussed in the preceding chapter).

1.   Assume the page-breaking algorithm is activated *and* that it is time for a
     page break. Here are the actions performed by the page-breaking algorithm
     under those circumstances.

     (a)   The page-breaking algorithm decides which part of the main vertical
           list should be printed on the current page. It removes that material
           from the current part of the main vertical list and stores it in box
           register 255. Any glue where the break occurred (the presence of glue
           at the break is the most likely of all cases) is discarded.
     (b)   Any insertion material the page-breaking algorithm expects the out-
           put routine to print on the current page is stored in the box registers
           associated with the various insertion classes.
     (c)   The penalty of the break point, as it was computed by the page-
           breaking algorithm, is stored in counter parameter \outputpenalty.

           Normally, the value stored in this register is irrelevant, as far as the
           output routine is concerned. This register however is necessary in some
           special cases as a way of communication between the page breaking
           algorithm and the output routine; see 33.6.1, p. 52, for details.

2.   The output routine is called. Here is a list of actions typically performed by
     the output routine:

     (a)   Adjust the page length if a flush bottom type of page layout was chosen
           (which is typically the case).
     (b)   Take any running head, the contents of box register 255, any box reg-
           isters containing insertion material and the running foot to compose
           a complete page. Then write the resulting page to the dvi file using
           \shipout.

           You might go back to the discussion of \shipout (29.8, p. III-531)
           which also discusses \deadcycles and \maxdeadcycles all of the men-

tioned control sequences are important in the context of output routines.

(c)   Increment the page number (usually stored in \count0).

33.3   Comments on Generating the Chapters on Output Routines

While generating the chapters on output routines in this series, naturally I wanted to include output generated by output routines presented in this volume into the text of this volume "directly," pasted in as *full page figures*. But to generate output with more or less two output routines simultaneously (the plain format output routine used to process this series *and* the output routine which is being discussed) is close to impossible and even if I had managed to do it, it would be quite complicated. Therefore I used an approach based on the merging of dvi files.

33.3.1   Merging DVI Files

To merge dvi files let me define the term of a *master* dvi *file* as a dvi file which, in the case of this chapter, contains the main body of the text. The figure pages where output generated by output routines presented in this chapter should be glued in (= overlaid with the text) are left empty, with the exception of figure caption, running head and page number.

A *pulled-in* dvi *file* is a dvi file which is first generated separately. In the case of this chapter, such pulled-in dvi files are generated by sample applications of output routines discussed in this chapter.

In the following description of how the complete dvi file is generated, assume for simplicity that there is only one pulled-in dvi file (you find a graphical explanation of this merging process in Figure 33.1 on the next page):

- First the master dvi file is generated.
- Next the pulled-in dvi file is generated in a separate TeX execution.
- Each "empty figure page" of the master dvi file is overlaid with one page of a pulled-in dvi file. The dvi file processor dvimerge performs this overlay.

    In effect, this step is identical to printing an empty figure page (step 1) and an example page of one of the output routines (step 2) on transparencies, overlaying them and then making a photocopy of both of them on a copy machine. Only here it is done by a program. For a more detailed description of this procedure see Bechtolsheim (1989).

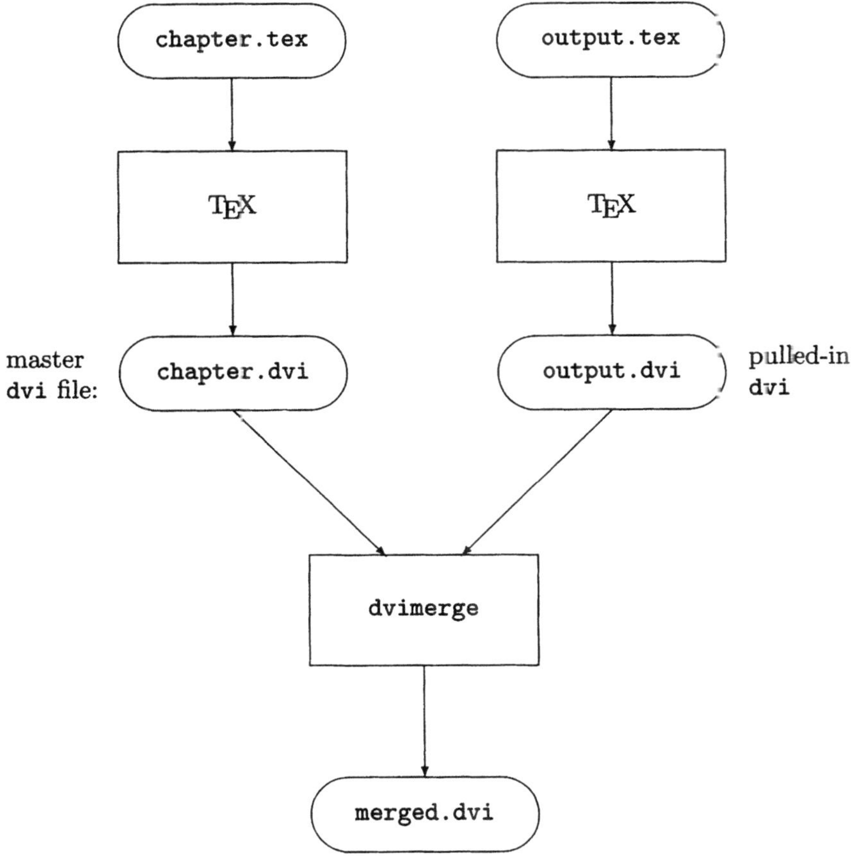

Figure 33.1. Explanation of dvimerge.

This approach of overlaying dvi files is used in other volumes of this series too, and it turned out to be a very useful thing to do under various circumstances.

The information about which page of which dvi file should be pulled-in to the master dvi file must be recorded in the master dvi file. This is done with a \special instruction (29.9, p. III-533). The \special instruction stores the following information in the master dvi file:

1. *File name* of the dvi file to be pulled-in.
2. *Page number* of the page of the pulled-in dvi file. The dvimerge program is able to select a specific page out of a pulled-in dvi file.
3. *Horizontal and vertical offset* that specifies the placement of the pulled-in dvi file in the master dvi file.
4. *Additional horizontal offset for even/odd numbered pages* by which pulled-in pages are shifted according to even and odd numbered pages of the master dvi file. There are two separate offsets for even and odd pages.

33.3.2   Dimensions in the Presented Output Routines

As mentioned before, sample output from the output routines discussed in this
and the following chapters are included as figures in the following chapter's text.
Therefore, the dimensions of the pages produced by these example applications of
output routines had to be adjusted (for instance, to leave room for figure captions
and page headers). Normally, the dimensions of the pages of these output routine
samples do *not* reflect the dimensions you would choose for a regular 8 1/2 in by
11 in page. Most likely when you use any of the presented output routines you
will have to adjust (= increase) the values of \hsize and \vsize. You should
in such a case first load the output routine you want to use using \InputD and
then change the dimensions. *Do not* change the original source code, please; see
5, p. xxxi.

33.3.3   Visible Boxes in Output Routines

In 9.3, p. I-318, I presented several macros which allow the user to print hboxes
and vboxes surrounded by rules. Also the related macros mark the base point by
a bullet and indicate the baseline by a dashed line. These macros are extremely
useful in the development of output routines because they allow one to visualize
the layout of a page easily. For instance, they make it easy to see where running
heads and footers are placed, the size of the page, etc. Therefore, many of the
example output routines use these macros, so when you develop your own output
routines, remember these macros are available.

You should set up your output routine in such a way that "with a flip of a
switch" all these rules disappear, so that after debugging your output routine you
have no problems eliminating these rules. The macro \EliminateRuledBoxes
comes in handy; see 9.3.12, p. I-342, for details. Because output routines form
implicit groups, calling \EliminateRuledBoxes inside an output routine does
not affect the use of the "ruled box macros" anywhere else; see 33.6.5, p. 54.

33.4   Basics of Output Routines

33.4.1   Setting Up an Output Routine, \output

In order to tell TeX which output routine you want to use, the routine must be
assigned to the token parameter, \output. There are two different approaches
to accomplish this.

1.   Assign the output routine you intend to use directly to \output.

```
1   \output = {
2       ...
3           \shipout\vbox{...}
4       ...
5   }
```

2. Define a macro that implements the output routine and then let \output trigger the execution of this macro.

```
1   \def\myoutput{
2       ...
3           \shipout\vbox{...}
4       ...
5   }
6   \output = {\myoutput}
```

The second approach has an advantage over the first when output routines have to be switched in the middle of a document (we will see an example of that later). Here is a setup which allows the switching of output routines.

Define a macro which contains the source of the *first* output routine.

```
1   \def\OutputOne{
2       ...
3   }
```

The same for the *second* output routine.

```
4   \def\OutputTwo{
5       ...
6   }
```

Now switch output routines.

```
7   \output = {\OutputOne}
8   ...
9   \output = {\OutputTwo}
10  ...
11  \output = {\OutputOne}
```

33.4.2   Logical / Physical Page

A *logical page* is what the page-breaking algorithm of TeX refers to as a page. Whatever is stored in box register 255 when the output routine is called is the logical page.

A *physical page* is a page as it is written to the dvi file by the output routine.

In most cases there is one logical page to one physical page: after the page-breaking algorithm calls the output routine, the logical page in box registers 255 becomes a physical page in the dvi file.

There are exceptions: double column output in TeX can be handled by generating each column separately as one logical page. The output routine, when called the first time, interprets the first logical page as a left column which is saved in a box register (no output to the dvi file occurs). At the time of the second call the saved left column and the current logical page (the right column) are combined into one physical page and written to the dvi file. Then the whole process starts all over again.

33.4.3   Page Dimensions

Figure 33.2 on the next page shows a standard 8 1/2 by 11 in page of TEX's plain format. The upper left corner of TEX's page is printed 1 in down and 1 in to the right of the upper left-hand corner of the paper. This means that the 1 in margins on the left, right, and top are *not* generated by TEX. As far as TEX is concerned, the page is not 8 1/2 in wide, but 6 1/2 in wide. The default of \hsize is 6.5 in, not 8.5 in.

Note that the device driver is responsible for printing TEX's page in such a way that the upper left corner of TEX's page is printed 1 in to the right and 1 in down from the upper left corner of the paper. If your driver does not place the page properly on the paper, then you can do this in two ways:

1. Many device drivers have an option that allow you to shift the output (see 2.5, p. I-11, item 2).
2. Dimension parameters \hoffset and \voffset allow shifting pages; see 33.6.4, p. 53, for details.

Observe that \vsize, the vertical size, does not include the height of the running head or footer line. \vsize is the vertical size of the text of a page only. The default of \vsize in the plain format is 8.9 in.

33.4.4   Left-Hand and Right-Hand Pages

In books or other documents, where the pages are printed on two sides, by convention, the left-hand side pages have even page numbers and the right-hand side pages have odd page numbers. In addition, these pages are normally shifted horizontally, for instance, towards the outside so that additional inside space for the book binding becomes available. This shifting of pages is another duty of the output routine. An output routine with this feature can be found in 34.4, p. 60.

In a document that is printed double-sided, chapters start on a right-hand side page (with an odd page number). It is therefore necessary to generate an empty left page, if the previous text ended on a right-hand side page. The following macro \NewPageRightHand checks whether the page number is even, after all preceding text was written out. If that is the case, a new empty page (with an even page number, of course) is generated.

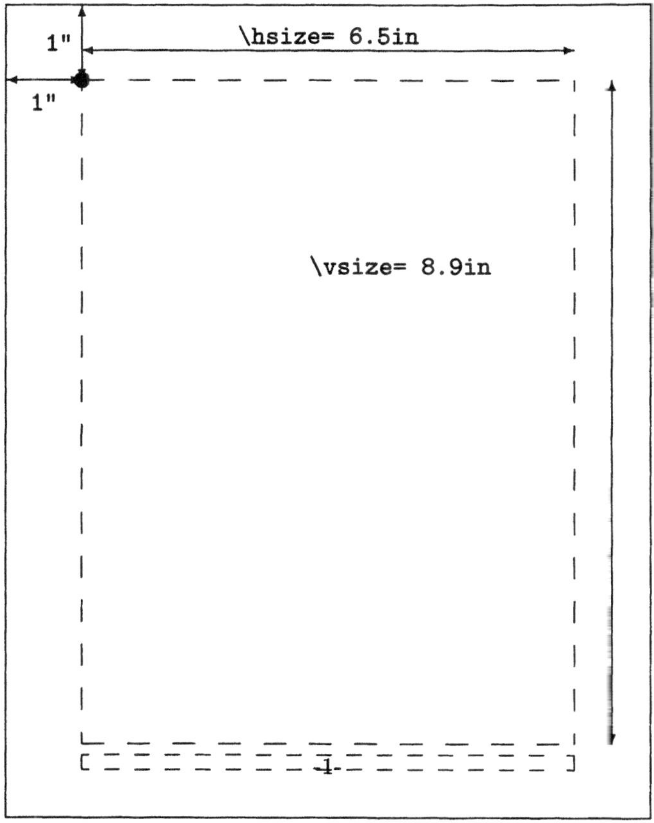

Figure 33.2. Standard page layout of an 8 1/2 in by 11 in page.

$$\mathcal{P}'\quad \bullet \texttt{ npright.tip } \bullet$$

```
15    \def\NewPageRightHand{%
```

Write out any remaining material (if there is any) by forcing a page break.

```
16        \vfill\supereject
17        \ifodd\count0
18        \else
```

The current page is an even page. Generate an empty page now. The empty hbox prevents the `\vfill\eject` from being ignored.

```
19            \hbox{}
20            \vfill\eject
21        \fi
22    }
```

The following macro `\ShouldBeRightHandPage`, when called, will determine whether the current page is a right hand page (odd page number). If this is true, nothing happens. If it is false, then `\NewPageRightHand` is called but not before an error message was issued.

```
23    \def\ShouldBeRightHandPage{%
24        \ifodd\count0
25        \else
26            \errmessage{\string\ShouldBeRightHandPage:
27                current page number is \the\count0.
28                Should be odd and NOT EVEN.}%
29            \NewPageRightHand
30        \fi
31    }
```

The definition of `\NewPageLeftHand` follows along the lines of the definition of `\NewPageRightHand`.

```
32    \def\NewPageLeftHand{%
33        \vfill\supereject
34        \ifodd\count0
35            \hbox{}
36            \vfill\eject
37        \fi
38    }
```

• End of <code>npright.tip</code> •

33.4.5   Portrait Mode and Landscape Mode

Usually a page is printed in *portrait mode*: the short side of the text page is at the top and bottom of the paper, the long side is at the left and right. In *landscape mode*, the page is rotated 90 degrees with the long side of the text page at the top and bottom of the paper.

TeX itself has no notion of the orientation of a page. A page in TeX has some horizontal and vertical size, and TeX does not care about the ratios of those values. The *device driver* is responsible for printing a page in landscape mode (not all drivers have this feature).

There are also cases where one would actually like to switch back and forth between landscape and portrait mode in the same document; this is beyond the scope of the discussions here. Again you need the cooperation of your driver.

33.5   Page Numbering, `\pageno`

Page numbering affects the output routine. In most cases, a simple page number is printed on the bottom of each page. Page numbers can also be found in running heads. For instance, in this series the page numbers are placed "on the outside."

The page number is stored in counter register 0 in plain TeX (\count0) and most other formats. This is a very useful convention (as we will see later), but not an absolute requirement.

The \countdef instruction (see 3.3.7, p. I-49) can be used to give any counter register a symbolic name. It is used in the plain format to assign the name \pageno to counter register 0: \countdef\pageno = 0 does just that. Therefore, in most cases \pageno instead of \count0 is written in this and the following chapters.

33.5.1   Page Numbers and \shipout

During the discussion of the \shipout instruction (29.8, p. III-531) it was mentioned that counter registers 0...9 are written to the dvi file. Therefore, by selecting counter register 0 to contain the page number, the page number is written to the dvi file. Programs like device drivers or dviselect can use this information to access a page or particular pages of a dvi file.

Counter registers 0 through 9 are also echoed on the terminal and into the log file (see for details 29.8, p. III-531), which is another reason for selecting counter register 0 for the page number: as the document's pages are written to the dvi file [1] [2] [3] ... appears on the screen, more than familiar to you by now.

33.5.2   Roman Page Numbers in the Plain Format

Roman page numbers are frequently used in the preliminary part of a document. It is used that way in this series (see the table of contents, acknowledgements and so forth). Of course, one of the macros presented in 3.4.1, p. I-68, can be used to generate roman page numbers. The approach of the plain format is different and is discussed now.

Remember that \romannumeral prints any value or a counter registers' content as a roman numeral. In the plain format, *negative page numbers* are not thought of as being negative page numbers (who would use negative page numbers in the first place?!) but as roman page numbers. So page number −3 in the plain format really means roman page number iii. The following Subsection discusses this in more details.

33.5.3   Plain Format Macros to Administer Page Numbers

Several macros of the plain format deal with page numbering (and roman page numbers in particular).

1. To use roman page numbers, one has to assign a negative value to \pageno
   (which is the same as \count0). For instance, at the beginning of the preliminaries of a book one would enter \pageno = -1. To switch back to arabic
   numerals simply write \pageno = 1. No special macro for that purpose exists.

2. The macro \advancepageno increments the page number by one. This macro
   is called by the output routine of the plain format. Because output routines
   form implicit groups the new value for the page number must be changed
   globally. This macro also must accommodate roman page numbers: in the
   case of negative (= roman) page numbers, the page counter must be *de-
   creased* rather than increased by one. Therefore, the definition of macro
   \advancepageno reads as follows:

```
1  \def\advancepageno{%
2      \ifnum\pageno < 0
3          \global\advance\pageno by -1
4      \else
5          \global\advance\pageno by 1
6      \fi
7  }
```

3. The term *folio* in typesetting means the place where the page number is
   printed on a page. Therefore the macro in the plain format which prints the
   current page number is called \folio. Here is the definition of this macro.
   Note how a negative page number is printed as a (positive) roman numeral.

```
1  \def\folio{%
2      \ifnum\pageno < 0
3          \romannumeral -\pageno
4      \else
5          \number\pageno
6      \fi
7  }
```

4. Page numbers in the plain format are printed in the footerline, centered. The
   relevant definitions of the plain format read as follows:

```
1  \newtoks\footline
2  \footline = {\hss\tenrm\folio\hss}
```

   Because \footline is a token register and not a macro you need to write
   \the\footline to actually extract the content from this register.

5. The \nopagenumbers macro eliminates printing page numbers:

```
1  \def\nopagenumbers{\footline = {\hfil}}
```

33.5.4   Double Page Numbers

Double page numbers like "3–5" (which might stand for Chapter 3, page 5)
can be handled in TeX by using two counter registers: for instance, use counter

register 0 for the chapter number, and counter register 1 for the page number within that chapter. Using \countdef in this case you might assign two names as follows:

```
1        \countdef\ChapterNo = 0
2        \countdef\PageNo = 1
```

Again, because of the particular choice of counter register indices used above, the following statements are true (see 34.4, p. 60, for an example):

1. Both counter registers are written to the dvi file.
2. Both counter registers are echoed on the terminal and in the log file when the output routine writes a page to the dvi file. For instance, the output, when chapter 3 is processed, will read as follows: [3.1] [3.2] [3.3].

33.5.5   Page Numbers as in "Page 3 of 12"

In legal documents, it might be desirable to print page numbers as in "page 3 of 12." Naturally, it is not until after a document has been processed that the page number of the last page is known, and therefore, there is no way to actually write pages to a dvi file until the whole document has been processed once. There are two ways to solve this problem:

1. Use a *two pass approach*. After the first pass the page number of the last page is written to a file. In the second pass the page number of the last page is known from the first pass.

   To find the page number of the last page, execute \supereject after the text is finished. \pageno now contains a page number that is one higher than the page number of the last page. This value can now be written to a file and can be used in the second path.
2. Use a *one pass approach* in which all pages are saved in box registers. Only after the last page is processed page numbers can be added to all the page box registers. Finally the pages can be written out to the dvi file.

   The obvious problem with this approach is that TeX may run out of memory trying to store all the pages in box registers. This approach will work only for short documents, whereas the first approach will always work.

33.6   Box Register 255 and the Output Routine

As mentioned before, the page-breaking algorithm will load the page to be printed (the *logical page* in this book) to box register 255. The output routine will then take the page out of this register, typically add the running head and footer, and then ship this page to the dvi file.

The output routine must empty box register 255, even if it does not use the current page (for instance, because the current page is being skipped). See 34.2, p. 58, for an example to explain this point.

33.6.1 \outputpenalty, \supereject

After TEX's page-breaking algorithm decides on the page break, it does not only load the current page into box register 255, but also stores a penalty value in counter parameter \outputpenalty. The value stored in this parameter is determined as follows:

- If the page break occurred at a penalty, this value's penalty is stored in this counter parameter.
- Otherwise the value of 10000 is stored in this parameter. Note that because a penalty of 10000 cannot occur at a break point it is possible to discriminate between a page break at a penalty and a page break at a glue.

This way the output routine has access, although to a *very* limited degree only, to the information that caused the page break. Let me now show cases where the content of \outputpenalty is actually used in the output routine:

1. The \supereject macro is a macro of the plain format that instructs TEX to print out all the remaining insertion material (this macro is usually invoked at the very end of a document or at the very end of major subdivisions of a document, such as at the end of every chapter). Here is the definition of this macro:

```
1    \def\supereject{%
2        \par                 % Force vertical mode.
3        \penalty -20000      % Force page break.
4    }
```

Note that as far as the page breaking algorithm is concerned a penalty of -20000 is equivalent to a penalty of -10000. Therefore, the \supereject will force an ordinary page break and otherwise has no special effect on the page breaking algorithm.

However, note that the value loaded into \outputpenalty will be -20000 and the output routine (at least in the case of the plain format) will now write-out all remaining insertions. This penalty value is a convention in the plain format and how the desired effect is actually achieved is discussed later. In this section we do *not* discuss what the output routine does with this information. For more details on \supereject see 35.4.7, p. 108.

2. 34.5, p. 66, shows an output routine where the penalty stored in \output-penalty is used to instruct the output routine to perform specific actions. A distinction has to be made between a page break programmed by the user (\vfill\eject), and one because of an overflow of a page.

33.6.2   Ending a Document, \end, \bye

The T_EX primitive \end instructs T_EX to terminate the processing of a document.
An error will be generated if any insertion material remains. Usually in the plain
format, the macro \bye is used defined as follows:

```
1   \outer\def\bye{%
2       \par
3       \vfill
4       \supereject
5       \end
6   }
```

The actions of \bye can be described as follows (\bye is defined using \outer
and therefore it cannot appear inside arguments of macros; see 21.6, p. III-179):

- Line 2: switch to vertical mode (if not yet in vertical mode).
- Line 3: allow the last page of a document to be short.
- Line 4: force a page break and also output all remaining insertion material
- Line 5: terminate T_EX's processing.

33.6.3   Vertical Material Generated in the Output Routine

The loading of box register 255 from the current page part of the main vertical
list normally does not remove everything from it, only most of it. For instance,
if a page break occurs in the middle of a paragraph then all lines which appear
on the next page (but not the current page) stay on the current page part of the
main vertical list, while the output routine is active.

Now observe that output routines can actually generate material themselves
(here I am not talking about material which is written to the dvi file). For
instance, if you wrote \centerline{This is fun} in a output routine *outside*
the material being written to the dvi file (using \shipout), this material will
be written to the main vertical list. This is the case in very few output routines
only; see 37.2, p. 166, for an example.

Let me now describe *where* this vertical material (generated in the output
routine) is placed with respect to the vertical material which is already on the
current page part of the main vertical list: this material is inserted *in front* of
the material that is already on the current page part of the main vertical list.

33.6.4   Shifting Pages Horizontally and Vertically, \hoffset and \voffset

Device drivers sometimes do not place their output exactly according to the
specifications previously set, with the reference point of the logical page 1 in

down and 1 in to the right of the corner of the paper (see 33.4.3, p. 46). It is often desirable to set the margins to a value different from the default values.

In this case, one can use the dimension parameter, \hoffset, to shift the whole page horizontally before the page is written to the dvi file. The page is obviously shifted to the right for positive values and to the left for negative values before writing the page to the dvi file. The default of this register is zero.

Similarly one can use \voffset to shift a page downwards for a positive dimension and upwards for negative dimensions.

These dimension registers should *not* be used by the output routine to shift left-hand and right-hand pages horizontally in double-sided documents because these registers should be reserved for the user to correct the placement of pages. See 33.4.4, p. 46.

33.6.5   Output Routines Form Implicit Groups

The execution of an output routine is performed inside an *implicit group* which is established automatically by TeX; see 19.4.10, p. III-106, for a discussion of implicit groups. This is quite useful, but there are a couple of cases where global changes must be triggered by the output routine. For example:

1. *Incrementing the page number.* It is the output routine which increments the page number in a document. Therefore \advancepageno (33.5.3, item 2, p. 50) contains \global before each \advance. See also 34.1, p. 57.
2. *Saving logical pages in box registers.* Some output routines save whole logical pages in box registers. In such cases, the saving of the logical page must be done globally. See 34.5.1, p. 72.

33.7   Page Layout and the Output Routine

The issue of page layout and output routines was already discussed in the previous chapter. Note that the page-breaking algorithm and output routine must be synchronized as far as the following two alternatives of page layout are concerned.

1. Bottom flush page layout (usually the case).
2. Ragged bottom page layout.

Occasionally you will see a message such as:

[1]   `Overfull \vbox (badness ...) has occurred while \output is active`

This message is the orthogonal counterpart to overfull lines which are reported when a paragraph is typeset (12.8.2, p. II-150), and insufficient stretchability or shrinkability is available. You need in particular to review the vertical glues on the current page part of the main vertical list or the amount of vertical glue inserted around headers, and so forth; see 32.4, p. 10, for a discussion of these glues.

33.8 Summary

In this chapter we learned:

- The page breaking algorithm and the output routine communicate via box registers (box register 255, for instance, contains the current page part of the main vertical list which should be printed on the current page) and via the \outputpenalty register.
- The merging of dvi files allows for the overlay of dvi files.
- Output routines can be tested easily using the ruled box macros presented in this series.
- An output routine is set up by assignment of the output routine code to \output.
- The output routine uses the \shipout primitive to write to a dvi file.
- The distinction between logical pages (pages as seen by the page breaking algorithm) and physical pages (pages as written to the dvi file by the output routine) was made.
- In the case of double-sided documents a distinction between left-hand and right-hand pages is necessary, because pages of such documents are usually shifted to the left or right.
- Two different page orientations, portrait and landscape mode, are known in typesetting and are supported by TeX. Unless a driver supports landscape mode, it is not possible to print a document in landscape mode.
- A page number is usually stored in counter register 0, because this way the page number is echoed on the terminal screen, as TeX processes a document and writes it to the dvi file.
- In the plain format roman page numbers are encoded as negative page numbers.
- Some documents use double page numbers, such as 3–4, which could mean, for instance, chapter 3, page 4 in that chapter.
- The \supereject macro is called to force writing out any held-over insertions.
- The \outputpenalty parameter contains the penalty associated with the current page at the most recent page break.
- The \end primitive instructs TeX to terminate processing a document. Usually the \bye macro is used which, among other things, first invokes \supereject.
- Output routines form implicit groups.
- Pages can be shifted by assigning values to \hoffset and \voffset. The output routine should not use these registers.

34
Some Simple Output Routines

To familiarize you with output routines we will discuss very simple output routines in this chapter. After two very basic output routines, I will show an output routine that surrounds each page printed by double rules. A routine with running heads that are different on left- and right-hand pages are also shown. We also discuss the distinction between logical and physical pages and show an output routine that collects multiple logical pages to print them together as one physical page.

After the discussion of the various output routines you will also find a discussion on how TeX allows for running heads through the use of marks.

34.1 Two Trivial Output Routines

Probably the most simple and still useful output routine you can have is the following:

```
1   \output = {\shipout\box255 }
```

This output routine simply takes the current logical page and writes it to the dvi file. No page numbers are printed or incremented, no running header or footer is printed. This output routine is automatically provided by TeX in case an output routine is not defined otherwise.

Now we will add code to increment the page number to the preceding output routine. It reads as follows:

```
1   \output = {
2       \shipout\box255
3       \global\advance\pageno by 1
4   }
```

Observe the use of \global when incrementing the page number because output routines form implicit groups; see 33.6.5, p. 54, for details.

34.2   An Output Routine for the Selective Printing of Pages

An output routine that allows the *selective printing* of pages of a document can be constructed based on the previous output routine. There are two ways to solve the problem of printing selected pages of a document:

1. Use `dviselect`, a TEX utility. This solution does *not* involve output routines. See 2.5, p. I-11, item 1, for details. One of the advantages of using `dviselect` is that it can be used with any output routine, and no changes to an already existing output routine are necessary.
2. Write a special output routine where pages, which are not to be printed, are *discarded by the output routine*. The rest of this Section discusses this approach.

For the following setup to work the user must initially load two counter registers, `\FirstPageNo` and `\LastPageNo`, with two page numbers. The output routine will only write those pages to the `dvi` file for which the page number falls within the range established by the above two numbers.

$$\mathcal{P} \quad \bullet \; \texttt{outpsel.tip} \; \bullet$$

Initialize the boundaries so all pages are printed by default by assigning very large page numbers.

```
15   \newcount\FirstPageNo            \FirstPageNo = -100000
16   \newcount\LastPageNo             \LastPageNo =   100000
```

The definition of the promised output routine begins here.

```
17   \output = {%
```

This output routine may discard pages and therefore it is *not* an error, if no output to the `dvi` file occurs. Therefore `\deadcycles` it set to zero each time the output routine is called. See 29.8, p. III-531, for more details on `\deadcycles`, etc.

```
18       \global\deadcycles = 0
```

If the current page's page number is too small, discard the page. In this case, note that box register 255 must be emptied; see 33.6, p. 51. Also a short message is printed.

```
19       \ifnum\pageno < \FirstPageNo
20           \setbox 255 = \box\voidbox
21           \NewLineMessage{Output routine discards page \the\pageno,
22               smaller than \string\FirstPageNo.}%
23       \else
```

Do the same if the current page number is too large.

```
24       \ifnum\pageno > \LastPageNo
25           \setbox 255 = \box\voidbox
26           \NewLineMessage{Output routine discards page \the\pageno,
```

```
27            larger than \string\LastPageNo.}%
28        \else
```

The current page number is within the specified range, so the page needs to be written to the **dvi** file. Use **\shipout** to do so.

```
29              \shipout\box255
30          \fi
31      \fi
```

Increment the page number.

```
32      \global\advance\pageno by 1
33  }
```

• End of <code>outpsel.tip</code> •

The default of the previously discussed output routine is, as already explained, to print all pages (with a "reasonable" page number). On the other hand, if you want to program this output routine to print pages 2...4, write

```
1  \FirstPageNo = 2
2  \LastPageNo = 4
```

before any output to the **dvi** file occurs.

34.3   Output Routine Generating Double Rules Around Pages

Next is an output routine where each page is surrounded by double rules. The instructions generating those double rules can be added easily to almost any output routine. The macros used to draw rules around every page are discussed in 9.3, p. I-318.

Fig. 34.1, p. 61, exhibits sample output from this output routine. Note that instead of the approach discussed here, you could have used **dvimerge** to overlay the requested rules; see 33.3.1, p. 42.

$\mathcal{P}$ • <code>ordbrule.tip</code> •

```
15  \InputD{box-mac.tip}              % 9.3.14, p. I-343.
16  \InputD{box-larg.tip}            % 8.3, p. I-298.
```

Set-up the page dimensions.

```
17  \hsize = 2.9in
18  \vsize = 6.0in
```

Set-up the output routine.

```
19  \output = {
20      \shipout\vbox{
21          \HboxR{%
22              \BoxLarger
```

```
23                    {\HboxR
24                      {\BoxLarger
25                         {\vbox to \vsize{\unvbox 255}}%
26                         {10pt}%
27                      }%
28                    }%
29                    {2pt}%
30             }%
31          }
32    }
```

• End of <code>ordbrule.tip</code> •

Here is an example using the preceding output routine (the output appears in Fig. 34.1 on the next page).

• <code>ex-out-double.tip</code> •

```
1    \input inputd.tip
2    \InputD{ordbrule.tip}                    % 34.3, p. 59.
```

Some text must be generated next.

```
3    \def\xx{%
4       This is a test. This is a test, and whether you like it or
5       not, this is {\it still\/} a test.
6    }
7    \def\yy{\xx\xx\xx\xx\par}
8    \yy\yy\yy\yy
9
10   \bye
```

• End of <code>ex-out-double.tip</code> •

34.4   An Output Routine with Running Heads

In the following example, we see another output routine with the following properties:

1. A running head is printed containing the current date, time, and the name of the TeX job (note that marks are necessary for more general running heads; see 34.7, p. 81).
2. A running foot is printed with *double* page numbers. For instance, page number 3–4 stands for page 4 of chapter 3.
3. *Crop marks* are printed on each page. Crop marks are reference points (outside the text area of a page) on the film produced by a phototypesetter
4. This output routine makes the distinction between left-hand and right-hand side pages (the assumption is that the document is printed double-sided). Pages are shifted horizontally: left pages (even page numbers) are shifted

This is a test. This is a test, and whether you like it or not, this is *still* a test. This is a test. This is a test, and whether you like it or not, this is *still* a test. This is a test. This is a test, and whether you like it or not, this is *still* a test. This is a test. This is a test, and whether you like it or not, this is *still* a test.

This is a test. This is a test, and whether you like it or not, this is *still* a test. This is a test. This is a test, and whether you like it or not, this is *still* a test. This is a test. This is a test, and whether you like it or not, this is *still* a test. This is a test. This is a test, and whether you like it or not, this is *still* a test.

This is a test. This is a test, and whether you like it or not, this is *still* a test. This is a test. This is a test, and whether you like it or not, this is *still* a test. This is a test. This is a test, and whether you like it or not, this is *still* a test. This is a test. This is a test, and whether you like it or not, this is *still* a test.

This is a test. This is a test, and whether you like it or not, this is *still* a test. This is a test. This is a test, and whether you like it or not, this is *still* a test. This is a test, and whether you like it or not, this is *still* a test. This is a test. This is a test, and whether you like it or not, this is *still* a test.

Figure 34.1. Double rule output routine, sample page.

to the right (away from the middle), and similarly right pages (odd page numbers) are shifted to the left (again away from the middle).

The output routine and an example of its application are presented as follows:

1. The file **orsimple.tip** (34.4.1 on this page) contains the output routine itself and the setup of some of the parameters, as, for instance, the amount by which odd and even pages are shifted horizontally.
2. The file **outsimple-ex.tip** contains the input by which the sample pages were generated. This file is reprinted in 34.4.2, p. 65.

    Observe that there are two choices of output routines that can be established by calling either **\OutputWithRules** or **\OutputWithoutRules**. If **\OutputWithRules** is used, macros **\HboxR**, **\VboxR**, and **\VtopR** are invoked to mark the placement of headers, footers, etc. If **\OutputWithoutRules** is called, this marking does not take place.
3. Figures 34.2–34.3, pp. 67–68, contain two pages of output generated with this output routine.

34.4.1   The Code of the Output Routine

Here is the source code of the output routine whose properties were just discussed:

$\mathcal{P}$ • orsimple.tip •

```
15   \InputD{box-mac.tip}                    % 9.3.14, p. I-343.
16   \InputD{pmtime.tip}                     % 3.3.8.2, p. I-51.
17   \InputD{graphmac.tip}                   % 9.2.1, p. I-314.
```

\LeftPageShift is a dimension register that holds the horizontal shift of left pages (even page numbers), **\RightPageShift** holds the horizontal shift of right pages (odd page numbers).

```
18   \newdimen\LeftPageShift      \LeftPageShift = -10pt
19   \newdimen\RightPageShift     \RightPageShift= -40pt
```

The macro **\HeadLine** prints a running header line. The current date and time, and the current job's name are printed.

```
20   \def\HeadLine{%
```

Here the differentiation between odd and even page numbers takes place.

```
21       \ifodd\PageNo
```

Odd page numbers (right-hand side pages): the date appears on the left side and the job name on the right side.

```
22           \HboxR to \hsize{%
23               \strut
24               \the\month/\the\day/\the\year, \PrintMilTime
```

```
25              \hfil
26              {\tt\jobname.tip}%
27          }%
28      \else
```

On left-hand pages it is just the opposite.

```
29          \HboxR to \hsize{%
30              \strut
31              {\tt\jobname.tip}%
32              \hfil
33              \the\month/\the\day/\the\year, \PrintMilTime
34          }%
35      \fi
36  }
```

Page numbering is set up as follows: \count0 contains the chapter number, \count1 the page number within each chapter.

```
37  \countdef\ChapNo = 0
38  \countdef\PageNo = 1
```

The macro \FootLine prints a running foot line. As mentioned before, the page number appears to the left for even numbered page and to the right for odd numbered pages.

```
39  \def\FootLine{%
40      \ifodd\PageNo
41          \HboxR to \hsize{%
42              \strut
43              \hfil
44              \bf \the\ChapNo--\the\PageNo
45          }%
46      \else
47          \HboxR to \hsize{%
48              \strut
49              \bf
50              \the\ChapNo--\the\PageNo
51              \hfil
52          }%
53      \fi
54  }
```

Box register \CropMarksBox is loaded with a box containing crop marks.

```
55  \newbox\CropMarksBox
```

Now compute the crop marks box. After the box has been built, its height, depth and width are set to zero. This way, when the box is printed, TeX's reference point does not move.

```
56  \setbox\CropMarksBox = \vbox{%
57      \offinterlineskip
```

The following positions of crop marks are in inches.

```
58      \SetScale{1in}
```

The thickness of crop mark lines is 0.5 pt.

```
59        \SetLineThickness{0.5pt}
```

The following crop mark positions may have to be adjusted by you. Print the upper left cross.

```
60        \DrawHLine(-0.8,0){0.2}
61        \DrawVLine(-0.7,-0.1){0.2}
```

Next print the upper right cross.

```
62        \DrawHLine(3.4,0){0.2}
63        \DrawVLine(3.5,-0.1){0.2}
```

Now print the lower left cross.

```
64        \DrawHLine(-0.8,-6.5){0.2}
65        \DrawVLine(-0.7,-6.6){0.2}
```

Finally print the lower right cross.

```
66        \DrawHLine(3.4,-6.5){0.2}
67        \DrawVLine(3.5,-6.6){0.2}
68  }
```

Set all the dimensions of the crop mark box to zero.

```
69  \ZeroBox{\CropMarksBox}
```

The definition of macro `\SimpleOutputRoutine` starts here (no output routine is set up at this point).

```
70  \def\SimpleOutputRoutine{%
```

Load `\dimen0` with the horizontal shift for the current page.

```
71        \ifodd\PageNo
72            \dimen0 = \RightPageShift
73        \else
74            \dimen0 = \LeftPageShift
75        \fi
```

Build the page and ship it out. First include crop marks. The current reference point is not moved because all dimensions of the crop mark box are zero.

```
76        \shipout\vbox{%
77            \offinterlineskip
78            \copy\CropMarksBox
```

Now print the page itself. The `\vskips` below control the spacing between running head and the text page, and the text page and the running foot. A bottom flush page layout is chosen.

```
79            \moveright\dimen0 \vbox{%
80                \HeadLine
81                \vskip 12pt
82                \VboxR to \vsize{%
83                    \unvbox 255
84                }
```

```
85                    \vskip 12pt
86                    \FootLine
87              }
88         }
```

Increment the page number. This must be done globally because of the implicit grouping in an output routine.

```
89         \global\advance\PageNo by 1
90   }
```

Now define the macro \OutputWithRules (no parameters). This macro sets-up the output routine just defined. Rules around certain items of the current page will be printed.

```
91   \def\OutputWithRules{%
92       \output = {\SimpleOutputRoutine}%
93   }
```

The macro \OutputWithoutRules sets-up the output routine so that no rules are printed around any of the entities, which are printed by the output routine. This is achieved by calling the \EliminateRuledBoxes macro before the macro \SimpleOutputRoutine is executed.

```
94   \def\OutputWithoutRules{
95       \output = {%
96           \EliminateRuledBoxes
97           \SimpleOutputRoutine
98       }
99   }
```

The default is the version of the output routine *without* rules.

```
100   \OutputWithoutRules
```

• End of orsimple.tip •

34.4.2   An Example Application of the Preceding Output Routine

Let me present now an example using the preceding output routine.

• ex-outsimple.tip •

```
1    \input inputd.tip
2    \InputD{orsimple.tip}                    % 34.4.1, p. 62
```

Initialize page numbering: start at chapter 3, page 1.

```
3    \ChapNo = 3
4    \PageNo = 1
```

Use the output routine with rules.

```
5    \OutputWithRules
```

Set-up the page layout.

```
 6   \hsize = 3.5in
 7   \vsize = 5.8in
 8   \baselineskip = 12pt
 9   \parskip = 10pt plus 2pt minus 1pt
10   \def\strut{\vrule height 8pt depth 4pt width 0pt}
11   \topskip = 10pt
```

Define the macros \spar and \xpar to generate sample paragraphs.

```
12   \def\spar{%
13       This is a sample paragraph. We just need some text to fill the
14       whole page. So keep your fingers crossed.
15   }
16   \def\xpar{%
17       \spar\spar\spar\spar\spar
18       \par
19   }
```

Generate paragraphs.

```
20   \xpar\xpar\xpar\xpar\xpar\xpar
21   \bye
```

• End of <code>ex-outsimple.tip</code> •

The output from the preceding example appears in Figs. 34.2–34.4, pp. 67–69.

34.5   Output Routine to Print Library Cards

I now show an output routine to print library cards. This is the first output routine where the distinction between logical and physical pages is important, because logical and physical pages are now different.

For this output routine, I assume that a library card is either one or two "logical pages" long which refers to either the front side, or the front and backside of a card. Because library cards are usually rather small, I print three "logical pages" as one physical page to save paper (it is assumed that the user later cuts the physical pages appropriately).

Some of the properties of this output routine are:

1. *Three logical* pages vertically stacked on *one physical* page are printed together.
2. The page layout is ragged bottom.
3. In case a library card extends beyond one logical page, it is necessary to indicate this at the bottom of the first logical page.

This output routine is presented as follows:

5/20/1993, 02:06 . . . . ex-outsimple.tip

This is a sample paragraph. We just need some text to fill the whole page. So keep your fingers crossed. This is a sample paragraph. We just need some text to fill the whole page. So keep your fingers crossed. This is a sample paragraph. We just need some text to fill the whole page. So keep your fingers crossed. This is a sample paragraph. We just need some text to fill the whole page. So keep your fingers crossed. This is a sample paragraph. We just need some text to fill the whole page. So keep your fingers crossed. This is a sample paragraph. We just need some text to fill the whole page. So keep your fingers crossed.

This is a sample paragraph. We just need some text to fill the whole page. So keep your fingers crossed. This is a sample paragraph. We just need some text to fill the whole page. So keep your fingers crossed. This is a sample paragraph. We just need some text to fill the whole page. So keep your fingers crossed. This is a sample paragraph. We just need some text to fill the whole page. So keep your fingers crossed. This is a sample paragraph. We just need some text to fill the whole page. So keep your fingers crossed. This is a sample paragraph. We just need some text to fill the whole page. So keep your fingers crossed.

This is a sample paragraph. We just need some text to fill the whole page. So keep your fingers crossed. This is a sample paragraph. We just need some text to fill the whole page. So keep your fingers crossed. This is a sample paragraph. We just need some text to fill the whole page. So keep your fingers crossed. This is a sample paragraph. We just need some text to fill the whole page. So keep your fingers crossed. This is a sample paragraph. We just need some text to fill the whole page. So keep your fingers crossed. This is a sample paragraph. We just need some text to fill the whole page. So keep your fingers crossed.

This is a sample paragraph. We just need some text to fill the whole page. So keep your fingers crossed. This

. . . . 3–1

Figure 34.2. Simple output routine example, a right-hand page.

ex-outsimple.tip.........................5/20/1993, 02:06

is a sample paragraph. We just need some text to fill the whole page. So keep your fingers crossed. This is a sample paragraph. We just need some text to fill the whole page. So keep your fingers crossed. This is a sample paragraph. We just need some text to fill the whole page. So keep your fingers crossed. This is a sample paragraph. We just need some text to fill the whole page. So keep your fingers crossed. This is a sample paragraph. We just need some text to fill the whole page. So keep your fingers crossed.

This is a sample paragraph. We just need some text to fill the whole page. So keep your fingers crossed. This is a sample paragraph. We just need some text to fill the whole page. So keep your fingers crossed. This is a sample paragraph. We just need some text to fill the whole page. So keep your fingers crossed. This is a sample paragraph. We just need some text to fill the whole page. So keep your fingers crossed. This is a sample paragraph. We just need some text to fill the whole page. So keep your fingers crossed. This is a sample paragraph. We just need some text to fill the whole page. So keep your fingers crossed.

This is a sample paragraph. We just need some text to fill the whole page. So keep your fingers crossed. This is a sample paragraph. We just need some text to fill the whole page. So keep your fingers crossed. This is a sample paragraph. We just need some text to fill the whole page. So keep your fingers crossed. This is a sample paragraph. We just need some text to fill the whole page. So keep your fingers crossed. This is a sample paragraph. We just need some text to fill the whole page. So keep your fingers crossed.

3-2

Figure 34.3. Simple output routine example, a left-hand page.

This is a sample paragraph. We just need some text to fill the whole page. So keep your fingers crossed. This is a sample paragraph. We just need some text to fill the whole page. So keep your fingers crossed. This is a sample paragraph. We just need some text to fill the whole page. So keep your fingers crossed. This is a sample paragraph. We just need some text to fill the whole page. So keep your fingers crossed. This is a sample paragraph. We just need some text to fill the whole page. So keep your fingers crossed.

This is a sample paragraph. We just need some text to fill the whole page. So keep your fingers crossed. This is a sample paragraph. We just need some text to fill the whole page. So keep your fingers crossed. This is a sample paragraph. We just need some text to fill the whole page. So keep your fingers crossed. This is a sample paragraph. We just need some text to fill the whole page. So keep your fingers crossed. This is a sample paragraph. We just need some text to fill the whole page. So keep your fingers crossed.

This is a sample paragraph. We just need some text to fill the whole page. So keep your fingers crossed. This is a sample paragraph. We just need some text to fill the whole page. So keep your fingers crossed. This is a sample paragraph. We just need some text to fill the whole page. So keep your fingers crossed. This is a sample paragraph. We just need some text to fill the whole page. So keep your fingers crossed. This is a sample paragraph. We just need some text to fill the whole page. So keep your fingers crossed.

This is a sample paragraph. We just need some text to fill the whole page. So keep your fingers crossed. This

Figure 34.4. Simple output routine example without ruled boxes.

1. The file **orcards.tip** contains some initial settings of some of TeX's parameters and the output routine itself (see 34.5.1 on this page).
2. The file **ex-outcards.tip** contains some sample source code. It can be found in 34.5.2, p. 73.
3. The output generated by those samples can be found in Figs. 34.5–34.7, pp. 77–79.
4. The log file generated by this output is reprinted in 34.5.3, p. 74.

34.5.1   The Code of the Output Routine

Here is the source code of this output routine:

$\mathcal{P}$ • orcards.tip •

```
15    \InputD{box-mac.tip}              % 9.3.14, p. I-343.
16    \InputD{modonead.tip}             % 3.3.9, p. I-52.
17    \InputD{nlm.tip}                  % 29.5.4, p. III-523.
```

Set-up page layout dimensions. A ragged bottom page layout was chosen.

```
18    \hsize = 3.0in
19    \baselineskip = 12pt
20    \topskip = 10pt plus 10pt
21    \vsize = 9\baselineskip
```

The page number is stored in \count0, as usual.

```
22    \countdef\PageNo = 0
```

Count all cards using the following counter.

```
23    \newcount\CardCount
24    \CardCount = 0
```

Define the macro \Card to generate the beginning of a card. This macro has one parameter, #1, which is the title of the card.

```
25    \def\Card #1{%
```

Terminate the preceding card.

```
26        \vfill\eject
```

Increment the card counter.

```
27        \advance\CardCount by 1
```

Print the card's beginning.

```
28        \noindent
29        {\bf #1}
30        \par
```

Save the card's name.

```
31        \def\CardTitle{#1}%
32    }
```

The output routine for printing library cards follows. Horizontal rules before and after each card are drawn. The dimension register, \RuleSpace, defines how much space is inserted before and after the text. The dimension stored in \BetweenRules defines how much vertical space is inserted between two consecutive rules.

```
33    \newdimen\RuleSpace        \RuleSpace = 10pt
34    \newdimen\BetweenRules  \BetweenRules = 20pt
```

Save the value for \baselineskip in dimension register \BaseLineSkipSave.

```
35    \newdimen\BaseLineSkipSave
36    \BaseLineSkipSave = \baselineskip
```

The following counter identifies the position of a logical page on a physical page. 0 is for the top logical page, 1 is for the middle logical page, and 2 is for the bottom logical page.

```
37    \newcount\PositionCount
38    \PositionCount = 0
```

The macro \HeadLine prints the running head line on each physical page. It has no parameters.

```
39    \def\HeadLine{%
40        \line{%
41            \strut
42            {\it Cards, page \the\PageNo}%
43            \hfil
44            \tt\jobname
45        }%
46    }
```

The macro \FootLine prints the footer line for each physical page.

```
47    \def\FootLine{%
48        \line{\strut*****\hfil*****}%
49    }
```

Declare three box registers to collect three logical pages.

```
50    \newbox\OutTopBox
51    \newbox\OutMidBox
52    \newbox\OutBotBox
```

\CardOutputRoutine, the output routine itself, begins here.

```
53    \def\CardOutputRoutine{%
```

Print an initial message each time the output routine is invoked.

```
54        \NewLineMessage{\string\CardOutputRoutine:
55            card number: \the\CardCount,
56            PositionCounter: \the\PositionCount,}%
57        \NewLineMessage{page number: \the\PageNo,
58            \string\outputpenalty: \the\outputpenalty.}%
```

Print the card title and "con't..." in the lower right-hand corner of a card, in case the text of this card continues on the next page. You must use a strut because otherwise the vertical spacing depends on the height and depth of the line generated now.

Make a distinction between an `\eject` and a `\supereject` on one side (condition below is true), and a penalty of > -10000 on the other side (condition below is false).

```
59          \ifnum\outputpenalty < -9999
60              \setbox0 = \hbox{\strut}
61          \else
62              \setbox0 = \line{%
63                  \strut
64                  \hfil
65                  \it
66                  \CardTitle\space con't\/\dots
67              }%
68          \fi
```

Save the individual logical pages. Remember that a ragged bottom page layout is used. Horizontal lines are added, before the three logical cards are printed out together to separate the three logical pages visually when printed.

```
69          \setbox 2 = \vbox{%
70              \offinterlineskip
71              \hrule
72              \vskip\RuleSpace
73              \VboxR to \vsize{\unvbox 255 \vfill}
74              \box0
75              \vskip\RuleSpace
76              \hrule
77          }
```

To save the top, middle, or bottom logical page, `\global` must be used because output routines form implicit groups.

```
78          \global\setbox
79              \ifcase\PositionCount
80                  \OutTopBox \or
81                  \OutMidBox \or
82                  \OutBotBox
83              \fi
84              = \box2
```

The following code is executed, when the last logical page of a physical page was just stored away. The complete physical page must now be shipped to the dvi file.

```
85          \ifnum\PositionCount = 2
86              \shipout\vbox{%
87                  \offinterlineskip
88                  \VboxR{
89                      \HeadLine
90                      \vskip\BaseLineSkipSave
91
```

```
 92                    \box\OutTopBox
 93                    \vskip\BetweenRules
 94                    \box\OutMidBox
 95                    \vskip\BetweenRules
 96                    \box\OutBotBox
 97
 98                    \vskip\BaseLineSkipSave
 99                    \FootLine
100              }
101           }
102        \fi
```

Was there a \supereject?

```
103        \ifnum\outputpenalty = -20000
104            \NewLineMessage{\string\supereject\space encountered.}
```

Yes, there was a \supereject. If one or two cards are left over, generate a blank card (a blank logical page) and restart the whole thing.

```
105        \ifnum\PositionCount < 2
106            \line{}
107            \vfill
108            \supereject
109        \fi
110     \fi
```

Increment by 1 (mod 3) the logical page position counter.

```
111        \ifnum\ModuloOneAdvanceNumCond{\PositionCount}{3} = 0
112            \global\advance\pageno by 1
113     \fi
114  }
```

Make the output routine that was just defined the current one.

```
115  \output = {\CardOutputRoutine}
```

● End of orcards.tip ●

34.5.2   The Source Code of the Example Application

Here is an example using the preceding output routine.

● ex-outcards.tip ●

```
 1   \input inputd.tip
 2   \InputD{orcards.tip}                    % 34.5.1, p. 70.
 3   \interlinepenalty = 34
```

Define two macros to generate library cards efficiently. \ShortCard generates a library card using one logical page, \LongCard generates a library card with two logical pages. Both macros have one parameter, #1, the title of a card.

```
 4   \def\ShortCard #1-%
```

```
 5      \Card{#1}%
 6      This is the example of a short card. The card's title is
 7      ''#1.'' We just print some sample text so we can show
 8      that our idea really works.
 9  }
10
11  \def\LongCard #1{%
12      \Card{#1}%
13      This is the example of a long card. The card's title is
14      ''#1.'' We just print some sample text so we can show
15      that our idea really works. But we need more text to fill
16      two logical pages or two cards.
17
18      This is fill text for a long card.
19      This is fill text for a long card.
20      This is fill text for a long card.
21      This is fill text for a long card.
22      This is fill text for a long card.
23      This is fill text for a long card.
24      This is fill text for a long card.
25      This is fill text for a long card.
26      This is fill text for a long card.
27      This is fill text for a long card.
28      This is fill text for a long card.
29  }
```

Finally some sample cards are generated.

```
30  \ShortCard{Alpha}
31  \ShortCard{Bravo}
32  \LongCard {Charlie}
33  \LongCard {Delta}
34  \ShortCard{Echo}
35  \bye
```

• End of <code>ex-outcards.tip</code> •

34.5.3   The Log File

This log file was generated when the example code of the previous Subsection
was executed.

• <code>ex-outcards.log</code> •

```
1  This is TeX, C Version 3.14 (...)
2  **&/usr/local/tex/lib/fmt/plain ex-outcards.tip
3  (ex-outcards.tip (inputd.tip
4  (namedef.tip
5  ) (inputdl.tip
6  \@InputDStream=\write0
7  ))
```

```
 8   (orcards.tip
 9   (box-mac.tip
10   (boxing7.tip
11   (boxing6.tip
12
13   (box-zero.tip
14   ) (boxing5.tip
15   (box-bul.tip
16   \@BulletBox=\box16
17   ) (box-bb.tip
18   \@BoxingBox=\box17
19   \@BaseLineLeaders=\box18
20   )
21   (box-thck.tip
22   \BoxRuleThickness=\dimen16
23   ))) (vcentx.tip
24   \@VcenterXBox=\box19
25   \@VcenterDimen=\dimen17
26   )
27   \@BoxRDimen=\dimen18
28   \@BoxRBox=\box20
29   \@BoxRNumber=\count26
30   ) (box-larg.tip
31   ) (emptybox.tip
32   ) (emprubox.tip
33   \@EmptyRuledBox=\box21
34   ) (boxrelim.tip
35   ) (linesr.tip
36   )) (modonead.tip
37   )
38   (nlm.tip
39   )
40   \CardCount=\count27
41   \RuleSpace=\dimen19
42   \BetweenRules=\dimen20
43   \BaseLineSkipSave=\dimen21
44   \PositionCount=\count28
45   \OutTopBox=\box22
46   \OutMidBox=\box23
47   \OutBotBox=\box24
48   )
49
50   \CardOutputRoutine: card number: 1, PositionCounter: 0,
51
52   page number: 1, \outputpenalty: -10000.
53
54   \CardOutputRoutine: card number: 2, PositionCounter: 1,
55
56   page number: 1, \outputpenalty: -10000.
57
58   \CardOutputRoutine: card number: 3, PositionCounter: 2,
59
```

```
page number: 1, \outputpenalty: 34. [1]

\CardOutputRoutine: card number: 3, PositionCounter: 0,

page number: 2, \outputpenalty: -10000.

\CardOutputRoutine: card number: 4, PositionCounter: 1,

page number: 2, \outputpenalty: 34.

\CardOutputRoutine: card number: 4, PositionCounter: 2,

page number: 2, \outputpenalty: -10000. [2]

\CardOutputRoutine: card number: 5, PositionCounter: 0,

page number: 3, \outputpenalty: -20000.

\supereject encountered.

\CardOutputRoutine: card number: 5, PositionCounter: 1,

page number: 3, \outputpenalty: -20000.

\supereject encountered.

\CardOutputRoutine: card number: 5, PositionCounter: 2,

page number: 3, \outputpenalty: -20000. [3]

\supereject encountered. )
Output written on ex-outcards.dvi (3 pages, 16548 bytes).
```

34.6   Specific Positioning Printing

Let us now examine the following problem. Assume the TeX source file you need
to print consists *only* of macro calls to the macro \PrintAtPosition where this
macro has the following four parameters:

- #1. The horizontal position (a dimension) of some text to print. This position
  is measured with respect to the zero point of TeX, (the zero point's print
  position is 1 in to the right and 1 in down from the upper left corner of the
  paper on which a document is printed.)
- #2. The vertical position (a dimension) of that text to print.
- #3. The text to print at the specified position.

Cards, page 1   `ex-outcards`

Alpha

This is the example of a short card. The card's title is "Alpha." We just print some sample text so we can show that our idea really works.

Bravo

This is the example of a short card. The card's title is "Bravo." We just print some sample text so we can show that our idea really works.

Charlie

This is the example of a long card. The card's title is "Charlie." We just print some sample text so we can show that our idea really works. But we need more text to fill two logical pages or two cards.

This is fill text for a long card. This is fill text for a long card. This is fill text for a long card. This is fill text for a long card. This is fill text for

Charlie con't...

`*****`   `*****`

Figure 34.5. Card example, page 1.

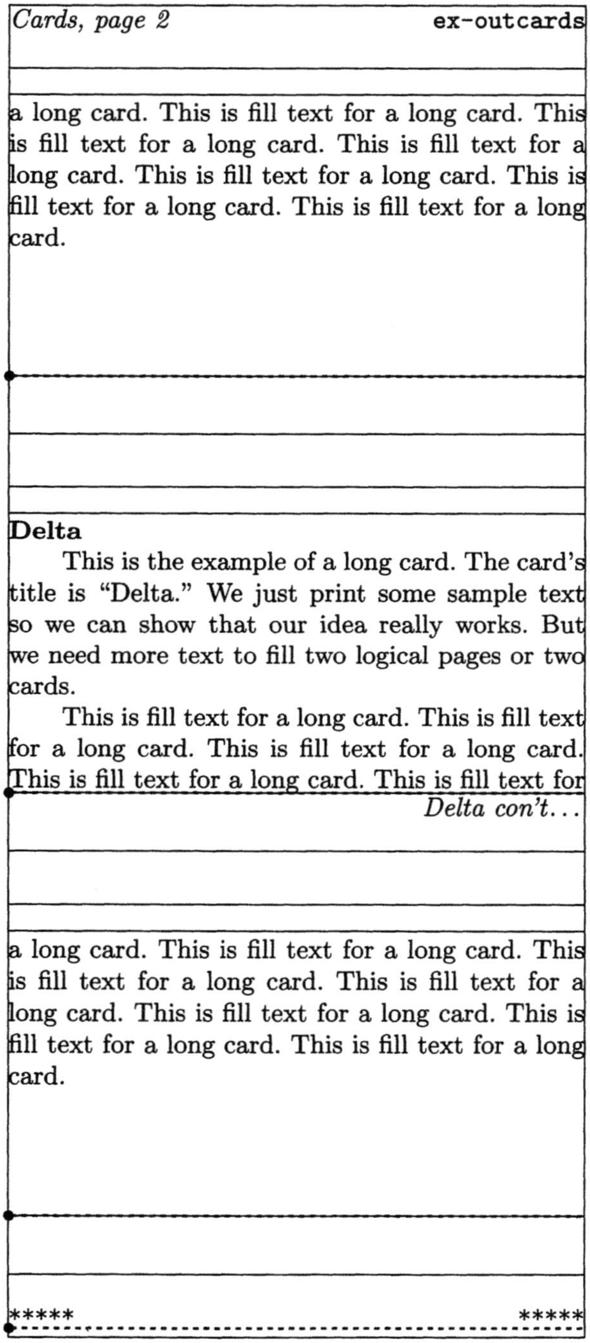

<pre>
 Cards, page 2   ex-outcards

 a long card. This is fill text for a long card. This
 is fill text for a long card. This is fill text for a
 long card. This is fill text for a long card. This is
 fill text for a long card. This is fill text for a long
 card.

 Delta
      This is the example of a long card. The card's
 title is "Delta." We just print some sample text
 so we can show that our idea really works. But
 we need more text to fill two logical pages or two
 cards.
      This is fill text for a long card. This is fill text
 for a long card. This is fill text for a long card.
 This is fill text for a long card. This is fill text for
                                    Delta con't...

 a long card. This is fill text for a long card. This
 is fill text for a long card. This is fill text for a
 long card. This is fill text for a long card. This is
 fill text for a long card. This is fill text for a long
 card.

 *****   *****
</pre>

Figure 34.6. Card example, page 2.

Cards, page 3   ex-outcards

Echo
    This is the example of a short card. The card's
title is "Echo." We just print some sample text so
we can show that our idea really works.

*****   *****

Figure 34.7. Card example, page 3.

- #4. A dimension register that is loaded as the result of the execution of this macro. This register contains the vertical position of the *bottom* of the box which was used by the macro to print #3.

Note that the position specified by #1 and #2 defines the *absolute* coordinates of text #3 to print. Usually there is *more than one call* to this macro per page.

1. The `\topskip` glue must be disabled, because it is undesirable to shift the current page down by the amount of `\topskip`.
2. The output must be printed in such a way that the reference point, initially (0, 0), does *not* move after a `\PrintAtPosition` call. This is achieved by using the macro `\ZeroBoxOut`.

First we will define macro `\PrintAtPosition`:

$$\mathcal{P}' \quad \bullet \; \texttt{atpos.tip} \; \bullet$$

```
15   \InputD{box-zero.tip}                        % 4.5.13, p. I-104.
16   \catcode`\@ = 11
```

The following box register is used to hold the text to be printed.

```
17   \newbox\@PrintAtPositionBox
```

The definition of macro `\PrintAtPosition` starts here.

```
18   \def\PrintAtPosition #1#2#3#4{%
19       \setbox\@PrintAtPositionBox = \hbox{%
```

Now the positioning takes place.

```
20           \hskip #1\relax
21           \lower #2\hbox{%
22               #3%
23           }%
24       }%
```

Contribute the box just generated to the current page without moving the reference point.

```
25       #4 = \dp\@PrintAtPositionBox
26       \ZeroBoxOut{\@PrintAtPositionBox}%
27   }
28   \catcode`\@ = 12
```

To use the preceding macro, you *must* first call `\SetUpPrintAtPosition` which will set up everything properly. This macro has no parameters. This macro turns off interline glue (`\offinterlineskip`), because otherwise TeX would insert interline glue after *each* call of `\PrintAtPosition`. Each item would then be regarded as a separate entity (in this case a zero dimension hbox). Therefore, without `\offinterlineskip`, the first text would be moved down by `\baselineskip`, the second text (on the same page) would be printed further down twice that amount, and so forth.

Note that this macro works with almost any output routine, particularly with the output routine of the plain format.

```
29    \def\SetUpPrintAtPosition{%
30        \nopagenumbers
31        \topskip = 0pt
32        \offinterlineskip
33    }
```

• End of <code>atpos.tip</code> •

The macros of this Section can be used in a variety of applications: I use this output routine in a `dvi` file processor that extracts positioning information from a `dvi` file, in order to place side notes in the margins of a document. Those side notes are actually processed in a *separate* TeX run, and they are placed in this separate run with the help of the preceding macro. The resulting `dvi` file is then overlaid with the `dvi` file of the main document to which these side notes belong. For further details see Bechtolsheim (1990a).

34.7   Marks

Let me now define *marks* in TeX. The purpose of marks in TeX is to generate the text in running heads, where the tex in these running head s based on the titles of chapters, headings of sections, and so forth. I will first explain the mark mechanism of TeX, before I show how the marking mechanism of TeX is applied to typeset running heads in this series.

Marks are in a certain sense similar to `\specials` in the sense that the associated information is *not* typeset, but is accessible in a different way. In the case of marks this information is not written to the `dvi` file but accessible using `\botmark`, `\topmark` and `\firstmark` (see below for details).

34.7.1   Generation of a Mark Using `\mark`

The primitive `\mark` is used to inform TeX about the text of a mark. For instance, `\mark{This is fun}` records the given text as a mark. What such a `\mark` is for will be discussed shortly. Notice that the text of a `\mark` is expanded right when the `\mark` is called, *not* when the text is extracted from `\topmark`, `\firstmark` or `\botmark` (see below).

34.7.2   Accessing Marks by `\topmark`, `\firstmark`, and `\botmark`

Marks can be accessed from inside the output routine and after a `\vsplit` operation. The application from inside the output routine is the most common one and will be discussed next. Further below marks and `\vsplit` are discussed.

When the output routine is invoked the following three macro-like control sequences are set up by TeX (you can think of these three control sequences as three parameterless macros defined by TeX):

1. \botmark. This control sequence expands to the text of the *most recent* mark encountered on the page that was just boxed.
2. \topmark. The value of \botmark of the preceding page.
3. \firstmark. The text of the *first* mark of the current page.

The following additional rules apply:

1. Initially, before any \mark command is issued, all three sequences \topmark, \firstmark and \botmark are set empty and therefore expand to nothing.
2. If there is no mark text encountered on the current page, all marks expand to \botmark of the preceding page.
3. All three values \topmark, \firstmark and \botmark are set on a global basis. They are not affected by TeX's grouping mechanism.
4. If you want to inspect any of the above \...marks, then use \show; for instance, \show\topmark will show you the current \topmark.

34.7.3   A Short Example

The following example shows how the various \...mark sequences evaluate. The assumption is that the following \mark commands were issued:

- Page 1: none.
- Page 2: \mark{AA}.
- Page 3: none.
- Page 4: \mark{BB} and \mark{CC}.
- Page 5: \mark{DD}.
- Page 6: none.

Thus the various \...mark sequences evaluate as shown in the following table:

Page number	\topmark	\firstmark	\botmark
1	null	null	null
2	null	AA	AA
3	AA	AA	AA
4	AA	BB	CC
5	CC	DD	DD
6	DD	DD	DD

34.7.3.1   Marks in \vsplit (\splitfirstmark, \splitbotmark)

Marks generated by \mark in a vbox, where this vbox is split using \vsplit, are handled similarly: use \splitfirstmark or \splitbotmark instead of \firstmark or \botmark to access those marks. Note that there is no "\splittopmark," because TeX has no notion of "the most recent \vsplit."

34.7.4   Using Marks for Printing Running Heads in Dictionaries

The application of marks is, as already indicated, the generation of running heads, as they are used in many types of documents. The generation of running heads for dictionaries is the easiest case, so it will be discussed here.

34.7.4.1   The Rules of the Game

First we need to establish the rules about running heads in *dictionaries* (see the Chicago Manual of Style, Chicago (1982), for details):

- For left-hand pages (even page numbers) use the *first* term that has its explanation actually beginning on that page. Discard therefore a term for which the explanation started on the preceding page but continued onto the current page.
- For right-hand pages (odd page numbers) use the last term of which the explanation began on that page (whether the explanation continues onto the next page or not is irrelevant).

34.7.4.2   A Macro for Mark Generation in Dictionaries

Next is the definition of a macro \DicEntry for the generation of \mark commands in a dictionary. To develop this macro (in connection with the rules about the contents of running heads just mentioned), let's assume the following scenario:

- Near the bottom of page 1 begins the explanation of term α, which continues onto page 2.
- After the explanation of α ended on page 2, term β is explained on page 2. This is therefore the first term that has its explanation begin on page 2.
- On page 3, towards the very end of this page, the explanation of term λ starts, the last term of page 3. It is irrelevant whether the explanation of λ continues onto page 4 or ends precisely on the bottom of page 4. The important assumption is that λ is the last term of page 3.

Now assume that a macro `\DicEntry` with one parameter, #1, the entry into the dictionary, is defined as follows (a call to this macro would be followed by the "explanation" of the dictionary entry):

$\mathcal{P}'$ • `dicentry.tip` •

```
15   \def\DicEntry #1{%
```

Finish preceding text.

```
16       \par
```

Use hanging indentation so that the dictionary term "sticks out."

```
17       \hangafter = 1
18       \hangindent = 5pt
19       \noindent
20       {\bf #1}%
```

Generate the mark here. The marks become part of the line containing the dictionary term.

```
21       \mark{#1}%
```

Space between the dictionary term and the following explaining paragraph is inserted here.

```
22       \hskip 1em plus .2em minus .2em
23       \ignorespaces
24   }
```

• End of `dicentry.tip` •

If you now recall the explanation of `\firstmark` and `\botmark` given before, then you will see that those two control sequences can be used to generate the running heads of left-hand and right-hand pages respectively. The next Subsubsection shows a modification of the plain format output routine with the running heads set up accordingly.

34.7.4.3   A "Dictionary Running Head"

The following modification of the plain format output routine includes running heads.

$\mathcal{P}$ • `rh-dict.tip` •

```
15   \InputD{box-mac.tip}                    % 9.3.14, p. I-343.
```

In the example here we generate small pages, where four small pages are printed on one page of this series. Typeset the text ragged right (this makes life easier, because of the small page width.)

```
16   \raggedright
17   \hsize = 11pc
18   \vsize = 19pc
19   \parskip = 3pt plus 2pt minus 2pt
```

Now set up the running head. This modifies the plain format's `\headline` macro.

```
20    \headline = {%
21        \ifodd\pageno
```

Odd pages (right hand page): use `\botmark`.

```
22            \RightlineR{\it\botmark}%
23        \else
```

Even pages: use `\firstmark`.

```
24            \LeftlineR{\it\firstmark}%
25        \fi
26    }
27    \EliminateRuledBoxes
```

• End of <code>rh-dict.tip</code> •

34.7.4.4   The Example Source Code

Here is the source code of a complete example using the previously presented
code.

• <code>ex-dicentry.tip</code> •

```
1    \input inputd.tip
2    \InputD{dicentry.tip}              % 34.7.4.2, p. 34.
3    \InputD{rh-dict.tip}               % 34.7.4.3, p. 34.
```

Start on page 2, an even page, so that the output has a page with an even page
number on the left and a page with an odd number on the right. All together
four mini-pages will be generated which are then subsequently glued together
and reprinted in one figure.

```
4    \pageno = 2
```

Now generate some sample text.

```
5    \DicEntry{Alpha} Alpha is a letter in the Greek alphabet,
6    and because we are using \TeX, we really should
7    write $\alpha$.
8
9    \DicEntry{Beta} Another one of the Greek letters. Imagine
10   there is a long explanation coming now, or imagine there is
11   none and all they give you is a reference to another dictionary.
12   Then, of course, you would ask yourself, why you bought this
13   dictionary.
14
15   \DicEntry{Gamma} Now, come on! Can't you come up with
16   something else besides Greek letter?!
17
18   \def\SillyText{Imagine there is an explanation of this
19       term here. Or imagine there is no such term explained.
20       Now that's something for an explanation, is it not? }
```

Echo 2

explanation of this term here. Or imagine there is no such term explained. Now that's something for an explanation, is it not?!

Echo 2   Imagine there is an explanation of this term here. Or imagine there is no such term explained. Now that's something for an explanation, is it not?!

Echo 3   Imagine there is an explanation of this term here. Or imagine there is no such term explained. Now that's something for an explanation, is it not?!

Echo 4   Imagine there is

2

Echo 7

an explanation of this term here. Or imagine there is no such term explained. Now that's something for an explanation, is it not?!

Echo 5   Imagine there is an explanation of this term here. Or imagine there is no such term explained. Now that's something for an explanation, is it not?!

Echo 6   Imagine there is an explanation of this term here. Or imagine there is no such term explained. Now that's something for an explanation, is it not?!

Echo 7   Imagine there is

3

Echo 8

an explanation of this term here. Or imagine there is no such term explained. Now that's something for an explanation, is it not?!

Echo 8   Imagine there is an explanation of this term here. Or imagine there is no such term explained. Now that's something for an explanation, is it not?!

Echo 9   Imagine there is an explanation of this term here. Or imagine there is no such term explained. Now that's something for an explanation, is it not?!

Echo 10   Imagine there is

4

Echo 11

an explanation of this term here. Or imagine there is no such term explained. Now that's something for an explanation, is it not?!

Echo 11   Imagine there is an explanation of this term here. Or imagine there is no such term explained. Now that's something for an explanation, is it not?!

5

Figure 34.8. Sample output, marks in a dictionary page.

```
21   \DicEntry{Echo} \SillyText
22   \DicEntry{Echo 2} \SillyText
23   \DicEntry{Echo 3} \SillyText
24   \DicEntry{Echo 4} \SillyText
25   \DicEntry{Echo 5} \SillyText
26   \DicEntry{Echo 6} \SillyText
27   \DicEntry{Echo 7} \SillyText
28   \DicEntry{Echo 8} \SillyText
29   \DicEntry{Echo 9} \SillyText
30   \DicEntry{Echo 10} \SillyText
31   \DicEntry{Echo 11} \SillyText
32   \bye
```

• End of `ex-dicentry.tip` •

34.7.5   Marks and Running Heads as Used in This Series

The rules of Springer-Verlag, the publisher of this series, require the use of chapter titles on left hand pages and section titles on the right hand pages. The Chicago Manual of Style (Chicago (1982), section 1.80) prescribes for this case that the *last* section title on a right-hand page is used for a running head. This arrangement is identical to the arrangement for dictionaries we just discussed.

For further details see 31.2.9, p. III-604, which contains the definition of macro \Section. 35.9.1, p. 122, discusses the generation of running heads.

34.8   Summary

In this chapter we learned:

- By default, TeX provides an output routine which does nothing else but ship the contents of box register 255 to the dvi file.
- An output routine is set up by assigning the proper code to \output.
- It is fairly easy to add an option to an output routine whereby only selective pages are printed.
- An output routine can shift pages horizontally depending on whether these pages are left- or right-hand pages. This is a feature used in printing double-sided documents. The same output routine also showed how running heads and running footers can change depending on whether a page is a left or right-hand page.
- An output routine where the distinction between logical and physical pages is important was presented (three logical pages are combined into one physical page and printed together).

- A macro was presented that allows text to be printed at arbitrary positions on the page.
- TEX's marking mechanism allows for the setup of running heads that depend on the current document's text. For instance, in dictionaries the first and the last term of a page can be listed in the running head. In a book section titles and chapter titles are printed in running heads.
- A mark is generated using the `\mark` command. The mark text is retrieved inside the output routine using `\botmark`, `\topmark` and `\firstmark`.
- The marking mechanism also works when boxes are split using `\vsplit`.

35
The Plain Format Output Routine and Insertions

In this chapter output routine and insertions of the plain format are discussed. It begins with a simplified version of the output routine, and gradually variations and extensions are introduced.

This chapter contains the first output routines with insertions. It is therefore important that you read this chapter carefully if you are interested in output routines with insertions. even if you are not interested in the output routine of the plain format.

35.1   A Brief Overview

Next let me present a mode detailed overview of the output routines discussed in this chapter:

1. We start with a simplified version of the plain format output routine without insertions; see 35.2, p. 91. We also discuss printing headers and footers in the plain format.
2. The original output routine (reformatted for better readability) can be found in 35.3, p. 96.
3. Next is a discussion of the insertions of the plain format. There are two insertion classes, one for footnotes and the other for top insertions.
4. We also discuss \midinsert. This instruction is of interest because of its flexibility. In addition, a modification of \midinsert is introduced; see 35.4.5, p. 104.
5. A modification of \topinsert to deal with oversized material is discussed in 35.5, p. 109.
6. A version of the plain format output routine is presented in 35.8, p. 120, where the current page part of the vertical list is written to the log file. This is very useful to trace problems with page layouts.
7. Finally, in 35.9, p. 121, we discuss an "extended plain format output routine" which was used for processing this series.

There are a number of macros for which different definitions are offered in this chapter. In order to keep things straight here is a brief overview which you may want to refer to later when reading this chapter:

- \headline, \footline. These macros are defined in 35.2.1 on the next page. These macros are *not* available in the output routine of this series of 35.9, p. 121.

  \nopagenumbers. This macro is defined in 35.2.2, p. 92, and its definition is *not* available in the output routine of this series (35.9, p. 121). Instead you need to call \SetPageLayout with the appropriate code of 35.9, item 4, p. 121. This macro is mentioned with macro \footline, because all macro \nopagenumbers does is to call \footline to set the footline to empty and therefore eliminating the printing of the page number.
- \makeheadline, \makefootline. These macros are used to *print* the headline and footline in the plain format's output routine. The definitions of these macros can be found in 35.2.3.2, p. 94, and 35.2.3.3, p. 95, respectively.
- \plainoutput. This is the plain format output routine. Three definitions of this macro are offered:

  - 35.2.3, p. 93, presents a *simplified version* of the plain format output routine.
  - 35.3, p. 96, presents the *default definition* of the plain format's output routine.
  -

- \pagebody. This macro is responsible for *printing the page body* (that is the page without headline and footline). The following definitions of this macro are offered:

  - 35.2.3.1, p. 93, presents the definition of this macro which goes with the simplified version of the plain format output routine.
  - 35.3, p. 96, presents the default definition of this macro.
  -

- \pagecontents. This macro produces the contents of a page (that it is directly used by \pagebody). The following definitions of this macro are offered:

  - 35.2.3.4, p. 95, presents the definition of this macro for the simplified plain format output routine.
  - 35.3, p. 96, presents the default definition of this macro.
  - 35.6.1, p. 113, presents a new definition of \pagecontents which is needed to use the redefined \endinsert and to display the vertical list of the page being printed (\@ShowPlainLists...).
- \footnote. This macro prints a footnote.

  - Its definition is reprinted in 35.4.1.3, p. 99.
  - A macro \FootNote generates automatically numbered footnotes for this series.

- \topinsert, \pageinsert, \midinsert.

- The *applications* of the preceding macros is shown in 35.4.3, p. 103, 35.4.4, p. 104, and 35.4.5, p. 104, respectively.
- The definitions of the preceding macros can be found in 35.4.6.2, p. 105.
- The real work is done in macro \endinsert.
  * The original definition of \endinsert can be found in 35.4.6.3, p. 106.
  * A different definition of \endinsert can be found in 35.6.2. This new definition accommodates the following extensions:
    + If oversized material is encountered in insertion material no extra empty pages are generated. Instead the material is artificially shortened. See 35.5.2, p. 110, for the relevant definition of \EndInsertTopInsFixed. This redefinition requires the use
    + A new \midinsert (which prevents the out of order printing of material) also needs to use this new version of \endinsert. See 35.7.2, p. 117, for detail.

- \supereject and \dosupereject. These macros are defined in 35.4.7, p. 108.

35.2   A Simplified Output Routine of the Plain Format

The main difference between the original output routine of the plain format and the following simplified output routine is that all insertions have been removed, so there are no footnotes, \topinserts, \midinserts, or \pageinserts (these instructions are explained shortly).

35.2.1   Running Header and Running Footer, \headline, \footline

The text of the running head and the running foot in the plain format are stored in *token registers* \headline and \footline respectively (note \headline and \footline are *not* macros). To print the contents of a token register, \the must be used (see 24.3, p. III-313). For instance, to print the headerline, a construct like \line{\the\headline}; to print the footline we use a construct like \line{\the\footline}; see 20.1.2, p. III-117.

Token register \headline is initialized as follows: \headline = {\hfil}, thus no running head is printed by default.

The default for the footline is \footline = {\hss\tenrm\folio\hss}. The footline is printed by \line{\the\footline} in the plain format output routine. The definition of \footline should remind you of the definition of \centerline (see 6.6.1, p. I-185). Also note that \folio prints the current page number (33.5.3, item 3, p. 50). In other words, the running footer of the plain format consists of a centered page number (unless changed by the user, of course).

Note that an explicit font change instruction, \tenrm (Computer Modern roman 10 pt), is part of the default setup of \footline. Why is this font change instruction necessary if the default font is that same font anyway? The reason is that the currently active font, when the output routine is actually called, might be a different font (for instance, an italics font may be active because a whole paragraph is typeset in italics, and the current page break occurred in the middle of this paragraph). And if that is true, the page number would also be printed in italics if it were not for the explicit font change as part of \footline.

You should therefore *always* include a font change instruction in case you define your own running head or running foot.

35.2.2 Suppressing the Printing of the Page Number, \nopagenumbers

The macro \nopagenumbers suppresses the printing of page numbers by clearing the contents of the \footline token register. The \nopagenumbers macro is defined as follows (no parameters):

```
1    \def\nopagenumbers{\footline = {\hfil}}
```

This macro should be applied carefully because the plain format does *not* define a macro that could be used to "turn back on" the printing of page numbers. The following explanation refers to perhaps the most typical application of \nopagenumbers (other than when a whole document is typeset without page numbers as might be the case with short memos or letters). Assume you only want to suppress the printing of page numbers for the title page of a document. The trick is simple: apply \nopagenumbers inside a group and *after* the title page has been written to the dvi file, terminate the group.

For example:

● ex-nopage.tip ●

Start a group. Turn off the printing of the page numbers.

```
1    {
2        \nopagenumbers
```

Print the title, then end this page.

```
3        \centerline{\bf Title}
4        ...
5        \vfill\eject
```

Turn on the printing of the page numbers again by terminating the previously started group.

```
6    }
```

One other small detail needs to be taken care of now: the page number must be reset to 1. The page number is incremented by the plain format output routine (that happens regardless of whether a page number is actually printed), and this page number change was done on a *global* basis. Note however, that if the title page is followed by the preliminaries of a document (such as table of contents and preface), roman page numbers are used. Do a \pageno = -1 in that case; see 33.5.3, p. 49, for details. Here we assume that we want to resume "ordinary" page numbering, using arabic page numbers.

```
7   \pageno = 1
8   ...
9   \bye
```

• End of ex-nopage.tip •

35.2.3   The Code of the Simplified Plain Format Output Routine

I will now present the code of the "simplified plain format output routine," our version of the plain format's output routine *without* any insertions. If you loaded the code below, then you would load this simplified output routine. Because this code has only an educational purpose but no practical one, the following source code file does *not* belong to the published files of this series.

Note that we call the definitions as given in the plain format "original definitions."

• op-simple.tip •

Define macro \plainoutput. This definition writes the current page to the dvi file and produces a page layout in accordance with the plain format's specifications. There is no change compared with the original definition.

```
1   \def\plainoutput{%
```

\shipout is TeX's primitive to write pages to the dvi file.

```
2       \shipout\vbox{%
```

The page which is being written consists of a running head, a page body, and a running foot. In the end increase the page number.

```
3           \makeheadline
4           \pagebody
5           \makefootline
6       }%
7       \advancepageno
8   }
```

35.2.3.1   The Definition of Macro \pagebody

The macro \pagebody actually generates the text page excluding header and footer. This definition of \pagebody is simplified compared with the original one, which can be found in 35.3, p. 96.

```
 9   \def\pagebody{%
10       \vbox to \vsize{%
```

The following instruction tells TEX that there is not a maximum on the depth of a page.

```
11           \boxmaxdepth = \maxdepth
12           \pagecontents
13       }%
14   }
```

35.2.3.2   The Definition of Macro \makeheadline

The macro \makeheadline prints the running head. This is the original definition.

```
15   \def\makeheadline{%
```

Create a zero height vbox.

```
16       \vbox to 0pt{%
```

Skip upwards by 22.5 pt. If you subtract from 22.5 pt the value of 8.5 pt (which is the height of the headline, see source code line 19), then this value is the distance from the origin of the page, *above* which the headline is printed. Next look at \topskip (the default in the plain format is 10 pt), which determines the positioning of the first line of the text on the page with reference to the zero point (by default that first line's baseline is therefore 10 points *below* the origin of the page).

Adding the preceding two values of 14 pt and 10 pt results in 24 pt, which is twice the default value of \baselineskip. Therefore seen the page as a whole (including header and footer line) there will be precisely one empty line between the header and the first line of text.

```
17       \vskip -22.5pt
18       \line{%
19           \vbox to 8.5pt{}%
20           \the\headline
21       }%
```

Skip down to the baseline of the zero height vbox just being built.

```
22       \vss
23       }%
```

There is no interline glue, so \topskip takes care of the proper spacing.

```
24       \nointerlineskip
25   }
```

The preceding discussion has assumed that the header's material is not of excessive height or depth. Regardless of how deep the material is, the header will be printed without any warning in the case of excessive depth because \vss will compensate for any depth. Note that if the height of the header's material is excessive (above 8.5 pt, which is the minimum height of the header's material caused by the strut-like construct \vbox to 8.5pt{}), the position of the top edge of the header will always stay the same at 22.5 pt above the zero point of the page. The higher the header's text the more the header's baseline will move downwards, and might (without a warning message) run into the page's text area. Note that in none of the cases just discussed is the placement of the text page itself affected in any way.

35.2.3.3   The Definition of Macro \makefootline

The macro \makefootline prints the running foot. By setting \baselineskip to 24 pt, the distance between the last line of the text page (assuming that the text page itself is completely filled) and the footer (measured from baseline to baseline) will be 24 pt. In other words, there is one empty line between the last line of the page and the running foot (assuming a single-spaced document), which is identical to the relation of header and first line of the text page. The following is the original definition of \makefootline.

```
26    \def\makefootline{%
27        \baselineskip = 24pt
28        \line{\the\footline}%
29    }
```

35.2.3.4   The Definition of Macro \pagecontents

The macro \pagecontents prints the text page and also takes care of the page layout. To understand the following macro definition, you need to see it in the context of the definition of \pagebody above, which encloses the vertical list generated by the following macro into a vertical box of length \vsize.

   This definition of the macro is simplified; for the original definition of this macro see 35.3 on the next page.

```
30    \def\pagecontents{%
```

Save the depth of the current page, then extract the current page from box register 255. Use \unvbox to allow the vertical glues of the main vertical list to regain their stretchabilities again.

```
31        \dimen@ = \dp 255
32        \unvbox 255
```

If a ragged bottom page layout is chosen, you need to insert \vfil glue so that the only place where the page length is adjusted is on the bottom of the current page (no other glue, for instance, none of the \parskip glues, are supposed to be stretched). The \kern below in effect sets the depth of the current page to zero.

```
33          \ifr@ggedbottom
34              \kern -\dimen@
35                  \vfil
36          \fi
37  }
```

One last point here. The question is why the preceding code (lines 34 and 35) does not simply read \vss. The advantage of the above solution is that an *oversized page* (a page too high) is still reported as being too long (an overfull vbox is reported), whereas if you had used \vss, then this glue would compensate for the excessive length of the page, regardless of the height of the current page.

• End of <code>op-simple.tip</code> •

35.3   The Ordinary Plain Format Output Routine

The TeX code of the original version of the plain format output routine presented below has been reformatted for improved readability. Nothing essential has been changed. To understand the following code you need to understand the output routine of the preceding section. You also need to know that there are *two* different insertion classes in plain TeX (details follow later):

1. \footins (for footnotes),
2. \topins (for topinsertions).

There are two main differences between the output routine of the preceding section and the output routine of the plain format.

First, look at the definition of macro \plainoutput in the plain format:

```
1   \def\plainoutput{%
2       \shipout\vbox{%
3           \makeheadline
4           \pagebody
5           \makefootline
6       }%
7       \advancepageno
8       \ifnum\outputpenalty > -20000
9       \else
10          \dosupereject
11      \fi
12  }
```

The main difference between the definition of \plainoutput and the definition of \PlainOutputSimplified discussed before is the test for \outputpenalty and the execution of \dosupereject if \outputpenalty is ≤ -20000.

Second, the definition of \pagecontents now changes quite a bit, because material from two insertion classes may need to be printed (note that in 35.6.1,

p. 113, yet another definition of **\pagecontents** is offered):

```
1     \def\pagecontents{%
```

If there is top-insertion material, print it.

```
2          \ifvoid\topins
3          \else
4              \unvbox\topins
5          \fi
```

The following code is as before.

```
6          \dimen@ = \dp255
7          \unvbox 255
```

Similar to the preceding code for top-insertion material, test for footnote insertion material and print it (preceded by a footnote rule). Do nothing, if there is no footnote insertion material.

```
8          \ifvoid\footins
9          \else
10             \vskip\skip\footins
11             \footnoterule
12             \unvbox\footins
13         \fi
```

Same as before.

```
14         \ifr@ggedbottom
15             \kern -\dimen@
16             \vfil
17         \fi
18     }
```

35.4   The Insertions of the Plain Format

In the preceding section, we learned that there are *two* different *insertion classes* in the plain format:

1. \footins (for footnotes).
2. \topins (for topinserts).

The plain format actually has four instructions which *generate* (or at least may generate) insertion material. These instructions are the following:

1. \footnote, generates material for insertion class \footins.
2. \topinsert, generates material for insertion class \topins.
3. \pageinsert, a special case of \topinsert.
4. \midinsert, either equivalent to a \topinsert or to a \vbox command.

A detailed discussion of those instructions follows now. Note that these four instructions do *not* belong to four *different* insertion classes!

35.4.1   Footnotes in the Plain Format, \footnote

Footnotes in the plain format are handled by the insertion class \footins. The macro to generate a footnote is \footnote, which has two arguments. The first argument, #1, is the footnote marker; frequently a superscript is used as a footnote marker (^{1}). The second argument of \footnote is the footnote itself. Here is a brief example of how a footnote can be generated: \footnote{^{1}}{Footnote text...}.

The plain format does *not* automatically number footnotes so you must keep track of the footnote numbers yourself. 35.4.2, p. 102, shows a macro for the automatic numbering of footnotes in documents.

Note that the definition of the \footnote macro below shows only one argument, the footnote marker. The footnote text is handled with a technique explained before in 19.4.15.3, p. III-112. The preceding statement that \footnote has *two* arguments is technically wrong, but as a practical point it does not matter much.

Footnotes must be discussed from two points of view: *generating* the footnote and *printing* the footnotes by the output routine. The generation of footnotes is discussed now. See 35.3 on the previous page for a discussion of the printing of footnotes.

Some additional discussion of footnote-related problems in output routines can be found in 36.3, p. 156.

35.4.1.1   The Beginning of the Footnote Processing Macros

The footnote-related macros of the plain format was reformatted for improved readability. The \interfootnotelinepenalty penalty (as the name suggests) used as interline penalty for footnotes.

● op-foot.tip ●

```
1   \catcode'\@ = 11
2   \newcount\interfootnotelinepenalty
3   \interfootnotelinepenalty = 100
```

35.4.1.2   Setting up the Insertion Class \footins

Declare a footnote insertion class \footins.

```
4   \newinsert\footins
```

The parameters of the insertion class \footins, are set up now: a space (corresponding to a \bigskip) must be reserved on every page (this space accounts for the footnote rule and space above it, between the text itself and the footnote). The magnification factor of this insertion class is 1. There is a maximum of 3 in of footnotes per page allowed.

```
5    \skip\footins = \bigskipamount
6    \count\footins = 1000
7    \dimen\footins = 8in
```

35.4.1.3   Defining Macro \footnote

Define the following footnote macro with one/two arguments:

- #1. The footnote marker, printed with the footnote text and the footnote.
- #2. The footnote text itself. Note that what appears to be a parameter to \footnote is really not, since the footnote text is absorbed by using the \aftergroup trick discussed in 19.4.15.3, p. III-112.

```
8    \def\footnote #1{%
9        \let\@sf = \empty
10       \ifhmode
```

If you are in horizontal mode, you need to save the space factor in \@sf and restore it later from \@sf (see the definition of \SaveSpaceFactor in 16.2.9, p. II-281, for a more detailed explanation).

```
11           \edef\@sf{\spacefactor = \the\spacefactor}%
12           \/%
13       \fi
```

Print the footnote marker itself here.

```
14       #1%
```

Restore the space factor now (if TeX was in horizontal mode when \footnote was called, otherwise do nothing).

```
15       \@sf
```

Call the \vfootnote macro. See below for its definition and meaning.

```
16       \vfootnote{#1}%
17    }
```

35.4.1.4   The Definition of \vfootnote

The macro \vfootnote continues processing a footnote. Its only parameter, #1, is the footnote marker, which was already an argument to \footnote. Note that until now, the footnote text has *not* yet been absorbed or looked at.

```
18    \def\vfootnote #1{%
```

Begin material for insertion class `\footins`.

```
19        \insert\footins\bgroup
20            \interlinepenalty = \interfootnotelinepenalty
```

Top baseline for broken footnotes.

```
21            \splittopskip = \ht\strutbox
22            \splitmaxdepth = \dp\strutbox
```

The following penalty strongly discourages TeX from splitting the text of one footnote across two pages.

```
23            \floatingpenalty = 20000
```

Note that a footnote may be generated from within some text when `\narrower` was issued. Therefore, during the generation of the footnotes, parameters like `\leftskip` and so forth are set back to their defaults. Because of the implicit grouping caused by `\insert` any of those changes will be undone after the footnote text was generated.

```
24            \leftskip = 0pt
25            \rightskip = 0pt
26            \spaceskip = 0pt
27            \xspaceskip = 0pt
```

The footnote marker appears left flush with the ordinary text margin, while the footnote text itself is printed with a small indentation on the left side.

```
28            \textindent{#1}%
```

A strut is inserted into the footnote text's first line to ensure correct vertical spacing. A strut is also inserted at the very end of the footnote text (see the definition of `\@foot` below).

```
29            \footstrut
```

The following `\futurelet`, together with the macro `\fo@t` (the "decision making macro") causes the execution of either `\f@@t` or `\f@t` to follow next. The `\f@@t` macro is executed if the footnote text is enclosed in curly braces (this is by far the most likely case), while the `\f@t` macro is executed if the footnote text consists of a single token, not enclosed in curly braces. See 23.4, p. III-252, for details on `\futurelet`.

```
30            \futurelet\next\fo@t
31    }
```

35.4.1.5   The Definitions of Macros \fo@t, \f@@t and \f@t

The definition of macro `\fo@t` comes next.

```
32    \def\fo@t{%
```

```
33        \ifcat\bgroup \noexpand\next
```

\next is the same as \bgroup, which in the sense used here, is the same as an opening curly brace.

```
34            \let\next = \f@@t
35        \else
```

No opening curly brace seen.

```
36            \let\next = \f@t
37        \fi
```

Here the execution of \f@t or \f@@t takes place, whichever is appropriate.

```
38        \next
39    }
```

The technique applied by \f@@t to process the footnote text is discussed in 23.1.8, p. III-242.

```
40    \def\f@@t{%
41        \bgroup
42        \aftergroup\@foot
43        \let\next =
44    }
```

As mentioned above, the following macro prints the footnote text if it consists of *one* single token.

```
45    \def\f@t #1{%
46        #1%
47        \@foot
48    }
```


35.4.1.6   The Definition of \@foot

The macro \@foot is called when the *end* of the footnote text is reached (its execution is either caused by the \@foot call contained in \f@t or by the \aftergroup of \f@@t).

```
49    \def\@foot{%
```

A strut is inserted to make the last line of the footnote text of the proper height. The following \vskip generates a small amount of additional vertical space between the current and the next footnote, if these two footnotes are printed on the same page.

```
50        \strut
51        \vskip 1pt
```

This \egroup matches the \bgroup of \@foot.

```
52        \egroup
53    }
```

35.4.1.7   The Definition of \footstrut

The \footstrut macro generates the footnote strut; see 7.4.1, item 3, p. I-236, for a discussion of a strut generated this way. Note that \splittopskip is the same as \ht\strutbox.

```
54    \def\footstrut{%
55        \vbox to \splittopskip{}%
56    }
57    \catcode'\@ = 12
```

• End of `op-foot.tip` •

35.4.2   Automatically Numbered Footnotes, \FootNote

I use the following macro \FootNote to number footnotes in this series automatically. This macro has one argument, #1, the text of a footnote. The footnote number is generated automatically (and reset to 1 at the beginning of every chapter). The printing of footnotes is done according to the guidelines of the publisher of this series.

$$\mathcal{P}'$$ • `ts-foot.tip` •

For the following macro source code to be usable, the output routine of 35.6, p. 113, must be loaded, which is the reason for the following \InputD call. This output routine is a variation of the plain format output routine (details are irrelevant at this point).

```
15    \InputD{op-pagec.tip}                    % 35.6.1, p. 113.
16    \catcode'\@ = 11
```

Declare a footnote counter and set it up so it is reset at the beginning of every chapter.

```
17    \NewCounter{FootNote}{\arabic}%
18        {\TheCounter{FootNote}}%
19        {\PrintCounter{FootNote}}
20    \AddCounterToResetList{FootNote}{ChapterNo}
```

The \FootNote macro is defined here.

```
21    \def\FootNote #1{%
```

Increment the footnote counter first (it was reset to zero at the beginning of the current chapter), so that the first footnote is numbered 1.

```
22        \StepCounter{FootNote}%
```

Now call the plain format's \footnote macro to generate the footnote. The footnote marker is first (use the current footnote number, raise it, and print it in a smaller font by using a superscript in TeX's math mode).

```
23        \footnote{$^{\PrintCounter{FootNote}}$}%
```

The following specifications (font size, line spacing) are according to the specifications of this series' publisher.

```
24          {%
25              \small
26              \baselineskip = 9pt
27              #1%
28          }%
29    }
```

The space between the body of the text and the beginning of the footnote text(s) is one empty line (12 pt), plus a thin line 5 pc wide (0.4 pt thick) and another 6 pt of vertical space, all together 18.4 pt.

```
30    \skip\footins = 18.4pt
```

The following macro causes the footnote(s) to be printed inside the output routine. It replaces the standard definition of this macro as given in 35.6, p. 113.

```
31    \def\@PrintFootnotePlain{%
```

Insert some vertical space between text page and footnote rule.

```
32        \vskip 12pt plus 2pt minus 1pt
```

Print the footnote rule.

```
33        \hrule width 5pc height 0.4pt depth 0pt
```

Insert some vertical space between the footnote rule and first footnote.

```
34        \vskip 6pt plus 1pt minus 0.5pt
```

Print the footnote(s).

```
35        \unvbox\footins
36    }
37    \catcode'\@ = 12
```

• End of `ts-foot.tip` •

35.4.3   Topinsertions, Applying \topinsert ... \endinsert

Besides the insertion class \footins, there is the insertion class \topins. Material for this insertion class can be generated using \topinsert, which is applied as follows: after \topinsert, the material to be inserted at the top is listed, and is terminated by \endinsert. For example,

```
1    \topinsert
2    $$
3        \vbox{
4            \halign{
5                ...
6            }
7        }
8    $$
```

```
9   \endinsert
```

TeX prints the material of a `\topinsert` at the *top* of the current page if there is sufficient space, or on top of the next page or following pages, depending on available space. If there is already material generated by `\topins` on the current page, the new material, if at all, is printed below the existing material. The details always depend on the specifics of a document, but note that the order of the material printed is always the same as the order of the material generated.

This insertion class is obviously intended for figures or tables in a document processed with the plain format.

35.4.4 Applying `\pageinsert` ... `\endinsert`

`\pageinsert` is identical to `\topinsert`, except that it is a full page `\topinsert`. The material of a `\pageinsert` is therefore also terminated by `\endinsert`. For instance,

```
1   \pageinsert
2       Material of a \pageinsert
3   \endinsert
```

and

```
1   \topinsert
2       \vbox to \vsize{
3           Material of a \pageinsert
4       }
5   \endinsert
```

are equivalent. If you assume, for example, that there is no other topinsert material pending, then TeX will fill up the current page with material following the `\pageinsert`. On the next page the material generated by `\pageinsert` will be printed. On the page after that page you will find regular text.

35.4.5 Applying `\midinsert` ... `\endinsert`

The `\midinsert` instruction is applied the same way the other two instructions just discussed are applied so that the material to be printed follows `\midinsert` and is terminated by `\endinsert`. Here is a very short example:

```
1   \midinsert
2       Material of a \midinsert
3   \endinsert
```

The main difference between `\topinsert` and `\midinsert` is that `\midinsert` first tests whether its material will fit on the page "right here" *without* causing a page break. If that is the case, the material is simply printed at the current location (imagine that in this case \midinsert and \endinsert have been removed from your text).

If, on the other hand, the material does *not* fit on the current page, the material is treated as if it were the material of a \topinsert; thus the \midinsert becomes in effect a \topinsert.

Obviously, \midinsert must determine the amount of available space on the current page. How the amount of available space can be determine is discussed in the context of macro \FreePageSpace (see 32.2.4, p. 6).

35.4.6   The Macros of the Insertion Class \topins

All three macros \topinsert, \pageinsert and \midinsert use insertion class \topins. The code for \topinsert, \pageinsert and \midinsert of the plain format are contained in the following source code file.

● pl-ins.tip ●

```
1    \catcode'\@ = 11
```

35.4.6.1   Setting Up Insertion Class \topins

Declare the insertion class \topins.

```
2    \newinsert\topins
```

Initialize the parameters of this insertion class: no additional vertical space is added when material of this insertion class is printed on the current page. The magnification factor is 1. There is no limit on the size of insertion class material on one page.

```
3    \skip\topins = 0pt
4    \count\topins = 1000
5    \dimen\topins = \maxdimen
```

Define two conditionals. \ifp@ge is true if a call to \pageinsert occurred. \if@mid is true if a \midinsert call occurred.

```
6    \newif\ifp@ge
7    \newif\if@mid
```

35.4.6.2   The Definitions of \topinsert, \pageinsert, and \midinsert

The definitions of macros \topinsert, \pageinsert, and \midinsert is next. All these macros (after setting two conditionals) call macro \@ins, which is at the heart of the matter and which does the processing for all three macros.

```
 8   \def\topinsert {\@midfalse\p@gefalse\@ins}
 9   \def\midinsert {\@midtrue \p@gefalse\@ins}
10   \def\pageinsert{\@midfalse\p@getrue \@ins}
```

The definition of macro `\@ins` begins here.

```
11   \def\@ins{%
12       \par
```

Start a group to keep any changes local. Then start to collect the insertion material in box register 0. Note that before this point, none of the insertion material of `\topinsert`, `\midinsert`, or `\pageinsert` was absorbed. The `\endinsert` by which the material is terminated, which belongs to one of the three insertion instructions, is called at the end of the material.

```
13       \begingroup
14       \setbox 0 = \vbox\bgroup
15   }
```

Actually to be truthful: it is not even `\@ins`, which does the real work, it is `\endinsert`, which we will discuss next.

35.4.6.3   The Original Definition of `\endinsert`

The `\endinsert` macro is defined next. The title of this subsubsection contains the word "original" because a different definition of `\endinsert` is offered in 35.6.2, p. 114.

```
16   \def\endinsert{%
```

The following `\egroup` finishes the vbox that was started by `\@ins`.

```
17       \egroup
```

Was the original instruction `\midinsert`?

```
18       \if@mid
```

Yes. Now test whether there is sufficient room on the current page: take the overall size of the material (i.e., height plus depth of box register 0), add a "fudge factor" (12 pt) and add `\pagetotal`. If this value exceeds `\pagegoal` ("available space") then the material will not fit on the current page and `\midinsert` is "redeclared" as a `\topinsert`. The approach taken below is along the lines of the computations of macro `\FreePageSpace` (see 32.2.4, p. 6).

```
19           \dimen@ = \ht0
20           \advance\dimen@ by \dp0
21           \advance\dimen@ by 12pt
22           \advance\dimen@ by \pagetotal
23           \advance\dimen@ by -\pageshrink
24           \ifdim\dimen@ > \pagegoal
25               \@midfalse
26               \p@gefalse
27           \fi
```

```
28          \fi
```

If \if@mid is still true at this point, then a \midinsert instruction was originally given *and* the material fits on the current page.

The \midinsert material is now printed and surrounded by vertical space. TeX is instructed that a page break right after the \midinsert material is a "good idea" by \goodbreak.

```
29          \if@mid
30              \bigskip
31              \box 0
32              \bigbreak
33          \else
```

Generate insertion material, insertion class \topins.

```
34              \insert\topins{%
```

The following penalty is inserted to have a legal break point between insertion material of the same class, which precedes this insertion material, and the current insertion material. The penalty indicates that TeX should try to keep the material together (if there is preceding material in the first place). The penalty value of 100 suggests that there is no great emphasis put on that.

```
35                  \penalty 100
```

No glue after a split \topins insertion is inserted and there is no limit on the depth of the material of the insertion. Breaking up such \topins material is regarded as harmless, but not encouraged either.

```
36                  \splittopskip = 0pt
37                  \splitmaxdepth = \maxdimen
38                  \floatingpenalty = 0
```

If the original instruction was \pageinsert, do the following now.

```
39                  \ifp@ge
```

Generate a zero depth box of this material.

```
40                      \dimen@ = \dp0
41                      \vbox to \vsize{
42                          \unvbox 0
```

The depth of the box generated here is reduced to zero by the insertion of negative kern in the amount of the depth of the material.

```
43                          \kern -\dimen@
44                      }%
45                  \else
```

The following code is executed to generate \topinsert or "converted" \midinsert material.

```
46                      \box 0
47                      \nobreak
48                      \bigskip
49                  \fi
```

```
50          }
51       \fi
```

The following \endgroup matches the \begingroup of \@ins. This ends the processing of \endinsert and regular text processing can now resume.

```
52       \endgroup
53    }
54    \catcode'\@ = 12
```

• End of <code>pl-ins.tip</code> •

35.4.7   The Definitions of \supereject and \dosupereject

Now let me discuss \supereject. This instruction, usually issued at the very end (it is part of the definition of macro \bye), makes sure that all held-over insertions are written out.

The macro \dosupereject will be defined first. This macro is called from within the output routine of the plain format, when a \supereject was encountered.

• <code>op-sup.tip</code> •

```
1    \def\dosupereject{%
```

Test whether insertions are held over (\insertpenalties contains the number of remaining insertions; see 32.7.6, p. 34).

```
2       \ifnum\insertpenalties > 0
```

There are insertions held over. Insert on the current page an empty hbox, then undo the effect of \topskip, which now leaves the full space on the page available for insertions. The \nobreak prevents the page break from occurring at the \kern, because a page break does occur at a kern if the kern is followed immediately by glue.

```
3          \line{}
4          \kern -\topskip
5          \nobreak
6          \vfill
```

\supereject is like \eject, except that a different penalty (of −20000) is inserted, which alerts the output routine of the plain format that all remaining over insertions need to be written.

```
7          \supereject
8       \fi
9    }
```

Next is the definition of \supereject. It is very straightforward. A penalty of −20000 instead of −10000 is used to allow the output routine to make a distinction between \eject and \supereject. The \par forces vertical mode.

```
10   \def\supereject{%
```

```
11        \par
12        \penalty -20000
13    }
```

• End of `op-sup.tip` •

35.5 Oversized Insertion Material

We will now discuss the handling of "oversized" material of the insertion class
`\topins`. First there is an example showing the problem of how TeX handles the
situation. Then a solution to this problem is offered.

The problem discussed in this section is not specifically limited to the plain
format, but is of interest to insertions in other output routines The solution,
though, is specific to the plain format's output routine.

35.5.1 An Example Showing the Problem

The following example generates oversized `\topinsert` material (material that
is too long). From a TeX user's point of view this is clearly an error. But this
is not the issue discussed here. Here we discuss what TeX is going to do when
presented with oversized insertion material.

Here is an example that demonstrates what happens.

• `ins-tool.tip` •

Write a short form of a "list of insertion file" to file `ins-tool.loi` This insertion
file is used to show on which pages the insertions are placed.

```
1    \newwrite\loiwrite
2    \immediate\openout\loiwrite = \jobname.loi
```

Define a macro to generate some oversized `\topinserts`. The macro has one
parameter, `#1`, a string that identifies the insertion and that is also written to
the insertion file.

```
3    \def\TooBigOfATopInsert #1{%
4        \topinsert
```

The insertion material generated is 10% larger than `\vsize` (incidentally the
problem discussed here would *not* occur if the user had used `\pageinsert` because `\pageinsert` generates material of length `\vsize`, regardless of the "real
length" of the material).

```
5            \vbox to 1.1\vsize{
6                This is insert ''#1.''
```

The list of insertions file is written here.

```
 7               \write\loiwrite{Insertion #1, page \the\pageno.}
 8               \vfill
 9            }
10         \endinsert
11   }
```

Generate four insertions and end the example.

```
12   \TooBigOfATopInsert{A}
13   \TooBigOfATopInsert{B}
14   \TooBigOfATopInsert{C}
15   \TooBigOfATopInsert{D}
16   \bye
```

• End of <code>ins-tool.tip</code> •

This is the reprint of the "insertion file list" generated by this example:

• <code>ins-tool.loi</code> •

```
1   Insertion A, page 2.
2   Insertion B, page 4.
3   Insertion C, page 6.
4   Insertion D, page 8.
```

What do you see? Every second page is empty! The material generated above would fit on 5 pages, but TeX really generates 9 pages: every oversized \topinsert is followed by one empty page.

35.5.2   Fixing the Problem

The problem in the preceding example can be fixed by forcing the length of the material to be less than \vsize. In addition, the macro source code below generates a warning message; the intention of the macro source code below is *not* to sweep this problem "under the user's rug," but to handle it more gracefully (by avoiding empty pages) and to alert the user of what happened.

The following source code results in a new definition of \midinsert. If you want to use it, simply load it and the new \midinsert macro definition is in effect.

$\mathcal{P}'$ • <code>topinfix.tip</code> •

```
15   \InputD{mspaces.tip}                      % 21.4.7, p. III-165.
```

A different definition of \endinsert stored in op-endin.tip must be loaded first, because it also accommodates another change discussed in this chapter (which required yet another modification); both changes can coexist peacefully and independently of each other.

```
16   \InputD{op-endin.tip}                     % 35.6.2, p. 114.
17   \InputD{maxmindi.tip}                     % 25.1.14, p. III-332.
18   \catcode'\@ = 11
19   \def\EndInsertTopInsFix{%
```

The following code is executed when the insertion material is treated as a "real `\topinsert`" (and not as a `\pageinsert`). This code is also executed in case of a converted `\midinsert`. First, the depth of the material is limited to 10 pt. If the depth is larger, a warning message is generated and the depth is set to 10 pt.

```
20        \ifdim\dp0 > 10pt
21            \wlog{\string\endinsert: \string\topinsert
22                material deeper than 10pt}%
23            \dp0 = 10pt
24        \fi
```

Now load the greater of zero and the current depth of the material into `\dimen0`.

```
25        \MaxDimen{\dimen0}{\dp0}{0pt}{}
```

The height of the insertion material is limited to $\verb|\vsize| - \verb|\dimen0| - \verb|\bigskip|$.

```
26        \dimen1 = \vsize
27        \advance\dimen1 by -\dimen0
28        \advance\dimen1 by -12pt
```

Print a warning message in case the material is too large.

```
29        \ifdim\ht0 > \dimen1
30            \wlog{\string\endinsert: \noexpand\topinsert
31                material too long (\the\ht0),}%
32            \wlog{\EightSpaces shortened to \the\dimen1.}%
```

Actually reduce the height of the material.

```
33            \ht0 = \dimen1
34        \fi
35    }
36    \catcode'\@ = 12
```

• End of <code>topinfix.tip</code> •

35.5.3   Repeating the Preceding Example

Let us repeat the preceding example with the new version of `\endinsert`, which invokes `\EndInsertTopInsFix`. Otherwise, the following source code is identical to the preceding example source code.

• <code>ins-tool2.tip</code> •

```
1   \input inputd.tip
2   \InputD{topinfix.tip}                        % 35.5.2, p. 110.
```

The insertion log file is written here.

```
3   \newwrite\loiwrite
4   \immediate\openout\loiwrite = \jobname.loi
5   \def\TooBigOfATopInsert #1{%
6       \topinsert
```

```
 7            \vbox to 1.1\vsize{%
 8                This is insert ``#1.''
 9                \write\loiwrite{Insertion #1, page \the\pageno.}%
10                \vfill
11            }%
12        \endinsert
13    }
14    \TooBigOfATopInsert{A}
15    \TooBigOfATopInsert{B}
16    \TooBigOfATopInsert{C}
17    \TooBigOfATopInsert{D}
18    \bye
```

• End of <code>ins-tool2.tip</code> •

The following insertion file list is generated now (compare this with the preceding insertion file):

• <code>ins-tool2.loi</code> •

```
1    Insertion A, page 1.
2    Insertion B, page 2.
3    Insertion C, page 3.
4    Insertion D, page 4.
```

Let me also reprint the log file so you can see the warning messages printed by \EndInsertTopInsFix.

• <code>ins-tool2.log</code> •

```
 1    This is TeX, C Version 3.14 (...)
 2    **&/usr/local/tex/lib/fmt/plain ins-tool2.tip
 3    (ins-tool2.tip (inputd.tip
 4    (namedef.tip
 5    ) (inputdl.tip
 6    \@InputDStream=\write0
 7    ))
 8    (topinfix.tip
 9    (mspaces.tip
10    ) (op-endin.tip
11    ) (maxmindi.tip
12    ))
13    \loiwrite=\write1
14    \endinsert: \topinsert material too long (707.52672pt),
15    shortened to 631.20255pt.
16    \endinsert: \topinsert material too long (707.52672pt),
17    shortened to 631.20255pt.
18    \endinsert: \topinsert material too long (707.52672pt),
19    shortened to 631.20255pt.
20    \endinsert: \topinsert material too long (707.52672pt),
21    shortened to 631.20255pt.
22
23    [1] [2] [3] [4] [5] .)
24    Output written on ins-tool2.dvi (5 pages, 664 bytes).
```

35.6   Modifying \pagecontents and \endinsert

Due to some changes I propose in this chapter, the plain format output routine
and related macros need to be modified (the preceding subsection discussed such
a change). Those changes are:

1. Handling oversized \topinserts requires a redefined \endinsert (as mentioned this was the subject of the discussion of the preceding subsection).
2. An improved \midinsert is discussed in 35.7, p. 116, which requires a new \endinsert.
3. A version of the plain format output routine that writes the main vertical list and if non-empty writes lists of the insertion material printed on a particular page into the log file, is discussed in 35.8, p. 120.
4. The output routine used for this series requires a modification to the printing of footnotes according to guidelines of this publisher (Guidelines (1990)); see 35.4.2, p. 102.

35.6.1   A Redefined \pagecontents

The changes to macro \pagecontents are made in the following file:

$$\mathcal{P}' \quad \bullet \text{ op-pagec.tip } \bullet$$

```
15   \InputD{namedef.tip}                          % 19.1.8, p. III-73.
16   \catcode'\@ = 11
17   \def\pagecontents{%
```

Note that \NameUse{@ShowPlainLists} expands to \relax, if \@ShowPlain-
Lists is *not* defined, which is equivalent to saying "if macro source code show-
pll.tip is not loaded" (see 35.8, p. 120). Macro \@ShowPlainLists is responsible for displaying the relevant box registers (box register 255 and the insertion
related box registers).

```
18       \wlog{\noexpand\pagecontents from op-pagec.tip called.}%
19       \NameUse{@ShowPlainLists}%
20       \ifvoid\topins
21           \wlog{\string\pagecontents: no topinserts.}%
22           \if\NameDefinedConditional{@TopInsertSize}%
23               \ifdim\@TopInsertSize < 20pt
24                   \global\@TopInsertSize = 0pt
25               \fi
26           \fi
27       \else
```

Similarly, \NameUse{@UpdateTopInsertSize} does nothing if \@UpdateTopIn-
sertSize is not defined, which is equivalent to saying that macro source code
new-midi.tip (see 35.7, p. 116) is not loaded.

```
28           \NameUse{ChangeBarPush}%
```

```
29          \NameUse{@UpdateTopInsertSize}%
30          \unvbox\topins
31          \NameUse{ChangeBarPop}%
32       \fi
33       \dimen@ = \dp 255
34       \unvbox 255
```

The printing of footnotes is changed in 35.4.2, p. 102. To accommodate this, the printing of footnotes is moved into a separate macro \@PrintFootnotePlain.

```
35       \ifvoid\footins
36       \else
37          \@PrintFootnotePlain
38       \fi
```

The following code is identical to the original definition of \pagecontents.

```
39       \ifr@ggedbottom
40          \kern -\dimen@
41          \vfil
42       \fi
43    }
```

Here is the definition of macro \@PrintFootnotePlain. This macro can be changed (and was changed to print this series; see 35.4.2, p. 102), for instance, if a different format for the printing of footnotes is required.

```
44    \def\@PrintFootnotePlain{%
45       \vskip\skip\footins
46       \footnoterule
47       \unvbox\footins
48    }
49    \catcode'\@ = 12
```

• End of op-pagec.tip •

35.6.2   An Extended \endinsert

Here is the source code for a changed \endinsert. Comments are not provided, unless the code is different from the original definition of \endinsert. The purpose of this extended \endinsert is to accommodate two changes done in this chapter:

1.   \midinsert was redefined; see 35.7, p. 116.
2.   \endinsert needed to be extended to shorten oversized \topinsert material; see 35.5.2, p. 110.

Here is the new source code of \endinsert.

$\mathcal{P}'$ • op-endin.tip •

```
15    \catcode'\@ = 11
```

```
16    \def\endinsert{%
17        \egroup
18        \if@mid
19            \dimen@ = \ht0
20            \advance\dimen@ by \dp0
21            \advance\dimen@ by 12pt
22            \advance\dimen@ by \pagetotal
23            \ifdim\dimen@ > \pagegoal
24                \@midfalse
25                \p@gefalse
26            \fi
27        \fi
```

In 35.7 on the next page a redefined \midinsert is discussed. If this redefined
\midinsert is to be in effect, \MidInsertFix must be invoked here, otherwise
nothing should happen here. \NameUse expands to \relax, if \MidInsertFix is
not defined.

```
28        \NameUse{MidInsertFix}%
29        \if@mid
30            \bigskip
31            \box 0
32            \bigbreak
33        \else
34            \insert\topins{%
35                \penalty 100
36                \splittopskip = 0pt
37                \splitmaxdepth = \maxdimen
38                \floatingpenalty = 0
39                \ifp@ge
40                    \dimen@ = \dp0
41                    \vbox to \vsize{
42                        \unvbox 0
43                        \kern -\dimen@
44                    }%
```

For the following call, see 35.7 on the next page for an explanation. Note: if
\MidInsertFixPage is undefined, nothing happens here.

```
45                    \NameUse{MidInsertFixPage}%
46                \else
```

If the macro \EndInsertTopInsFix is defined, then the length of oversized material based on \topinsert is reduced. See 35.5.2, p. 110, for details.

```
47                    \NameUse{EndInsertTopInsFix}%
48                    \NameUse{MidInsertFixTop}%
```

From here on it's as usual.

```
49                    \box 0
50                    \nobreak
51                    \bigskip
52                \fi
53            }
```

```
54        \fi
55        \endgroup
56    }
57    \catcode'\@ = 12
```

• End of `op-endin.tip` •

35.7   Improving \midinsert

In the preceding subsection we already "advertised" a modified \midinsert. To justify any change to \midinsert, we need to discuss a potential problem with \midinsert. Here is a brief example to that effect:

1. Assume \vsize is 8 in.
2. Assume 5 in worth of material has been contributed to the current page (\pagetotal is 5 in) with no insertion material (\pagegoal is still 8 in, the value of \vsize). In effect there is still 8 in − 5 in = 3 in free space left on the current page.
3. Now assume a \midinsert instruction generates material 4 in long to print figure A. This material does *not* fit on the current page, where only 3 in are left. Therefore TeX converts this \midinsert into a \topinsert.
4. Another \midinsert follows *immediately* to print figure B. This time, however, the material is only 2 in long. In other words, this material *does* fit on the current page. The \midinsert therefore is printed at the current location and is *not* converted into a \topinsert.

Note that in the end the material of both \midinserts is printed in *reverse order* compared with the order in which this material was *generated*. In general this is *undesirable*, because figure B would be printed before figure A.

35.7.1   The Strategy of Fixing \midinsert

Macro \midinsert can be modified to print material always in the generated order. Here is the general strategy:

1. A dimension register \@TopInsertSize is introduced. This register contains the length of any \topinsert material that was *generated, but not yet written* to the dvi file by the output routine. If you think about insertions as *queues*, (queues that hold material to be placed later on a page) then this register contains the length of the \topins queue (the term *length* as it is used here does *not* refer to the number of elements in the queue, but to the length of the material held in that queue).

This register is updated as follows:

(a)   Each time insertion material of the class \topins is generated, \@Top-InsertSize is *incremented* by the sum of the height and depth of the material. It is incremented regardless of whether the original instruction was a \topinsert, \pageinsert or a "converted \midinsert."

(b)   Each time the output routine writes \topins material to the dvi file, \@TopInsertSize is *decremented* by the length of material written out to the dvi file (in other words, it is decremented by the height plus depth of \box\topins).

2.   \@TopInsertSize can now be used by the redefined \midinsert macro (defined below) to determine whether the \topins queue is empty or not. In other words, regardless of whether the material fits on the current page, if \@TopInsertSize is greater than zero, a \midinsert must be converted to a \topinsert. Otherwise, if \@TopInsertSize is zero, the new \midinsert and the old \midinsert behave in the same way, so if the material fits onto the current page, it will be placed there.

35.7.2   The Source Code of the New \midinsert

In the altered definition of \midinsert, the following two changes are triggered when the macro source code below is loaded:

1.   \endinsert is changed, because a redefined \endinsert causing \MidInsertFix to be invoked. Macro \MidInsertFix is defined below.

2.   \pagecontents is changed, since a redefined \pagecontents is loaded, which invokes \MidInsertFixPage. Again, macro \MidInsertFixPage is defined below.

If you want to use this new version of \midinsert, simply load the following macro source file and the necessary redefinitions will take place.

$\mathcal{P}'$ • new-midi.tip •

```
15   \InputD{op-pagec.tip}                    % 35.6.1, p. 113.
16   \InputD{op-endin.tip}                     % 35.6.2, p. 114.
17   \catcode'\@ = 11
```

Save the length of the \topins-related insertion material held over and not yet printed in the following dimension register.

```
18   \newdimen\@TopInsertSize
19   \@TopInsertSize = 0pt
```

For debugging purposes the following macro \ReportTopInsertSize is quite useful: it writes the current value of \@TopInsertSize plus some additional comment (defined by #1, the only argument of the macro) into the log file.

```
20   \def\ReportTopInsertSize #1{%
```

```
21        \wlog{\string\@TopInsertSize: "#1"}%
22        \wlog{\EightSpaces The value of \noexpand\@TopInsertSize is
23            \the\@TopInsertSize}%
24    }
```

The following macro \MidInsertFix is called by the redefined \endinsert.

```
25    \def\MidInsertFix{%
```

Check whether the \topins insertion queue still contains material. If that is the case, the material *must not* be printed in place.

```
26        \ReportTopInsertSize{\string\MidInsertFix}%
27        \ifdim\@TopInsertSize > 0pt
```

There is remaining material of the insertion class \topins, and therefore this \midinsert must become a \topinsert.

```
28            \@midfalse
29            \p@gefalse
30        \fi
31    }
```

The following macro \MidInsertFixPage adds the length of \pageinsert material to the queue-length register \@TopInsertSize.

```
32    \def\MidInsertFixPage{%
33        \ReportTopInsertSize{\string\MidInsertFixPage[1]}%
34        \global\advance\@TopInsertSize by \vsize
35        \ReportTopInsertSize{\string\MidInsertFixPage[2]}%
36    }
```

The following macro \MidInsertFixTop adds the length of any \topinsert material or any "converted" \midinsert material to the queue-length register \@TopInsertSize.

```
37    \def\MidInsertFixTop{%
38        \ReportTopInsertSize{\string\MidInsertFixTop[1]}%
39        \global\advance\@TopInsertSize by \ht0
40        \global\advance\@TopInsertSize by \dp0
41        \ReportTopInsertSize{\string\MidInsertFixTop[2]}%
42    }
```

The following macro is called by the new \pagecontents macro (as defined in 35.6, p. 113); the size of the material of the insertion class \topins must be subtracted from \@TopInsertSize.

```
43    \def\@UpdateTopInsertSize{%
44        \ReportTopInsertSize{\string\@UpdateTopInsertSize[1]}%
45        \global\advance\@TopInsertSize by -\ht\topins
46        \global\advance\@TopInsertSize by -\dp\topins
47        \ReportTopInsertSize{\string\@UpdateTopInsertSize[2]}%
48    }
49    \catcode`\@ = 12
```

• End of new-midi.tip •

35.7.3 Showing That It Works

Here is a very brief example that shows that everything works. We use the dimensions as specified in the introductory example which tried to justify the modification to \midinsert.

• <code>ex-new-midi.tip</code> •

```
1    \input inputd.tip
2    \InputD{new-midi.tip}                      % 35.7.2, p. 117.
```

Next we need an insertion list file so we will know where the material material is printed.

```
3    \newwrite\loiwrite
4    \immediate\openout\loiwrite = \jobname.loi
```

For simplicity, use a nice round value for \vsize.

```
5    \vsize = 8in
```

Define a macro which generates a \midinsert, with the size specified as parameter #1, and some identification text, #2.

```
6    \def\MidInsertPlay #1#2{%
7        \midinsert
8          \vbox to #1{%
9              This is an insert.
10             \write\loiwrite{Insertion #2, page \the\pageno.}%
11             \vfill
12         }
13       \endinsert
14   }
```

Fill the first page with 5 in of material.

```
15   \vbox to 5in{}
```

Generate two \midinserts (the "long one" first).

```
16   \MidInsertPlay{4in}{The long one}
17   \MidInsertPlay{2in}{The short one}
18   \bye
```

• End of <code>ex-new-midi.tip</code> •

The insertion list file generated by the preceding source file reads as follows (note the correct order of the two insertions).

• <code>ex-new-midi.loi</code> •

```
1    Insertion The long one, page 2.
2    Insertion The short one, page 2.
```

35.8   Dumping the Page's Main Vertical List into the Logfile, \ShowPlainLists...

I now present a source code file which applies the following changes to the output routine of the plain format.

1. The contents of the vertical list of the main page (that is, of box register 255) is logged in the log file.
2. The contents of the box register, which contains the text of footnotes to be printed on the current page, is logged in the log file. If there are no footnotes on a page, a short message to this effect is written to the log file.
3. The contents of box register \topins (which contains the insertion material of the insertion class of the same name) is logged into the log file. If there is no insertion material of this class, a short message to this effect is written.

Note that all of the above log operations display the various lists with depth 1, but have "unlimited" length so that the full lists are logged.

The following TeX source code can be used by you as follows: first load this file using \InputD (see 28.4.3, p. III-467). In this way the processing sketched above can be enabled by \ShowPlainListstrue, whereas \ShowPlainLists-false, which is also the default, disables the listed features.

$$\mathcal{P}'\quad \bullet \text{ showpll.tip } \bullet$$

```
15   \InputD{shboxes.tip}                      % 4.5.15, p. I-111.
```

The following macro source code loads a redefined \pagecontents macro that invokes \@ShowPlainLists.

```
16   \InputD{op-pagec.tip}                     % 35.6.1, p. 113.
17   \catcode'\@ = 11
```

Define the \ifShowPlainLists conditional now.

```
18   \newif\ifShowPlainLists
19   \ShowPlainListsfalse
```

Now define the macro \@ShowPlainLists which is called by the new definition of \pagecontents of 35.6, p. 113.

```
20   \def\@ShowPlainLists{%
```

First write to the log file the vertical list of the main page (box register 255).

```
21       \ifShowPlainLists
22           \wlog{*** \string\@ShowPlainLists: main vertical list ***}%
23           \wlog{*** Page number (\string\count0): \the\count0
24               \space***}%
25           \ShowBoxDepthOne{255}%
```

Do the same with footnote material. Print a short message, if there is no footnote material for the current page.

```
26       \ifvoid\footins
```

```
27              \wlog{\string\@ShowPlainLists: no footnotes.}%
28          \else
29              \wlog{*** \string\@ShowPlainLists: footnote box ***}%
30              \ShowBoxDepthOne{\footins}%
31          \fi
```

Do the same with material related to insertion class \topins.

```
32          \ifvoid\topins
33              \wlog{\string\@ShowPlainLists: no topinserts.}%
34          \else
35              \wlog{*** \string\@ShowPlainLists: top inserts ***}%
36              \ShowBoxDepthTwo{\topins}%
37          \fi
38          \wlog{*** \string\@ShowPlainLists: end dump of
39              page: \the\count0 \space ***}%
40      \fi
41  }
42  \catcode'\@ = 12
```

• End of `showpll.tip` •

For an application of this macro see 10.9.2, p. II-29.

35.9   Output Routine and Related Macros for This Series

At this point we will study the output routine and other related macros for this series. In particular, changes concerning the page layout (running heads) were made compared with the plain format's original output routine.

The following is a summary of those changes that affect the output routine or the page layout:

1. Page numbers are handled differently in that there is a counter PageNo declared, based on the counter macros of 3.4, p. I-68. See also 30.7.8, p. III-562.
2. A special macro for the automatic generation of footnote numbers is used; see 35.4.2, p. 102.
3. According to 30.7.5.1, p. III-555, and 30.7.5.4, p. III-559, certain macros must be treated as "robust macros." This also affects the output routine.
4. Page numbers are handled by storing a *page layout code* in counter register \@PageLayoutCode. This register can have the following values (note that whether the page numbers are printed as arabic numbers or roman numerals is *not* determined by those codes, but by the \NewCounter or \Reassign-Counter call for counter PageNo).

    (a)   0. The page layout works as in the plain format. In particular you can use \nopagenumbers to suppress page numbering. This value is also the default.

(b)   1. No page numbers and running heads are printed. This applies tc all pages until changed.

(c)   2. Same as 1 for the current page. But after this page has been printed, a switch to mode 3 for the following page (and pages) is executed.

(d)   3. This page layout code is for the preliminaries and endmatter of this series: left-hand and right-hand pages contain a description of the current part of the document (such as table of contents) and the page number (the page number is positioned to the very left on left-hand pages and to the very right on right-hand pages).

(e)   4. This page layout code is identical to page layout code 1 except that a change to page layout code 5 occurs after the current page has been printed.

(f)   5. This page layout is largely identical to the page layout code 3. The main difference is that the current chapter heading is printed on left-hand pages and the current section heading on right-hand pages.

35.9.1   The Output Routine Used in This Series

Here is the source code of the output routine used in this series. Some other additional macro definitions are also given.

$\mathcal{P}'$ • ts-outpu.tip •

```
15    \InputD{namedef.tip}              % 19.1.8, p. III-73.
16    \InputD{showpll.tip}              % 35.8, p. 120.
17    \InputD{new-midi.tip}             % 35.7.2, p. 117.
18    \InputD{topinfix.tip}             % 35.5.2, p. 110.
19    \InputD{nathd.tip}                % 8.1.3, p. I-264.
20    \InputD{rangetst.tip}             % 25.1.9, p. III-328.
```

Writing a document such as this series and of this size I found it useful to let TeX help me point out those pages with bad page breaks. In other words I wanted to have access to the information about the quality of page breaks *without* actually looking at the pages (actually *correcting* bad page breaks usually requires looking at them, of course).

The following conditional, if set to true, causes the parameters related to page breaks to be written to a special log file. By default this feature is disabled. Call macro \WriteToPageLogFile to enable the feature.

```
21    \newif\ifWritePageLog
22    \WritePageLogfalse
```

Get a special stream for the page log file.

```
23    \newwrite\PageLogStream
```

Now define macro \WriteToPageLogFile.

```
24    \def\WritePageLogFile{%
25        \immediate\openout\PageLogStream = \jobname.plog
```

```
26        \WritePageLogtrue
27    }
```

Save the badness of a specific box in the following box register.

```
28    \newcount\BadnessSave
```

Sometimes (to see where a page break went wrong) it is useful to see the main vertical list of a page of this book. By default though, we don't look at it.

```
29    \ShowPlainListsfalse
30    \catcode'\@ = 11
```

Now define macro \SetPageLayout which will set up the page layout according to the codes just specified. This macro has one parameter #1, the layout code.

```
31    \newcount\@PageLayoutCode
32    \def\SetPageLayout #1{%
33        \global\@PageLayoutCode = #1
34        \CheckRange{\@PageLayoutCode}{0}{5}%
35            {\string\SetPageLayout: }
36    }
```

The following call determines the default page layout.

```
37    \SetPageLayout{0}%
```

The left running head normally contains the chapter's heading and the right running head usually contains the current section heading. For certain parts of this series, for instance, for the first page of every chapter or for certain pages of the front matter (title pages, dedication page), it is empty.

We define both running heads as empty initially.

```
38    \def\@LeftRunningHead{}
39    \def\@RightRunningHead{}
```

The following macro \NewPageRightHandSpecial is invoked at the end of every chapter and at the end of other major subdivisions. It causes an empty left-hand page (even page number) to be generated, if the current text ended on a right-hand page.

```
40    \newcount\@SavedPageLayoutCode
41    \def\NewPageRightHandSpecial{%
42        \vfill
43        \supereject
44        \ifodd\count0
45            \wlog{\string\NewPageRightHandSpecial: no empty
46                page to generate (\string\count0 = \the\count0)}%
47        \else
```

Current page number is even.

```
48            \hbox{}
```

No running head on an empty right-hand page! Save the old layout code, print the empty right-hand page and restore the layout code.

```
49          \@SavedPageLayoutCode = \@PageLayoutCode
50          \SetPageLayout{1}%
51          \wlog{\string\NewPageRightHandSpecial: empty page
52             generated,
53             (\string\count0 = \the\count0).}%
54          \vfill
55          \eject
56          \SetPageLayout{\@SavedPageLayoutCode}%
57       \fi
58    }
```

Pages can be shifted to the left depending on whether the current page number is odd or even.

```
59    \newdimen\OddPagesHorizontalShift
60    \newdimen\EvenPagesHorizontalShift
61    \newdimen\CurrentPageShift
```

The redefinition of the plain format output routine for this series begins here.

```
62    \def\plainoutput{%
```

First dump box 255 and the insertion boxes into the log file if this has been requested.

```
63       \@ShowPlainLists
```

Running heads must be expanded now. Note that these expansions are local to this output routine because output routines in general are executed inside implicit groups.

```
64       \edef\@LeftRunningHead{\@LeftRunningHead}%
65       \edef\@RightRunningHead{\@RightRunningHead}%
```

Set the running head and running foot to empty at first, if the plain format layout is *not* in effect.

```
66       \ifnum\@PageLayoutCode = 0
67       \else
68          \headline = {}
69          \footline = {}
70       \fi
```

The page layout is next.

```
71       \ifcase\@PageLayoutCode
```

If the page layout code is 0, the plain format's defaults apply, so you should leave everything as it is.

```
72       \or
```

If the page layout code is 1, no running head is printed. This has been already set up, so no additional TeX code needs to be executed.

```
73       \or
```

The page layout code is 2, that is no running head and foot. Next page uses page layout code 3.

```
74              \global\@PageLayoutCode = 3
75          \or
```

The page layout code is 3. Execute \@PageLayoutCodeThree now.

```
76              \@PageLayoutCodeThree
77          \or
```

The page layout code is 4, next time it should be 5.

```
78              \global\@PageLayoutCode = 5
79          \or
```

The page layout code is 5. Set-up running heads now.

```
80              \ifodd\count0
81                  \headline = {%
```

Use the proper size and type face.

```
82                      \small\rm
83                      \hfil
```

The use of \botmark to generate the proper running head is explained in 34.7.
See also 31.2.9, p. III-604.

```
84                      \botmark
85                      \hskip 18pt
86                      \PrintCounter{PageNo}%
87                  }%
88              \else
89                  \headline = {%
90                      \small\rm
91                      \PrintCounter{PageNo}%
92                      \hskip 18pt
93                      \@LeftRunningHead
94                      \hfil
95                  }%
96              \fi
```

The end of all legal page layout codes is reached. Generate an error message if a
layout code > 5 is found.

```
97          \else
98              \errmessage{\string\plainoutput: \string\@PageLayoutCode
99                  out of range.}%
100         \fi
```

Get the complete page together in box register 4 and save its badness in \Bad-
nessSave.

```
101         \setbox 4 = \vbox{%
102             \pagebody
103         }%
```

Quality control: various parameter values are written to a special log file.

```
104         \ifWritePageLog
```

```
105            \immediate\write\PageLogStream{%
106                Part name: \CurrentPartName, Page \the\count0
107            }%
108            \immediate\write\PageLogStream{%
109                Page \the\count0:
110                stretch: \the\pagestretch,
111                shrink: \the\pageshrink,
112                outputpenalty: \the\outputpenalty
113            }%
114            \immediate\write\PageLogStream{%
115                pagefilstretch: \the\pagefilstretch,
116                pagefillstretch: \the\pagefillstretch
117            }%
118            \immediate\write\PageLogStream{%
119                ht 255: \the\ht255,
120                dp 255: \the\dp255
121            }%
122            \NaturalHeight{\dimen0}{255}%
123            \NaturalDepth{\dimen1}{255}%
124            \immediate\write\PageLogStream{%
125                Natural height: \the\dimen0,
126                Natural depth:  \the\dimen1
127            }%
```

Compute the difference between the ideal and the actual height and report this
value.

```
128            \dimen2 = \vsize
129            \advance\dimen2 by -\dimen0
130            \immediate\write\PageLogStream{%
131                Ideal height: \the\vsize,
132                Difference:   \the\dimen2
133            }%
```

Build the page in box register 4. Report the badness, the height and the depth
of the box being generated.

```
134            \immediate\write\PageLogStream{%
135                badness: \the\BadnessSave,
136                height: \the\ht4,
137                depth: \the\dp4
138            }%
```

Write an empty line into the protocol file so that the output from various pages
is separated.

```
139            \immediate\write\PageLogStream{}%
140        \fi
```

Robust macros should *expand* while the page is built, because the running head
might contain calls to otherwise robust macros. The page is built in box register 5.

```
141        \setbox 5 = \vbox{%
142            \makeheadline
143            \box 4
```

```
144        \makefootline
145      }%
```

On the other hand, during the following \shipout *robust macros* should *not* expand.

```
146      \@MakeRobustMacros
```

Compute the amount by which the pages are shifted horizontally depending on the current page number (odd or even) and then do it.

```
147      \ifodd\count0
148          \CurrentPageShift = \OddPagesHorizontalShift
149      \else
150          \CurrentPageShift = \EvenPagesHorizontalShift
151      \fi
152      \wlog{Redefined \string\plainoutput (ts-outpu.tip):
153          Shifting: shift amount is \the\CurrentPageShift\space\space
154          (page is \the\pageno).}%
155      \setbox 6 = \vbox{%
156          \moveright\CurrentPageShift \box5
157      }%
158      \shipout\box6
```

The following call of \advancepageno calls the version of this macro as given in 30.7.8, p. III-562.

```
159      \advancepageno
```

The following is identical to the standard plain format output routine.

```
160      \ifnum\outputpenalty > -10000
161      \else
162          \dosupereject
163      \fi
164    }
```

35.9.2   The Definition of \@PageLayoutCodeThree

Processing page layout code 3 is done by the following macro. This layout is also used somewhere else; therefore the settings below are stored in a separate macro rather then included directly into the preceding code.

```
165  \def\@PageLayoutCodeThree{%
166      \ifodd\count0
167          \headline = {%
```

The type size, font and other specifications are set according to the publisher's guidelines.

```
168              \small\rm
169              \hfil
170              \@RightRunningHead
```

```
171                \hskip 18pt
172                \PrintCounter{PageNo}%
173            }%
174        \else
175            \headline = {%
176                \small\rm
177                \PrintCounter{PageNo}%
178                \hskip 18pt
179                \@LeftRunningHead
180                \hfil
181            }%
182        \fi
183    }
184    \catcode'\@ = 12
```

• End of `ts-outpu.tip` •

35.9.3   Figures and Tables Using \topinsert

In the macros used to generate figures and tables for this series, I found using a modified \topinsert sufficient.

$\mathcal{P}'$ • `ts-float.tip` •

```
15    \InputD{dblarg.tip}              % 23.4.4.2, p. III-258.
16    \InputD{centpar.tip}             % 25.1.15, p. III-333.
17    \InputD{futlet.tip}              % 23.4.3, p. III-256.
18    \InputD{compst.tip}              % 25.1.17.1, p. III-334.
```

Loading the following code ensures that insertions that are too long are shortened to a little less than \vsize so that no extra blank pages are generated by TeX. See 35.5.2, p. 110, for a discussion of this issue.

```
19    \InputD{topinfix.tip}            % 35.5.2, p. 110.
20    \catcode'\@ = 11
```

Declare two counters, one for figures and the other for tables. Both are reset at the beginning of each chapter. Figure and table numbers are formed by the pair chapter number and serial number within the chapter.

```
21    \NewCounter{FigureNo}{\arabic}%
22        {\TheCounter{ChapterNo}.\TheCounter{FigureNo}}%
23        {\TheCounter{ChapterNo}.\TheCounter{FigureNo}}%
24    \NewCounter{TableNo}{\arabic}%
25        {\TheCounter{ChapterNo}.\TheCounter{TableNo}}%
26        {\TheCounter{ChapterNo}.\TheCounter{TableNo}}%
27    \AddCounterToResetList{FigureNo}{ChapterNo}
28    \AddCounterToResetList{TableNo}{ChapterNo}
```

In general, handling of captions goes as follows: the \Caption macro is a macro with one argument, the text of the caption. In case a different caption text is needed for the list of figures or tables from the caption text being printed, this macro can be invoked with one optional argument preceding the regular argument. The optional argument, enclosed in square brackets, determines the entry for the list of figures or tables (the regular argument then determines what is printed in the caption). Here is the definition of \Caption.

```
29    \def\Caption{\DblArg{\@Caption}}
```

The default definition of \@Caption is to print an error message, because the correct definition of \@Caption is provided by \BeginFigure or \BeginTable below. Thus captions may only be defined within a figure or table environment.

```
30    \def\@Caption [#1]#2{%
31        \errmessage{\string\@Caption: caption out of place, caption
32            text = #1.}%
33    }
```

A figure is enclosed in \BeginFigure and \EndFigure. The figure is saved in a vbox similar to the way \topinsert does it. The \BeginFigure macro has one optional argument (enclosed in square brackets). If this optional argument is present (a p), a \pageinsert, instead of a \topinsert, is used.

```
34    \def\BeginFigure{%
35        \DoFutureLet{\ifx}{[}{\@BeginFigure}{\@BeginFigure[t]}%
36    }
```

The following conditional is set to true, if \@BeginFigure's parameter is correct.

```
37    \newif\if@FigureInsertOk
```

Now define macro \@BeginFigure which has one argument, [#1], enclosed in square brackets. This parameter can be either a t, in which case a \topinsert is used below, or it can be a p, in which case a \pageinsert is generated below.

```
38    \def\@BeginFigure [#1]{%
39        \begingroup
```

Test whether the argument is correct and set up the proper \...insert call.

```
40        \def\@UseThisInsert{}%
41        \@FigureInsertOkfalse
42        \if\StringsEqualConditional{#1}{t}%
43            \def\@UseThisInsert{\topinsert}%
44            \@FigureInsertOktrue
45        \fi
46        \if\StringsEqualConditional{#1}{p}%
47            \def\@UseThisInsert{\pageinsert}%
48            \@FigureInsertOktrue
49        \fi
50        \if@FigureInsertOk
51        \else
52            \errmessage{\string\@BeginFigure: illegal argument
53                "#1." Use [t] instead.}%
```

If the code is wrong, use a \topinsert.

```
54            \def\@UseThisInsert{\topinsert}%
55        \fi
```

Clear out the figure caption text.

```
56        \gdef\FigureCaptionText{}%
57        \gdef\FigureCaptionTextLof{}%
```

Increment the figure number counter.

```
58        \StepCounter{FigureNo}%
```

Define \Label to refer to the current figure number.

```
59        \def\Label ##1{\@Label{##1}{\RefCounter{FigureNo}}{0}}%
```

The following definition of \@Caption handles captions of figures. The first argument (enclosed in square brackets) determines the list of figures' entry and the second one determines the text printed with the figure.

```
60        \def\@Caption [##1]##2{%
61            \gdef\FigureCaptionTextLof{##1}%
62            \gdef\FigureCaptionText{##2}%
63            \wlog{\noexpand\@Caption (figures), caption text saved:
64                        ##2.}%
65        }%
```

Now start to collect the figure itself.

```
66        \setbox 0 = \vbox\bgroup
67    }
```

The \EndFigure macro which will be defined now, first terminates the vbox that was opened by \BeginFigure.

```
68    \def\EndFigure{%
69        \egroup
```

Box register 0 now contains the complete figure. The caption (if any is specified) is added now, before \topinsert or \pageinsert is actually called. The macro \@UseThisInsert expands to either \topinsert or \pageinsert.

```
70        \@UseThisInsert
71            \box0
```

Print a caption text, if one was provided.

```
72            \if\EmptyStringConditional{\FigureCaptionText}%
73            \else
```

The following \bigskip determines the distance between the figure itself and the caption below it.

```
74                \bigskip
```

If the figure caption is short enough, it is printed centered on one line, otherwise it is printed as an ordinary paragraph.

```
75                \CenterOrParagraph{%
76                    \small
77                    Figure~\PrintCounter{FigureNo}.
78                    \FigureCaptionText
79                }%
```

Write the necessary information to the auxiliary file of the current part, so a list
of figures can be generated later.

```
80                  \WriteToAuxSpecial{lof}{1}{\PrintCounter{FigureNo}}%
81                     {\FigureCaptionTextLof}{\PrintCounter{PageNo}}%
82           \fi
```

End the \topinsert or \pageinsert started above.

```
83        \endinsert
```

End the \begingroup of \BeginFigure.

```
84        \endgroup
85    }
```

Tables are next and are done in the same way. The caption of a table appears
on the *top* of the table though! The text below is less documented because it is
largely identical to the text of \BeginFigure. First define macro \BeginTable:

```
86    \def\BeginTable{%
87        \begingroup
```

Clear out the table caption text.

```
88        \gdef\TableCaptionText{}%
89        \gdef\TableCaptionTextLot{}%
90        \StepCounter{TableNo}%
91        \def\Label ##1{\@Label{##1}{\RefCounter{TableNo}}{0.}}%
92        \def\@Caption [##1]##2{%
93           \gdef\TableCaptionText{##2}%
94           \gdef\TableCaptionTextLot{##1}%
95           \wlog{\noexpand\@Caption for tables: ##2}%
96        }%
97        \setbox 0 = \vbox\bgroup
98    }
```

The \EndTable macro is defined first; it terminates the vbox, which was
used to collect the table.

```
99    \def\EndTable{%
100       \egroup
```

Box register 0 now contains the complete table. The caption is added first, then
\topinsert is called.

```
101       \topinsert
102          \if\EmptyStringConditional{\TableCaptionText}%
103          \else
104             \smallskip
105             \CenterOrParagraph{%
106                \small
107                Table~\PrintCounter{TableNo}.
108                \TableCaptionText}%
109             \WriteToAuxSpecial{lot}{1}{\PrintCounter{TableNo}}%
110                {\TableCaptionTextLot}{\PrintCounter{PageNo}}%
```

The following \smallskip causes some space between the caption and the table body itself to be inserted.

```
111                    \smallskip
112            \fi
113            \box0
114        \endinsert
115        \endgroup
116  }
117  \catcode'\@ = 12
```

• End of `ts-float.tip` •

35.10   Summary

- The running head and footer of a page are determined by the contents of the **\headline** and \footline token registers in the plain format. By default the running head is empty and the footer contains a centered page number.
- The printing of page numbers can be suppressed with **\nopagenumbers**.
- Macros **\makeheadline** and **\makefootline** print the running head and the footer. These macros act, when called by **\pagecontents**, the macro which builds the page (this macro is the central macro of the plain format output routine).
- The plain format has two insertion classes. Insertion class **\footins** is used for footnotes and footnotes are generated with the **\footnote** command. While you are responsible for numbering footnotes yourself, it is very easy to design a macro numbering footnotes automatically. The insertion class **\topins** is the other insertion class, used when macros **\topinsert** or **\pageinsert** are called. A modified **\topinsert** dealing with oversized material was also discussed.
- **\midinsert** is like **\topinsert**, if some material cannot fit on a page. On the other hand, if there is sufficient room for the material, then it is placed right in the current location. A modified version of **\midinsert** was discussed, preventing a reversal in the order of insertion material.
- A version of the output routine of the plain format was discussed where the current page's vertical list is written to the log file.
- The output routine used for this series, which is a modification of the plain format output routine, was also presented. One of the modifications was the addition of the writing of a protocol file with certain quality control parameters to this file. Macros for dealing with figures and tables were also presented.

36
Output Routines with Insertions

In this chapter we will continue our discussion of output routines with insertions. The previous Chapter on the plain format output routine already demonstrated output routines with insertions.

The purpose of an insertion is to allow the user to enter certain entities, such as a table or a footnote, together with regular text. However, the output of insertion material does not occur at the location where the regular text is printed, but somewhere else, and that location is determined by the insertion class (and is usually different between different insertion classes).

36.1 An Output Routine Setup for the Development of an Index

The following output routine is useful for printing a document while an index for this document is developed. In 29.1, p. III-489, I already discussed briefly how an index is generated for a document processed with TeX:

1. The \write instruction (28.5, p. III-472) is used to write index expressions and page numbers to an index file.
2. This index file is sorted by an index sorting program.
3. The sorted index file is read back into the document to be printed, typically at the end of a document.

Now let us discuss how you can *simplify the development process* of an index. There is of course the possibility of generating the index and then going through every entry of the index and checking all the page numbers. This is a very tedious process, and if you think about it for a moment there is a much easier way to do it.

It is much more convenient to check the index on a page-by-page basis. Simply print all index terms that would refer to a page in the margins of that page. Then you can verify very easily whether you forgot any relevant index terms for a specific page.

Let me discuss an implementation strategy here. An insertion class \indexins will be defined below. Because all index terms are printed in the margins, the page layout itself is *not* influenced by this; index terms printed in the margin do *not* take away space from the current page and therefore they have *no* influence on the page layout (we assume that the margin is sufficiently long to allow the printing of all index terms of one page, which is a reasonable assumption). By setting the magnification factor for this insertion class to zero TeX is told that an insertion of that class does *not* take up any space on the current page (see also 32.7.3, p. 30).

36.1.1   The Output Routine for Index Printing

Here is the source code of an output routine used to print index terms in the margins of a document.

𝒫   • index-or.tip •

```
15    \InputD{box-mac.tip}                    % 9.3.14, p. I-343.
16    \InputD{setstrut.tip}                    % 7.4.3.1, p. I-239.
```

Define insertion class \indexins. The magnification of this insertion class is set to zero (see the top of this page for an explanation of this). There is no limit on the number of index terms printed per page and overflow is actually possible (in practice this probably never occurs).

```
17    \newinsert\indexins
18    \dimen\indexins = \maxdimen
19    \skip\indexins = 0pt
20    \count\indexins = 0
```

Open the index file for writing out index information. I did not choose the usual file extension for index files of this series (**idx**) to avoid conflicts with processing this example; instead I used the extension **idxx**.

```
21    \newwrite\indexwrite
22    \openout\indexwrite = \jobname.idxx
```

Define macro \index now. This macro has one parameter, #1, an index term.

```
23    \def\index #1{%
```

The macro first writes the index term out to the index file.

```
24        \write\indexwrite{\noexpand\ix{#1}{\the\pageno}}%
```

Now an insertion containing the current index entry is generated. The index term will be printed in a small font (5 pt roman).

```
25        \insert\indexins{%
26            \baselineskip = 7pt
```

```
27              \ComputeStrut
28              \hbox{\fiverm\MyStrut #1}
29          }%
30      }
```

The following macro \indexx has one parameter, #1, an index term that is very similar to the macro \index presented before. The main difference is that the term is not only inserted into the index but also printed in the text itself.

```
31      \def\indexx #1{%
32          \index{#1}%
33          #1%
34      }
```

Now define the output routine for printing the index terms out in the margins. This output routine is straightforward. It prints, right to left, the current text page, some horizontal white space of 20 pt, and the box register containing the index terms of the current page. See 41.5, p. 355, for a discussion of \valign.

```
35      \output = {%
36          \shipout\hbox-{%
37              \valign{
38                  #\vfil
39              \cr
40                  \VtopR{\unvbox255}\cr
41                  \hbox to 20pt{}\cr
42                  \BoxR\indexins\cr
43              }
44          }
```

Increment the page number and the output routine is finished.

```
45          \global\advance\pageno by 1
46      }
```

• End of `index-or.tip` •

36.1.2   An Example Application of the Preceding Output Routine

For an example application of this output routine we have the following source code:

• `ex-indexins.tip` •

Load the necessary macros.

```
1   \input inputd.tip
2   \InputD{index-or.tip}                    % 36.1.1, p. 134.
```

Change the page layout dimensions.

```
3   \hsize = 3.0in
4   \vsize = 6.0in
```

Generate some text including some index terms:

```
5        This is now some text, of the new index insertion output
6    routine. Let's index the term ''\indexx{spring},'' also
7    ''\indexx{fall},'' and also ''\indexx{summer}.''
8    One could index some more terms, if one wanted. There is
9    ''\indexx{winter},'' a season we missed so far.
10   \par
```

Of course, one may index a word not appearing in the text, as is done next.

```
11   \index{seasons}
```

Here is another paragraph. Now refer to an index term given before. Note that this implies that the index term will be printed twice out in the right-hand margin of the current text page (which is hardly a problem).

```
12       More index terms, and of course there is the
13   possibility, that the same term is indexed again. Like, for instance,
14   ''\indexx{winter}''. One of the disadvantages of this form of index,
15   and we want to be honest\index{honesty} about it, is the fact
16   that the index listing out in the margin is {\it not\/}
17   sorted\index{unsorted index}.
18
19       Ok, that is \indexx{good enough}. Actually it is
20       \indexx{real good enough}.
21   \bye
```

• End of <code>ex-indexins.tip</code> •

The above input generates output printed in Fig. 36.1 on the next page. This index file is generated by the preceding example:

```
1    \ix {spring}{1}
2    \ix {fall}{1}
3    \ix {summer}{1}
4    \ix {winter}{1}
5    \ix {seasons}{1}
6    \ix {winter}{1}
7    \ix {honesty}{1}
8    \ix {unsorted index}{1}
9    \ix {good enough}{1}
10   \ix {real good enough}{1}
```

36.1.3   A Different Approach Solving the Same Problem

A different solution using a `dvi` file processor (see 33.3.1, p. 42) can be implemented easily: the index terms to be written into the margins of the text pages of the document would be typeset in a separate TeX execution and then overlaid on top of the main document pages using **dvimerge**. This has the following advantages (compared to the approach presented above, based on the output routine):

> This is now some text, of the new index insertion output routine. Let's index the term "spring," also "fall," and also "summer." One could index some more terms, if one wanted. There is "winter," a season we missed so far.
>
> More index terms, and of course there is the possibility, that the same term is indexed again. Like, for instance, "winter". One of the disadvantages of this form of index, and we want to be honest about it, is the fact that the index listing out in the margin is *not* sorted.
>
> Ok, that is good enough. Actually it is real good enough.

Figure 36.1. Sample output for the output routine that supports the development of an index for some document.

1. No changes to any output routine are necessary. Note that the above output routine is really only a sketch of changes which you have to apply to whatever output routine you actually use.
2. The index expressions can be sorted (on a page-by-page basis) before they are printed out in the margins, and duplicates can be removed.
3. For the final printing of the document, where the marginal index terms are omitted, one simply skips the step of executing `dvimerge`. In the approach based on the output routine the document needs to be reprocessed with a different definition of `\index`, which thus generates no `\insert` calls, or the output routine must be changed to discard marginal notes.

36.2   An Output Routine with Flexible Figure Caption Placement

The next output routine is rather interesting[1]. The page layout (and therefore the output routine) distinguishes between left-hand and right-hand pages. In the case of the problem below this has quite broad implications. Instead of lengthy explanations I recommend that you look at the output of a sample application of the following output routines as it appears in Figs. 36.2–36.3, pp. 150–151. The page layout can be briefly summarized as follows:

1. The running heads of left-hand and right-hand pages are different.

---

[1] The problem leading to this output routine was, in a slightly more complicated fashion, proposed by Lia Mitchell from the Los Alamos National Laboratory, Los Alamos, New Mexico.

2. The running head and the footer lines are wider than the text page itself. The text page is shifted horizontally all the way to the right for left-hand pages, and all the way to the left for right-hand pages, within the limits established by the running head and footer.

3. Figures are printed *differently* on left-hand and right-hand pages: figure captions are always printed on the outside. To get TeX to place the figure captions on the correct side of a figure will be the main focus of the following discussion.

36.2.1   Sketching Two Different Solutions

Figures have their captions be printed differently, either on the left or on the right of the figure itself. The problem of deciding *where* the caption of a figure should be printed is interesting, because it is *not* possible to decide on the proper figure caption placement at the time the figure is *generated*: at this time it is *unknown* whether a figure will be later *printed* on a page with an even or an odd page number.

The problem below does *not* allow for figure placement using `\midinsert`, but along the lines of `\topinsert` only.

There are two solutions to this problem.

1. Use a *two pass approach*. Using a two pass approach is conceptually very easy and also very reliable. During the *first pass* captions are placed with a "don't care" attitude. During this pass a file is written, called the *figure placement file*, which is a reduced form of a "list of figure file." This figure placement file contains the page numbers of the pages on which the figures were finally printed.

    During the *second pass* these page numbers are read in from the figure placement file, and then depending on whether the read-in page number is even or odd, the caption is generated properly, at the time the figure is *generated*. This approach therefore does not really involve the output routine, because the captions and the figures are all generated properly.

    An implementation of this solution is discussed in 36.2.4, p. 141, but, of course, this solution is intellectually by far not as challenging as the following one pass solution.

2. Use a *one pass approach*. An insertion class is defined which is used to contain all figures with their captions *always* on the *left side*. This therefore delivers the correct placement of the captions for left-hand pages (even page numbers) *only*.

    In parallel a "shadow insertion queue" is maintained. This shadow insertion queue is *not* an insertion in the TeX sense. This shadow queue is maintained by the macros presented below. It contains *all* figures with their captions on the *right* side.

    The processing done by the output routine is obviously different depending on whether the current page number is odd or even.

(a)   If the current page number is *odd*, the material from the box associated with the figure insertion class is printed. The same amount of material (material of the same vertical length) will be removed from the shadow box queue, because the same figure or figures with their captions on the right-hand side are in this queue and are not needed.

(b)   If the current page number is *even*, the box containing the figures with their captions on the left side (the box from the insertion class) is discarded and the same amount of material is taken from the shadow insertion queue and printed.

This solution is slightly more complicated than the first approach, but obviously has the advantage that one pass suffices.

36.2.2   Presenting the Solutions

Both of the above solutions are presented in the following locations:

1.   TeX code common to both output routines is presented 36.2.3 on this page.
2.   The first output routine, based on a *two pass* approach, can be found in 36.2.4, p. 141.
3.   The second output routine, using *one pass* only, can be found in 36.2.5, p. 145.
4.   The TeX source code of an example application of both output routines can be found in 36.2.6, p. 148.
5.   In Figs 36.2–36.5, pp. 150–153, you find sample output generated by these output routines.

36.2.3   Code Common to Both Output Routines

Here is the source code common to both output routines.

𝒫 • ola-comm.tip •

```
15   \InputD{box-mac.tip}                    % 9.3.14, p. I-343.
```

Set \topskip and \vsize.

```
16   \topskip = 10pt
17   \vsize = 6.0in
```

Define the horizontal shift of even and odd pages.

```
18   \newdimen\LeftPageShift
19   \LeftPageShift =  0.25in
20   \newdimen\RightPageShift
21   \RightPageShift= -0.25in
```

Define various "horizontal sizes" now. \hsize is the width of each text page.
\PageWidth is the width of the "whole page," the width of running heads and
footers. \DiffWidth is the difference of \PageWidth and \hsize.

```
22    \hsize = 24pc
23
24    \newdimen\PageWidth
25    \PageWidth = 29pc
26
27    \newdimen\DiffWidth
28    \DiffWidth = \PageWidth
29    \advance\DiffWidth by -\hsize
```

Define macro \PageLine, similar to \line, but using \PageWidth instead
of \hsize.

```
30    \def\PageLine{\hbox to \PageWidth}
```

Define \SetHeader, a macro to set up the running head. The text of the
headerline is the macro's only parameter #1. This macro in turn defines two
macros \LeftHeader and \RightHeader used by the output routine later to
actually print the running head. Then the running head is initialized.

```
31    \def\SetHeader #1{
32        \def\RightHeader {\PageLine{\it\hfil #1}}
33        \def\LeftHeader  {\PageLine{\it #1\hfil}}
34    }
35    \SetHeader{Some Header}
```

Define a similar macro, \SetFooter with one parameter (#1), as text of the
footer line. Macros \RightFooter and \LeftFooter are defined by \SetFooter
and called by the output routine to print the footer line. Also a default footer
line is set up.

```
36    \def\SetFooter #1{
37        \def\RightFooter {%
38            \PageLine{%
39                \hfil
40                \it #1%
41                \hskip 0.5in
42                \bf \the\pageno
43            }%
44        }
45        \def\LeftFooter {%
46            \PageLine{%
47                \bf \the\pageno
48                \hskip 0.5in
49                \it #1%
50                \hfil
51            }%
52        }
53    }
54    \SetFooter{Some Footer}
```

Now define an insertion class called \FigureIns. The usual magnification factor of 1.0 is used for this insertion class. The maximum space a figure can occupy is the current page length. An additional space of 12 pt is reserved on a page if there is at least one figure on a page.

```
55    \newinsert\FigureIns
56    \count\FigureIns = 1000
57    \dimen\FigureIns = \vsize
58    \skip\FigureIns = 12pt
```

Now define macro \PrepareFigureBox that generates a sample figure box. This macro loads box register 0 with a box to print the figure, and box register 1 with a box to print the caption of the figure. This macro has three parameters:

- #1. The figure number.
- #2. The height of the figure box.
- #3. The caption's text.

```
59    \def\PrepareFigureBoxes #1#2#3{%
```

Load box register 0 now as discussed above.

```
60        \setbox 0 = \VboxR to #2{
61            \vfil
62            \centerline{\tt Figure~#1}
63            \vfil
64        }
```

Load box register 1 now as discussed above.

```
65        \setbox 1 = \VboxR{
66            \hsize = \DiffWidth
67            \raggedright
68            \noindent
69            \strut #3
70        }
71    }
```

• End of <code>ola-comm.tip</code> •

36.2.4   The Source Code of the Two Pass Output Routine

The source code of the first output routine (based on a two pass approach) begins here.

$\mathcal{P}$ • <code>ola-2p.tip</code> •

```
15    \InputD{namedef.tip}                    % 19.1.8, p. III-73.
16    \catcode`\@ = 11
```

Define the file name of the figure placement file.

```
17    \def\FigurePlacementFileName{\jobname.fip }
```

Counter `\Fco` is used to number the figures internally.

```
18    \newcount\Fco
```

Define streams for reading and writing the figure placement file.

```
19    \newread\FigurePlacementStreamIn
20    \newwrite\FigurePlacementStreamOut
```

Now define macro `\ReadTheFigurePlacementFile` which opens the figure placement file for input, reads in the whole file, and then closes the figure placement file. The figure placement file is then opened again for output.

```
21    \def\ReadTheFigurePlacementFile{%
22        \Fco = 1
23        \openin\FigurePlacementStreamIn = \FigurePlacementFileName
24        \ActuallyReadFigurePlacementFile
25        \immediate\openout\FigurePlacementStreamOut =
26            \FigurePlacementFileName
27    }
```

Now define macro `\ActuallyReadFigurePlacementFile`. This macro actually reads in the figure placement file.

```
28    \def\ActuallyReadFigurePlacementFile{%
```

Initial EOF: file was not there in the first place.

```
29        \ifeof\FigurePlacementStreamIn
30            \closein\FigurePlacementStreamIn
31            \let\ReadInFigurePlacementFileNext = \relax
32        \else
```

Take care of the `\par` appended at the end of a file read in using `\read`; see 28.2.4, p. III-452, for a discussion of this issue. This `\par` is in effect skipped, and the next time this macro is executed the preceding `\ifeof` will evaluate to true.

```
33            \read\FigurePlacementStreamIn to \ALineFromTheFile
34            \def\ParMeansDone{\par}%
35            \ifx\ALineFromTheFile\ParMeansDone
36            \else
```

The current figure serial number (found in counter `\Fco`) is used to define a macro where "Fig-\the\Fco" is the token used for the name of this macro. In other words, macros `\Fig-1`, `\Fig-2`, and so forth are defined. The replacement text of these macros is the page number of where this figure was printed during the first run.

```
37                \NameEdef{Fig-\the\Fco}{\ALineFromTheFile}
38                \advance\Fco by 1
39            \fi
40            \let\ReadInFigurePlacementFileNext =
41                \ActuallyReadFigurePlacementFile
42        \fi
43        \ReadInFigurePlacementFileNext
44    }
```

The end of this macro definition is reached now.

Another figure counter will be needed later.

```
45    \newcount\FBTwoCount
46    \FBTwoCount = 1
```

The following conditional is used to save the information of whether a caption should be printed on the left of a figure or right of a figure.

```
47    \newif\ifLeftSideCaption
```

Define macro \FBTwoPass to generate a sample figure. This macro has the following parameters:

- #1. The figure number.
- #2. The height of the box.
- #3. The figure caption.

```
48    \def\FBTwoPass #1#2#3{%
```

This sets up box register 0 to hold the figure, and box register 1 to hold the caption.

```
49        \PrepareFigureBoxes{#1}{#2}{#3}
```

Generate the insertion.

```
50        \insert\FigureIns{%
```

Write out the new placement information (that is page number) to the figure placement file.

```
51           \write\FigurePlacementStreamOut{\the\pageno}
```

Get placement information for the current figure from the figure placement file previously read-in. Note that there may be no such figure placement information available for a figure, for instance, because the very first processing of the current document occurs and there was no figure placement file to begin with.

```
52           \if\NameDefinedConditional{Fig-\the\FBTwoCount}%
```

There is figure placement information available for this figure because macro \Fig-x, where x is the number of the current figure, is defined.

```
53               \wlog{Figure \the\FBTwoCount:
54                     \NameUse{Fig-\the\FBTwoCount}}%
55               \expandafter\ifodd\NameUse{Fig-\the\FBTwoCount}%
56                   \LeftSideCaptionfalse
57               \else
58                   \LeftSideCaptiontrue
59               \fi
60           \else
```

It is unknown where the current caption should be placed because no macro \Fig-x is defined. Place the caption arbitrarily, on the left side.

```
61               \LeftSideCaptiontrue
```

```
62          \fi
```

Now actually generate the figure and the caption on the correct side of the figure (if this is the second pass of the processing of a document, the placement of the caption is correct).

```
63          \ifLeftSideCaption
64            \hbox{%
65              \valign{%
66                ##\vfil
67              \cr
68                \copy 1\cr
69                \copy 0\cr
70              }%
71            }%
72          \else
73            \hbox{%
74              \valign{%
75                ##\vfil
76              \cr
77                \copy 0\cr
78                \copy 1\cr
79              }%
80            }%
81          \fi
82        }
```

Increment the figure number counter.

```
83        \advance\FBTwoCount by 1
84  }
```

\OutCaptionGameTwoPass is a macro which should be assigned to \output so that the two pass output routine is used.

```
85  \def\OutCaptionGameTwoPass{%
```

Clear box registers 0 and 1.

```
86        \setbox0 = \box\voidb@x
87        \setbox1 = \box\voidb@x
```

Load dimension register 0 with the horizontal shift distance of the text with respect to the page as established by header and footer. Also set up the running heads now.

```
88        \ifodd\pageno
89          \dimen0 = 0pt
90          \let\Header = \RightHeader
91          \let\Footer = \RightFooter
92        \else
```

Even page number.

```
93          \dimen0 = \DiffWidth
94          \let\Header = \LeftHeader
95          \let\Footer = \LeftFooter
```

```
96        \fi
```

Build the page and ship it out. Note that after the second pass the placement of the figure caption is correct.

```
 97       \shipout\vbox{%
 98           \hrule height 1pt
 99           \vskip 5pt
100           \Header
101           \vskip 12pt
102           \ifvoid\FigureIns
103           \else
104               \box\FigureIns
105               \vskip\skip\FigureIns
106           \fi
107           \moveright\dimen0 \BoxR 255
108           \vskip 12pt
109           \Footer
110       }
```

Increase the page number.

```
111       \global\advance\pageno by 1
```

Deal with \supereject.

```
112       \ifnum\outputpenalty > -20000
113       \else
114           \dosupereject
115       \fi
116   }
117   \catcode'\@ = 12
```

• End of <code>ola-2p.tip</code> •

One concluding remark: it is *you* who must make sure that a document is processed at least twice. If you want to minimize processing you can save the figure placement file at the beginning of a run and compare it with the new figure placement file which is written during the TeX execution. Only if the two files are different would you need to reprocess the document.

36.2.5 The Code of the Second Output Routine (One Pass Approach)

The code of the second output routine which uses a one pass approach is:

$\mathcal{P}$ • <code>ola-1p.tip</code> •

```
15   \InputD{shboxes.tip}                    % 4.5.15, p. I-111.
16   \catcode'\@ = 11
```

The following box register is used to maintain the shadow insert on queue.

```
17   \newbox\InsertOtherBox
```

Define a macro \FBOnePass with the same parameters as \FBTwoPass (needless to say, this macro implements the second approach):

- #1. The figure number.
- #2. The height of the figure.
- #3. The caption of the figure.

```
18    \def\FBOnePass #1#2#3{%
19        \PrepareFigureBoxes{#1}{#2}{#3}
```

Box register \InsertOtherBox contains the same material, but this time the captions are printed on the right side of the figures.

```
20        \setbox\InsertOtherBox = \vbox{%
21            \offinterlineskip
```

Get the previous content of the box, add a break point (\penalty 0), and then add the new figure and its caption.

```
22                \unvbox\InsertOtherBox
23                \penalty 0
24                \hbox{%
25                    \valign{%
26                        ##\vfil
27                    \cr
28                        \copy 0\cr
29                        \copy 1\cr
30                    }%
31                }
32        }
```

The insertion class \FigureIns contains the figures with their caption always on the left side.

```
33                \insert\FigureIns{%
34                    \hbox{%
35                        \valign{%
36                            ##\vfil
37                        \cr
38                            \copy 1\cr
39                            \copy 0\cr
40                        }%
41                    }
42                }
43        }
```

The new output routine \OutCaptionGameOnePass starts here. Again a macro is defined which must be assigned to \output.

```
44    \def\OutCaptionGameOnePass{%
```

Clear box register 0.

```
45        \setbox0 = \box\voidb@x
```

Load \dimen0 with the horizontal shift of the text page itself. Define \Header
and \Footer properly, depending on whether the current page is an odd or an
even page. Box register 0 below is loaded with the material to be printed in case
there is figure insertion material on the current page. Box register 1 is used to
discard certain material (note that both box registers can be used safely because
of the implicit grouping which occurs inside output routines). \dimen1 is loaded
with the height of the insertion material (if there is any).

```
46        \ifodd\pageno
47            \dimen0 = 0pt
48            \let\Header = \RightHeader
49            \let\Footer = \RightFooter
50            \ifvoid\FigureIns
51            \else
```

On odd pages we use the material from the shadow box queue and discard the
material from the box register associated with the insertion class \FigureIns.

```
52                \dimen1 = \ht\FigureIns
53                \setbox1 = \box\FigureIns
54                \setbox0 = \vsplit\InsertOtherBox to \dimen1
55            \fi
```

Deal with even page numbers now.

```
56        \else
57            \dimen0 = \DiffWidth
58            \let\Header = \LeftHeader
59            \let\Footer = \LeftFooter
60            \ifvoid\FigureIns
61            \else
```

For pages with even page numbers the material from the insertion class is used,
and the same amount of material from the shadow queue must be discarded.

```
62                \dimen1 = \ht\FigureIns
63                \setbox1 = \vsplit\InsertOtherBox to \dimen1
64                \setbox0 = \box\FigureIns
65            \fi
66        \fi
```

From now on it's downhill: build the page and ship it out.

```
67        \shipout\vbox{%
```

The top part of the final page is built now.

```
68            \hrule height 1pt
69            \vskip 5pt
70            \Header
71            \vskip 12pt
```

The figure or figures are printed now, followed by a small vertical space between figures and the following text. Remember that box register 0 contains the
material to be printed.

```
72            \ifvoid 0
```

```
73              \else
74                  \box 0
75                  \vskip\skip\FigureIns
76              \fi
```

Now the text page itself is printed, shifted by the appropriate amount to the right.

```
77              \moveright\dimen0 \BoxR 255
78              \vskip 12pt
79              \Footer
80          }
```

Advance the page number.

```
81          \global\advance\pageno by 1
```

The end of this output routine is near.

```
82          \ifnum\outputpenalty > -20000
83          \else
84              \dosupereject
85          \fi
86      }
87  \catcode'\@ = 12
```

• End of <code>ola-1p.tip</code> •

36.2.6   The Source Code of the Example Applications of the Preceding Output Routines

In our example using the preceding two output routines, the output generated by the two output routines will be, of course, identical.

• <code>ola-exam.tip</code> •

First the necessary macros are loaded.

```
1   \input inputd.tip
2   \InputD{ola-comm.tip}           % 36.2.3, p. 139.
3   \InputD{ola-2p.tip}             % 36.2.4, p. 141.
4   \InputD{ola-1p.tip}             % 36.2.5, p. 145.
```

Set-up the page layout. The values of \hsize and \vsize were set when ola-common.tip was read-in.

```
5   \parskip = 6pt plus 2pt minus 1pt
```

Now define macro \SPar to generate a sample paragraph *and* a figure. This macro has two parameters:

- #1. The figure number.
- #2. The height of the figure to be generated by this macro.

```
 6    \def\SPar #1#2{%
 7        This is a sample paragraph. We just need some text to fill the
 8        whole page. So keep your fingers crossed. This is a sample
 9        paragraph to insert Figure~#1. Let us write a little more text,
10        so the figure can be inserted and there is also some text going
11        with it. You have fun, and a good time too!
12
13        \fb{#1}{#2}{Some caption here (Figure~#1, height = #2).}
14        \par
15    }
```

Now define macro \Examples which generates some paragraphs and figures, using the \SPar macro just defined.

```
16    \def\Examples{
17        \SPar{1}{1.0in}
18        \SPar{2}{1.3in}
19        \SPar{3}{0.8in}
20        \SPar{4}{1.0in}
21        \SPar{5}{0.8in}
22        \SPar{6}{0.8in}
23        \SPar{7}{1.6in}
24        \vfill\supereject
25    }
```

First set up everything to use the first output routine and generate some examples.

```
26    \output = {\OutCaptionGameTwoPass}
27    \let\fb = \FBTwoPass
28    \SetHeader{Two Pass Output Routine}
29    \SetFooter{\tt\string\OutCaptionGameTwoPass}
30    \ReadTheFigurePlacementFile
31    \Examples
```

Now set up the second output routine and generate some examples. The example pages generated now start with page number 101.

```
32    \pageno = 101
33    \output = {\OutCaptionGameOnePass}
34    \let\fb = \FBOnePass
35    \SetHeader{One Pass Output Routine}
36    \SetFooter{\tt\string\OutCaptionGameOnePass}
37    \Examples
```

This is the end of the examples.

```
38    \end
```

• End of <code>ola-exam.tip</code> •

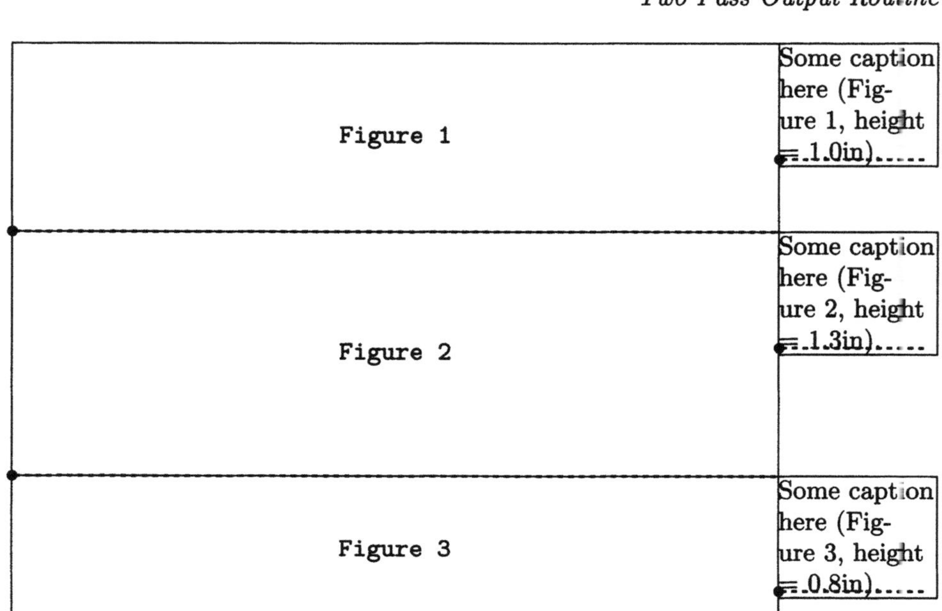

This is a sample paragraph. We just need some text to fill the whole page. So keep your fingers crossed. This is a sample paragraph to insert Figure 1. Let us write a little more text, so the figure can be inserted and there is also some text going with it. You have fun, and a good time too!

This is a sample paragraph. We just need some text to fill the whole page. So keep your fingers crossed. This is a sample paragraph to insert Figure 2. Let us write a little more text, so the figure can be inserted and there is also some text going with it. You have fun, and a good time too!

This is a sample paragraph. We just need some text to fill the whole page. So keep your fingers crossed. This is a sample paragraph to insert Figure 3. Let us write a little more text, so the figure can be inserted and there is also some text going with it. You have fun, and a good time too!

`\OutCaptionGameTwoPass`   1

Figure 36.2. Output, two pass approach, alternating placement of figure captions, page 1

Two Pass Output Routine

<table>
<tr><td>Some caption here (Figure 4, height = 1.0in).....</td><td>Figure 4</td></tr>
<tr><td>Some caption here (Figure 5, height = 0.8in).....</td><td>Figure 5</td></tr>
<tr><td>Some caption here (Figure 6, height = 0.8in).....</td><td>Figure 6</td></tr>
</table>

This is a sample paragraph. We just need some text to fill the whole page. So keep your fingers crossed. This is a sample paragraph to insert Figure 4. Let us write a little more text, so the figure can be inserted and there is also some text going with it. You have fun, and a good time too!

This is a sample paragraph. We just need some text to fill the whole page. So keep your fingers crossed. This is a sample paragraph to insert Figure 5. Let us write a little more text, so the figure can be inserted and there is also some text going with it. You have fun, and a good time too!

This is a sample paragraph. We just need some text to fill the whole page. So keep your fingers crossed. This is a sample paragraph to insert Figure 6. Let us write a little more text, so the figure can be inserted and there is also some text going with it. You have fun, and a good time too!

This is a sample paragraph. We just need some text to fill the whole page. So keep your fingers crossed. This is a sample paragraph to insert Figure 7. Let us write a little more text, so

2   `\OutCaptionGameTwoPass`

Figure 36.3. Output, two pass approach, alternating placement of figure captions, page 2

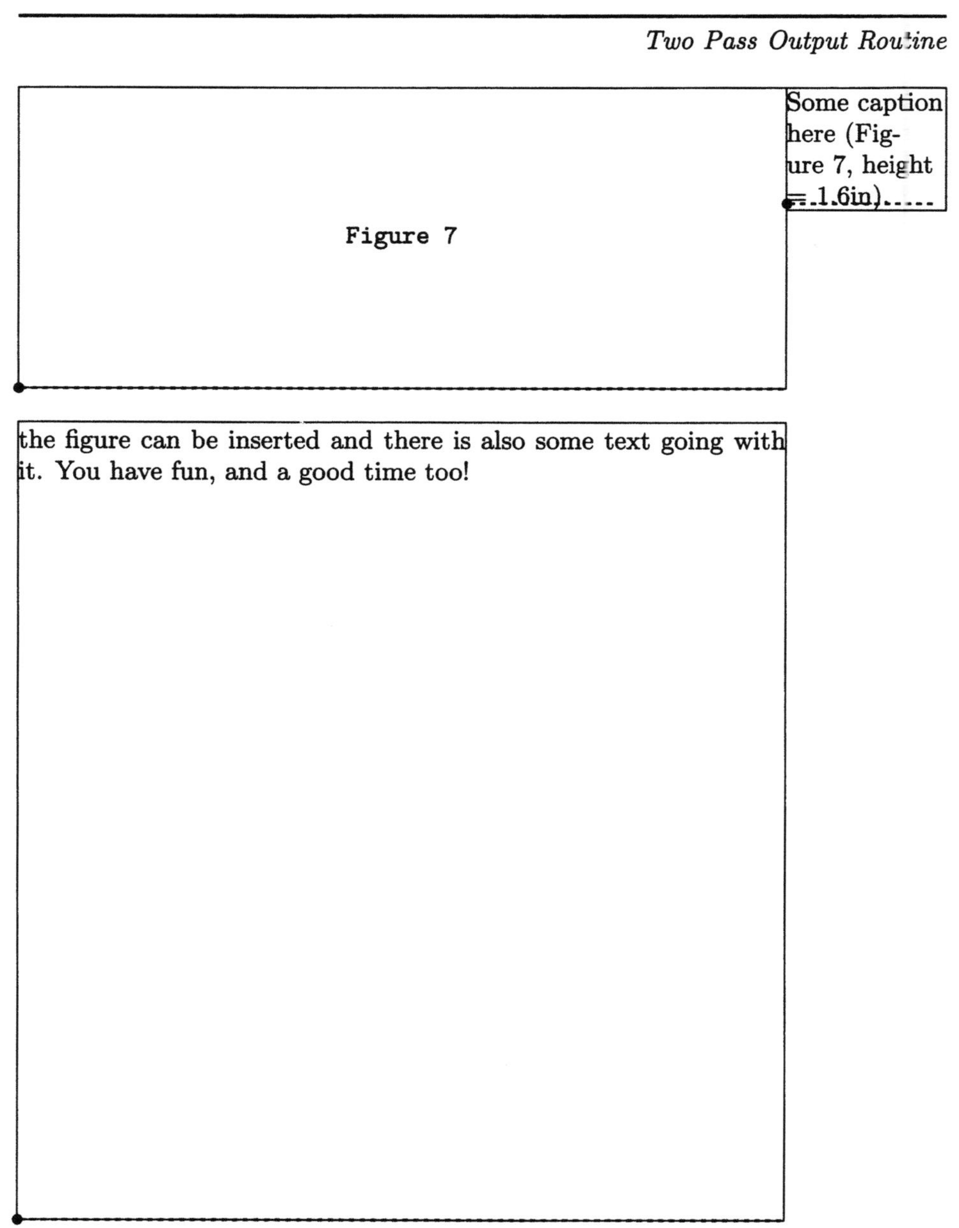

Figure 36.4. Output, two pass approach, alternating placement of figure captions, page 3

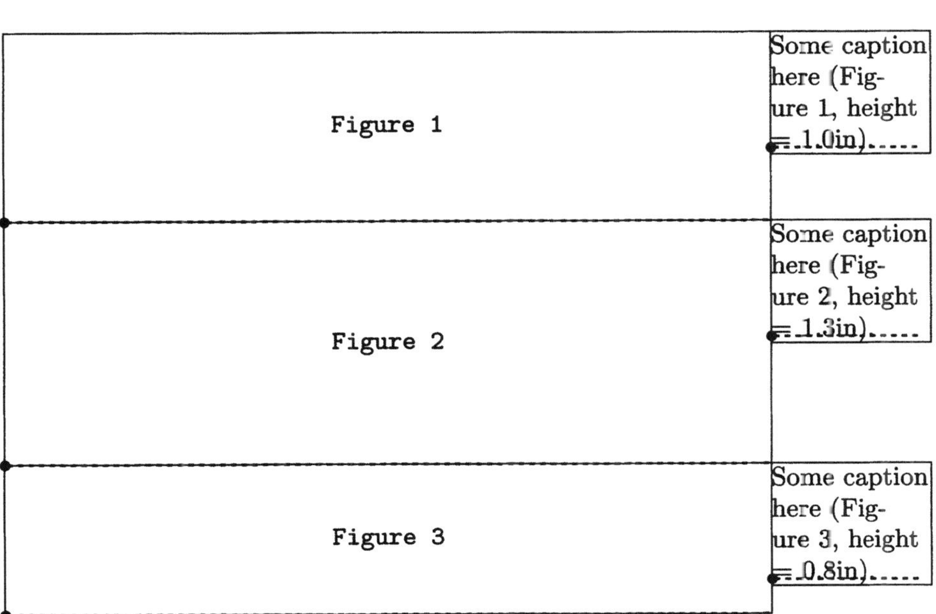

Figure 36.5. Output, one pass approach, alternating placement of figure captions, page 101

One Pass Output Routine

<table>
<tr><td>Some caption here (Figure 4, height =1.0in).....</td><td>Figure 4</td></tr>
<tr><td>Some caption here (Figure 5, height =0.8in).....</td><td>Figure 5</td></tr>
<tr><td>Some caption here (Figure 6, height =0.8in).....</td><td>Figure 6</td></tr>
</table>

This is a sample paragraph. We just need some text to fill the whole page. So keep your fingers crossed. This is a sample paragraph to insert Figure 4. Let us write a little more text. so the figure can be inserted and there is also some text going with it. You have fun, and a good time too!

This is a sample paragraph. We just need some text to fill the whole page. So keep your fingers crossed. This is a sample paragraph to insert Figure 5. Let us write a little more text. so the figure can be inserted and there is also some text going with it. You have fun, and a good time too!

This is a sample paragraph. We just need some text to fill the whole page. So keep your fingers crossed. This is a sample paragraph to insert Figure 6. Let us write a little more text. so the figure can be inserted and there is also some text going with it. You have fun, and a good time too!

This is a sample paragraph. We just need some text to fill the whole page. So keep your fingers crossed. This is a sample paragraph to insert Figure 7. Let us write a little more text. so

102   \OutCaptionGameOnePass

Figure 36.6. Output, one pass approach, alternating placement of figure captions, page 102

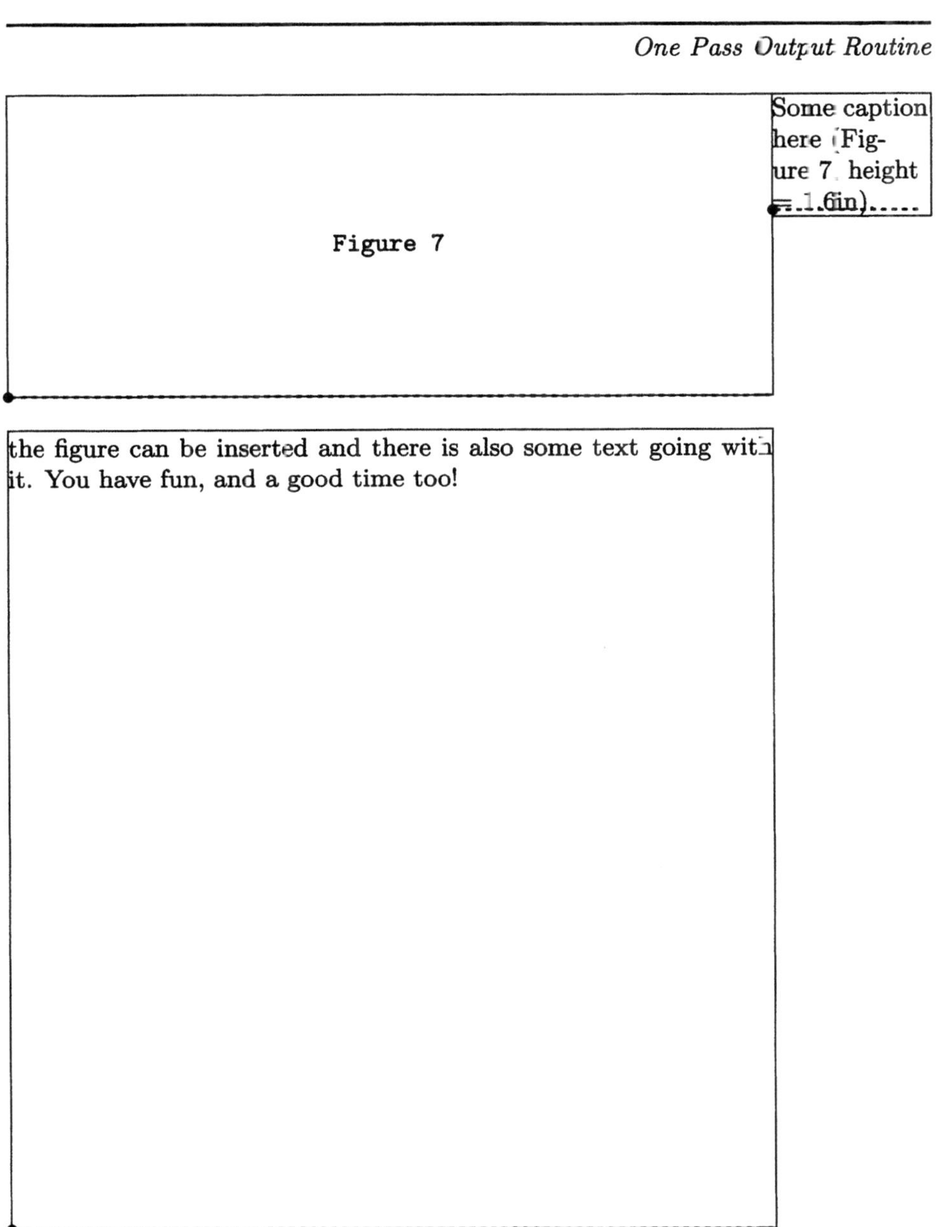

Figure 36.7. Output, one pass approach, alternating placement of figure captions, page 103

36.3 Footnote-Related Problems in the Design of Output Routines

36.3.1 Footnote Numbering Reset on Every Page

One problem encountered in the design of output routines is how to automatically number footnotes in such a way that the numbering of these page numbers is reset at the beginning of *every page*, rather than at the end of every chapter. The easiest way to solve this problem is to use a two pass approach that generates a footnote numbering file very similar to the figure placement file of the preceding output routine with variable caption placement of figures. You then use the information from this footnote placement file to determine footnote numbers properly.

I will not discuss an implementation, because it is very easy to derive from the implementation of the two pass approach of the output routine of the preceding Section (see 36.2.1, item 1, p. 138).

36.3.2 Multiple Footnote References

There is another footnote problem that occurs occasionally and is quite difficult to solve:

- Assume *multiple references* occur on the same page to the *same footnote text*.
- Where multiple references to the same footnote text occur, this footnote text is to be printed only once.

This is a very difficult problem to solve in TeX, because it requires a more powerful insertion concept than is provided.

This is an outline for a solution to a somewhat restricted problem. Assume a certain small set of acronyms is used in a document, and each acronym should be printed in the text with a footnote, that explains the meaning of the acronym. Needless to say, we want to print only one footnote, per acronym, per page, regardless of how often the acronym is used on a page.

Here is how this problem can be solved. For *each* acronym **A** do the following:

1. Create a *separate insertion class* for this acronym. Let us call the insertion class **A** for acronym **A**.
2. Set the magnification factor of this insertion class to zero. This means that regardless of how many insertions (footnote generation requests for the specific acronym) appear on a page, no space will be reserved. (Note that this cannot be the complete solution because if it were no space for the actual footnote explaining the acronym would be reserved).

3. Use the glue register of the insertion class **A** to reserve the appropriate amount of space on the current page for printing the explanation of the acronym. This corrects the space allocation problem that was just pointed out.

4. Modify the output routine used in the format you are using as follows: add a test which tests whether the box register associated with insertion class **A** is void or not.

    (a) If it is void, nothing needs to be done, because there was no reference to acronym **A** on the current page.

    (b) If it is *not* void, then there was obviously at least one, and potentially more, reference to this acronym. How many references occurred does not matter. In fact, the contents of the box register associated with this insertion class is discarded.

         Note however that TeX did reserve the proper amount of space, because the glue register associated with insertion class A was set properly.

Obviously, while the above solution is a little cumbersome for a large number of acronyms, it can be set up, for a small number very easily.

36.4   Limitations of the Insertion Concept

There are some limitations to the concept of insertions in TeX. Those limitations, for instance, are responsible for why insertions are not used to implement floating bodies (figures and tables) in LaTeX (insertions are used only for footnotes in LaTeX).

Insertions in TeX implement a *queuing mechanism*, but the user has no real control over that queue. Such detailed control over the queue is necessary, though, to overcome the following problems:

1. It is *not* possible to specify the maximum *number* of insertion pieces which are placed into the insertion box register by the page breaking algorithm and therefore printed on one page. For instance, it is *not* possible to specify something like "under all circumstances don't give me more than three pieces of \topins-related material."

2. It is also *not* possible to specify a threshold dimension x for an insertion class, so that if material *in excess* of x is placed on one page, no other material (in particular no regular text) is printed on that page. For instance, typesetting specifications of publishers frequently contain statements like 'if a page with a figure or with figures has room for less than 5 lines of text, then print no text (pring the figure or figures only) on that page."

3. The last problem (that can be worked around with a lot of trickery) is the following: assume you would like to place figures on the top or the bottom

of pages and mix this throughout a particular document. Two separate insertion classes \botins and \topins cannot be used, because first of all *you* don't want to be bothered with placement decisions (top versus bottom) and second of all the numbering of those figures should be linear, regardless of whether the figure is placed on the bottom or on the top.

In other words, you need to use *one* insertion class, and the output routine needs to decide where material is supposed to be placed. Yes, it can be done in TeX, but the power of the insertion mechanism is insufficient for a straightforward solution.

36.5 Summary

In this chapter we learned:

- A specialized insertion class can be used to help in the development of an index of a document: index terms referenced on pages are printed in the margins of the text.
- Two different output routines were discussed to print figures with captions where the captions are placed differently depending on whether the page number of the current page is even or odd. One approach is based on using two passes (during the first pass the correct placement of figure captions is ignored and during the second pass the information from the first pass is used to place captions properly). The other approach uses a "shadow insertion queue" in which figures and captions are maintained and placed opposite to the placement in the regular insertion queue. The output routine subsequently picks out the material from the correct queue.
- Two problems of numbering footnotes were discussed. In order to have footnotes numbered in such a way that the numbering starts at one at the beginning of each page, a two pass approach can be used. Also we discussed how multiple footnote references (where a certain footnote text is referenced more than once from the same page) can be handled.
- We discussed some of the inherent limitations of the insertion concept of TeX.

37
Double Column Output Routines

Two double column output routines will be presented in this chapter beginning with an explanation of their basic principles. There are essentially two different ways to handle double column output:

1. *Treat left- and right-hand columns as two separate logical pages.* When the output routine is called, then box register 255 either contains a "left column logical page" (in that case box register 255 is simply saved in another box register with no output to the dvi file) or a "right column logical page" (in this case the previously saved left column and the right column contained in box register 255 are now written together out to the dvi file). See 37.1 on this page for such an output routine.

2. *Generate left and right columns as a single column of double page length and then split this double long column into two columns using* \vsplit. In other words, the splitting of left and right columns is not done "automatically" by the page breaking algorithm as was the case in the previous approach, but the double long column is split into two columns by the output routine. A double column output routine using this approach can be found in 37.2, p. 166. The version of this output routine discussed at the given reference offers an additional feature in that it allows switching back and forth between single and double column output.

37.1   Double Column Output Routine I

The double column output routine presented in this section follows the first approach outlined in the introduction to this Chapter. In an output routine presented previously (the output routine to print library cards, see 34.5, p. 66) the concept of logical and physical pages was already explained In the output routine presented below there are two logical pages (the two columns) which are combined into one physical page, before being written out to the dvi file.

The definition of `\supereject` for this output routine must be changed compared to the definition of `\supereject` in the plain format. First of all, there are no insertions supported by the output routine presented below. More importantly, at the end of each job it must be checked whether the very last logical page was a left column (with no right column) or not. If the very last page was indeed a left column, then an empty right column must be generated, so that the left column is actually written out to the `dvi` file and not lost.

The following output routine is presented along the following lines:

1. 37.1.1 on this page contains the source of the output routine. It also contains the definition of macro `\SetUpDC` which must be called to set up this output routine.
2. 37.1.2, p. 163, contains the source code of an example using this output routine.
3. Fig. 37.1, p. 164, and Fig. 37.2, p. 165, contain output generated by the example source code.

37.1.1 Source Code of the Output Routine

Let us now examine the source code of the output routine.

$$\mathcal{P} \quad \bullet \text{ out2-or.tip } \bullet$$

```
15    \catcode'\@ = 11
```

`\@LeftColumnBox` is a box register to hold the left column of a page to be printed by this output routine. `\@RightColumnBox` is a box register to hold the right column of a page being printed by this output routine.

```
16    \newbox\@LeftColumnBox
17    \newbox\@RightColumnBox
```

`\@PageWidth` is the horizontal width of the physical page, including the space between the two columns. Note that `\hsize` must be set to the width of one column.

```
18    \newdimen\@PageWidth
```

The following macro `\SetUpDC` must be called initially to set up the dimensions when using this output routine. This macro should be called before any text is generated. This macro has three parameters:

- #1. The overall width of the physical page when the two logical pages are glued together.
- #2. The width of each column. Therefore there will be $\#1 - 2 * \#2$ horizontal space between the left and right column.
- #3. The page height (`\vsize`).

```
19    \def\SetUpDC #1#2#3{%
```

```
20        \@PageWidth = #1
21        \hsize = #2
22        \vsize = #3
23    }
```

The following macro generates a full length line. It does not use \hsize because \hsize is reduced to the width of one column in double column mode.

```
24    \def\@PageLine{\hbox to \@PageWidth}
```

The following macro generates the running head. You might want to change this macro to generate a different running head.

```
25    \def\Header{%
26        \@PageLine{%
27            \it Some Header Text\hfil
28            \PrintCurrentMonth\space \the\day, \the\year,
29            Page~\the\pageno
30        }%
31    }
```

The page layout of this output routine is by default ragged bottom. \topskip is set as discussed in 32.3.2, p. 9.

```
32    \topskip = 10pt plus 60pt
```

\@ColumnMode is used to keep track of whether the current column is a left (L) or right (R). The very first column is a left column.

```
33    \let\@ColumnMode = L
```

The definition of \supereject changes as mentioned before.

```
34    \def\supereject{%
35        \eject
```

The last column has been taken care of. If the current column is a right column, then the preceding column was a left column, stored in box register \@Left-ColumnBox, but which was not yet written to the dvi file.

```
36        \if\@ColumnMode R
```

This is indeed true, the very last column was a left column. Therefore generate an empty right column, which will be written out to the dvi file together with the retained left column.

```
37            \hbox{}
38            \vfill\eject
39        \fi
```

No special action was necessary, if the last column was a right-hand column.

```
40        \end
41    }
```

Here is where the output routine starts.

```
42    \output = {
43        \if\@ColumnMode L
```

The left column is simply saved. Note the use of `\global` because output routines form implicit groups.

```
44              \global\setbox\@LeftColumnBox = \vbox to \vsize{
45                  \unvbox 255
46                  \vfil
47              }
```

Force the width of this column to be precisely `\hsize` so that putting together the left and right columns will work correctly later. Also switch to "right column mode." When the right column is encountered, it is first stored in `\@Right-ColumnBox`.

```
48              \global\wd\@LeftColumnBox = \hsize
49              \global\let\@ColumnMode = R
50          \else
51              \setbox\@RightColumnBox = \vbox to \vsize{
52                  \unvbox 255
53                  \vfil
54              }
```

The width of the right column is again forced to be `\hsize` (which it will be in most cases anyway). Then the left and right columns are glued together and written out to the `dvi` file.

```
55              \wd\@RightColumnBox = \hsize
56              \shipout\vbox{
57                  \Header
58                  \vskip 24pt
59                  \@PageLine{%
60                      \box\@LeftColumnBox
61                      \hfil
62                      \box\@RightColumnBox
63                  }
64              }
```

The page number (counting physical pages) is incremented by one and the output routine switches back to "left column modus."

```
65              \advancepageno
66              \global\let\@ColumnMode = L
67          \fi
68      }
69      \catcode'\@ = 12
```

• End of <code>out2-or.tip</code> •

37.1.2   An Example Application

Here is the source code of an example application of the preceding output routine.

𝒫   • <code>ex-out2.tip</code> •

```
15    \input inputd.tip
16    \InputD{out2-or.tip}                    % 37.1.1, p. 160.
17    \InputD{samplepa.tip}                   % 27.1.3.3, p. III-402.
18    \InputD{alldate.tip}                    % 26.23, p. III-395.
```

Initialize everything.

```
19    \SetUpDC{29pc}{13pc}{35pc}
```

Use a ragged right paragraph layout because of the narrow columns.

```
20    \raggedright
```

Now generate some text for the first page.

```
21    \SamplePar{A-1}{3}
22    \SamplePar{A-2}{2}
23    \SamplePar{A-3}{4}
24    \SamplePar{A-4}{3}
25    \SamplePar{A-5}{2}
```

Perform an explicit "column break" now.

```
26    \vfill\eject
```

The following text appears as the fourth column (left column of the second physical page) in the document.

```
27    \SamplePar{A-6}{4}
```

It's all over!

```
28    \bye
```

• End of `ex-out2.tip` •

Identification of this paragraph: *A-1. Sample paragraph 1, with 3 sentences.* So here we go, and when you check the number of sentences, then note that these first two sentences do *not* count. This is one of the many sentences this macro generates, to be more specific it is sentence number 1 of 3. This is one of the many sentences this macro generates, to be more specific it is sentence number 2 of 3. This is one of the many sentences this macro generates, to be more specific it is sentence number 3 of 3.

Identification of this paragraph: *A-2. Sample paragraph 2, with 2 sentences.* So here we go, and when you check the number of sentences, then note that these first two sentences do *not* count. This is one of the many sentences this macro generates, to be more specific it is sentence number 1 of 2. This is one of the many sentences this macro generates, to be more specific it is sentence number 2 of 2.

Identification of this paragraph: *A-3. Sample paragraph 3, with 4 sentences.* So here we go, and when you check the number of sentences, then note that these first two sentences do *not* count.

This is one of the many sentences this macro generates, to be more specific it is sentence number 1 of 4. This is one of the many sentences this macro generates, to be more specific it is sentence number 2 of 4. This is one of the many sentences this macro generates, to be more specific it is sentence number 3 of 4. This is one of the many sentences this macro generates, to be more specific it is sentence number 4 of 4.

Identification of this paragraph: *A-4. Sample paragraph 4, with 3 sentences.* So here we go, and when you check the number of sentences, then note that these first two sentences do *not* count. This is one of the many sentences this macro generates, to be more specific it is sentence number 1 of 3. This is one of the many sentences this macro generates, to be more specific it is sentence number 2 of 3. This is one of the many sentences this macro generates, to be more specific it is sentence number 3 of 3.

Identification of this paragraph: *A-5. Sample paragraph 5, with 2 sentences.* So here we go, and when you check the number of sentences, then note that these first two sentences do *not* count.

Figure 37.1. Double column output routine, sample page 1.

Some Header Text *May 20, 1993, Page 2*

This is one of the many sentences this macro generates, to be more specific it is sentence number 1 of 2. This is one of the many sentences this macro generates, to be more specific it is sentence number 2 of 2.

Identification of this paragraph: *A-6. Sample paragraph 6, with 4 sentences.* So here we go, and when you check the number of sentences, then note that these first two sentences do *not* count. This is one of the many sentences this macro generates, to be more specific it is sentence number 1 of 4. This is one of the many sentences this macro generates, to be more specific it is sentence number 2 of 4. This is one of the many sentences this macro generates, to be more specific it is sentence number 3 of 4. This is one of the many sentences this macro generates, to be more specific it is sentence number 4 of 4.

Figure 37.2. Double column output routine, sample page 2.

37.2   A Single / Double Column Output Routine

This section introduces an output routine which allows the user to switch back and forth between single and double column page output. No footnotes or other types of insertions are provided. The code for this output routine was initially derived from Platt (1985) and extended considerably.

The output routine macros presented in this Section, when in double column mode, use the second approach described in the introduction to this chapter. A double length (with respect to the available vertical space on the current page) column is generated which is then split into two halves, the left and right column, using \vsplit. Note that this split operation is initiated by the output routine, after the page breaking algorithm has delivered a double length column.

In a certain sense the macros presented in this Section *also* incorporate aspects of the *previous* output routine because one has the possibility of explicitly triggering a "column break" between the left and the right column. If you do so, then the output routine below saves the left column in a box register and waits for the right column to "arrive." This approach is obviously more or less identical to the way the previously presented output routine works, which collects left and right columns separately.

Also note that the code below does not only contain an output routine but also other related macros.

As you will see shortly it is actually incorrect to talk about a *single* output routine. There are all together *three* output routines presented in the macros below and the macros will switch back and forth between these three output routines.

This Section of the series, besides demonstrating double column output, also discusses the following items:

- The macros show that *switching output routines* is definitely a useful thing to do.
- The output routine macros below show how an output routine writes material *back onto the main vertical list.*
- The output routine macros show manipulations of \pagegoal and \vsize from within an output routine.

37.2.1   Macros To Be Called by the User

A variety of macros are called by the user to control the page layout. These macros will be discussed in detail later, but here is a brief overview:

1. \SetUpDSC is an initialization macro and is discussed in 37.2.7.2, p. 170.
2. \BeginDoubleColumns starts double column output. See 37.2.7.8, p. 176, for further details.

3. \EndDoubleColumns is a macro to end double column output and to switch back to single column mode. See 37.2.7.8, p. 176, for details.
4. \eject. The \eject macro to force a usual page break or column break had to be redefined; see 37.2.7.5, p. 173. It is actually not this macro that changes, but the actions of the various output routines in case of an \eject are quite complicated.
5. \DSCHeader, \DSCFooter are header and footer macros discussed in 37.2.7.8, p. 182.
6. The \bye also had to be redefined; see 37.2.7.8, p. 183.

37.2.2   A Quick Rundown

The following is a brief explanation of how the output routine macros work. If you want to understand the details then there is no way around digging into the source code that appears in 37.2.7, p. 169.

The macros initially set up a *single* column output routine. This output routine is pretty much a standard type of output routine. In case of a page overflow in single column mode, the output routine will write the material of box 255 together with running head and footer to the dvi file.

When a switch to double column output occurs (\BeginDoubleColumns), the material collected so far of the current page part of the main vertical list is stored away in a box register called \@PageSoFar. Although this box register under the current circumstances can only contain single column material this is not necessarily always true: if repeated switching back and forth between single and double column modus takes place ("repeated" refers here to one and the same page), then \@PageSoFar may actually also contain double column material.

Before a switch to double column output occurs, a new value for \vsize is computed. To do so the vertical size of the material collected so far in box register \@BoxOfPageSoFar (zero if empty) is subtracted from the overall height of the page (\@ComputeVsizeForDoubleColumns). This value therefore reflects the amount of vertical space that is still available on the current page. This value is then multiplied by two, because taking left and right columns together makes twice this amount be available for double column output (remember that splitting up double column output into a left- and a right-hand column is handled by the output routine, and not by the page breaking algorithm).

Now assume that the following double column material fills up the remainder of the current page. When the double column output routine is called it will split the long column into a left and right column and write it out together with any material contained in \@BoxOfPageSoFar (if the current physical page did not contain any single column material, then this box register is of course empty). After this, the value of \vsize for double column output must be recomputed, because \@BoxOfPageSoFar is now empty for sure (it may not have been empty before). Now twice the regular page height is available for double column output.

Now assume that the user switches back to single column output by calling \EndDoubleColumns. This means that the material for the left and right column must be computed by splitting the current material in half. This case is not too different from the case where a regular "column overflow" occurs (previous paragraph), except for the fact that the two columns are now simply packed together with any other material of \@BoxOfPageSoFar and returned to the main vertical list. They are *not* written out to the dvi file because there is still room left on the current physical page. Now a switch to single column output occurs with any previous single or double column material already on the main vertical list. In other words, in a single column mode page break that might occur later, any preceding single or double column material is already on the main vertical list.

The macros also offer the possibility of explicitly generating a column break between a left and right column. If this occurs, then the output routine does not need to split the current column into a left and right column. Thus, there is no more the notion of a double length column that must be split to derive two single columns.

37.2.3   The Page Layout of the Presented Macros

With the initialization call to \SetUpDSC you have a choice of a bottom flush (parameter 5 of \SetUpDSC must be set to 0 in this case) or a ragged bottom (parameter 5 must be set to 1 then) type of page layout. This page layout applies to single column and double column text.

37.2.4   Column Break Computations

When the double column output routine presented below is called, the \outputpenalty can have the following different values (this explanation must be read together with the source code in 37.2.7 on the next page to be completely understandable):

1. \@EjectPenalty (-10000). This is the standard penalty for the case of a forced column break in double column mode and a standard page break in single column mode. This penalty (or for that matter \eject) must only occur in a left column, i.e. to indicate an explicit switch over from the left column to the right column. In case this penalty occurs in a right column, an error is reported. The assumption is that if \eject is used to switch from left to right column, the right column is terminated by switching back to single column mode.
2. \@BalancePenalty (-10001). This penalty is inserted by the \EndDoubleColumns macro to indicate the end of double column mode to the double column output routine. This is referred to as case A below.

3. The last case is a penalty > -10000 as it is encountered during a regular page break because of an overflow of the current page part of the main vertical list.

37.2.5   Column Break Cases in the Double Column Output Routine

The following cases of double column page breaks can be identified (again the following explanation should be read in connection with the later presented source code presented later):

- *Case A.* A \EndDoubleColumns is encountered. Balancing is initiated, generating a left and right column of approximately the same length. The macros below will make the left column at least as long as the right column and both columns the same length as much as possible.
- *Case B.* A *regular column break* occurs because of an overflow of the main vertical list. Again the two columns are balanced. Subsequently the two columns together with any remaining material in \@BoxOfPageSoFar (which might be empty) are written out to the dvi file.
- *Case C.* An *explicit column break* occurs (triggered by the user through an \eject, in the left of two columns). This means that the notion of a double length column, which will be split into a left- and right-hand column, does not make sense anymore. Instead there is now a left column that is saved, while output is regarded as part of the right column.

  After this case (case C) has happened, assuming correct input, either case D or case E will happen.
- *Case D.* The *right column overflows.* The material in box register 255 determines the right column (the left column was previously saved as part of the processing that relates to case C).
- *Case E.* A call to \EndDoubleColumns is encountered. Again the material in box register 255 determines the right column (the left column was previously saved).

  An explicit column break in the right column (\eject) is illegal. You must use \EndDoubleColumns to terminate the current column and then \BeginDoubleColumns to resume double column output.

37.2.6   On the Presentation of the Macros

The double column output macros are presented as follows:

1. 37.2.7 on this page contains the output routine itself.
2. 37.2.8, p. 184, contains the example input.
3. Figures 37.3 to 37.6 on pages 189–192 contain example output.

37.2.7   The Output Routine Macros Source Code

This section presents the source code of all the macros related to the output routine.

37.2.7.1   Initial Definitions

First some initial macros are loaded.

$$\mathcal{P} \quad \bullet \ \texttt{out-ds.tip} \ \bullet$$

```
15   \InputD{box-mac.tip}              % 9.3.14, p. I-343.
16   \InputD{rangetst.tip}             % 25.1.9, p. III-328.
17   \InputD{maxmindi.tip}             % 25.1.14, p. III-332.
18   \InputD{prot.tip}                 % 29.6.1, p. III-525.
19   \InputD{vtbox.tip}                % 8.1.4, p. I-265.
20   \InputD{nathd.tip}                % 8.1.3, p. I-264.
21   \InputD{shiftudb.tip}             % 7.5.7, p. I-254.
22   \catcode'\@ = 11
```

Now some registers will be defined.

   The following is the value of \hsize for single column output and at the same time overall page width for double column output.

```
23   \newdimen\@PageWidth
```

The width of a column in double column mode.

```
24   \newdimen\@ColWidth
```

The amount by which the left column is shifted to the right and the right column to the left with respect to the margins as established by the output in single column modus.

```
25   \newdimen\@ColIndent
```

Height of the page (single and double column output).

```
26   \newdimen\@PageHeight
```

Number of columns (2 or 1) of currently valid page layout.

```
27   \newcount\@DSCCurNumberOfColumns
```

Page layout code: 0 is bottom and top flush, 1 is ragged bottom. This applies to single and double column mode.

```
28   \newcount\@PageLayoutCodeDSC
```

Debugging? 0 if no, 1 if yes.

```
29   \newcount\@DSCDebugging
```

37.2.7.2   The Definition of Macro \SetUpDSC

The macro \SetUpDSC must be called to initialize everything. It is mainly responsible for setting up the page layout. It has the following six parameters:

- #1. The width of the whole page when using the single column mode.
- #2. The width of a column in double column mode.
- #3. The distance by which the left column is moved to the right with respect to the ordinary right-hand margin as established by the single column modus. The right-hand column is shifted by the same amount to the left. Obviously $2 * (\#2 + \#3)$ must be not greater than \hsize (or #1).
- #4. The page height.
- #5. The page layout (0 for top and bottom flush, 1 for ragged bottom). This code applies to single and double column output. When you switch frequently back and forth between single and double column output on one page, then using a ragged bottom type of page layout makes adjusting the page length much easier for TeX, because in ragged bottom page layout a page's length need not be adjusted.
- #6. A debugging flag. If 1, then a protocol file is written and various boxes of the output are printed with rules surrounding them using the ruled box macros of 9.3. If 0, then no protocol file is written and no rules are printed.

If you change \baselineskip for text being generated by the output routines that are presented here, then you should do so before calling \SetUpDSC.

```
30    \def\SetUpDSC #1#2#3#4#5#6{%
31        \@DSCDebugging = #6
32        \CheckZeroOneRange{\@DSCDebugging}%
33            {\string\SetUpDSC: debugging code wrong}
34        \ifnum\@DSCDebugging = 0
35            \ProtWritefalse
36        \else
37            \ProtWritetrue
38        \fi
```

A protocol file is written. This is set up.

```
39        \InitProtWrite
40        \WriteProtocol{0}{\string\SetUpDSC: begin}%
```

All the parameters need to be saved.

```
41        \@PageWidth = #1%
42        \@ColWidth = #2%
43        \@ColIndent = #3%
44        \@PageHeight = #4%
45        \@PageLayoutCodeDSC = #5%
46        \CheckZeroOneRange{\@PageLayoutCodeDSC}%
47            {\string\SetUpDSC: page layout code wrong}%
48        \hsize = \@PageWidth
49        \dimen0 = 2\@ColIndent
```

```
50        \advance\dimen0 by 2\@ColWidth
51        \ifdim\dimen0 > \hsize
52            \errmessage{\string\SetUpDSC: Initial values of
53                \string\hsize, \noexpand\@ColWidth and
54                \noexpand\@ColIndent do not make sense.}%
55        \fi
56        \vsize = \@PageHeight
```

Other initializations. Start out in single column modus.

```
57        \@DSCCurNumberOfColumns = 1
58        \@SetSingleColumnOutput{As part of \string\SetUpDSC}%
```

Page layout decisions are made here.

```
59        \ifcase \@PageLayoutCodeDSC
60            \topskip = 10pt                 % Top / bottom flush
61        \or
62            \topskip = 10pt plus 50pt    % Ragged bottom
63        \else
64            \errmessage{\string\SetUpDSC: illegal
65                page layout parameter.}%
66        \fi
67        \WriteProtocol{0}{\string\SetUpDSC: end}%
68    }
```

37.2.7.3   Restoring the Original Output Routine

The main purpose of the output routine presented here is to be able to switch
between single and double column output. Printing this series this output routine is only used for the index chapters. The following macro \UnDoDSC resets
the output routine to the original plain format output routine. Note that the
redefined \eject and \bye macros do *not* pose any problems and will work with
the output routine of the plain format.

```
69   \def\UnDoDSC{%
70       \global\output = {\plainoutput}%
71   }
```

37.2.7.4   Registers Used

Here is the declaration of some registers (box and dimension registers) used by
the output routine macros.

1. \@BoxOfPageSoFar. If in double column modus save here what should be
    printed on the current page together with the double column material cur-
    rently being accumulated.
2. \@DSCLeftColumnBox. Save the left column in this box register.

3. \@DSCLeftNaturalHeight. Natural height of the left column.
4. \@DSCRightColumnBox. Save the right column here.
5. \@DSCRightNaturalHeight. Natural height of the right column.
6. \@DSCTempBox. A temporary box register.

```
72    \newbox\@BoxOfPageSoFar
73    \newbox\@DSCLeftColumnBox
74    \newdimen\@DSCLeftNaturalHeight
75    \newbox\@DSCRightColumnBox
76    \newdimen\@DSCRightNaturalHeight
77    \newbox\@DSCTempBox
```

37.2.7.5   Auxiliary Macros and Other Definitions

Next we present some auxiliary macros needed for the implementation of the
output routine. The macro \@EliminateRulesConditional is called to eliminate
rules generated by macros like \HboxR:

```
78    \def\@EliminateRulesConditional{%
79        \ifnum\@DSCDebugging = 0
80            \EliminateRuledBoxes
81        \fi
82    }
```

The following conditional's default is false. The conditional is set to true by
the double column output routine upon a case C type of page break.

```
83    \newif\if@CaseCPageBreak
```

Some penalty definitions. Note that the following definition of \eject is identical
to the definition of \eject used with the plain format output routine except for
the fact that its definition refers to macro \@EjectPenalty.

```
84    \def\@EjectPenalty{-10000 }
85    \def\eject{\penalty\@EjectPenalty}
86    \def\@BalancePenalty{-10001 }
```

The following macro generates a full length line. It does not use \hsize
because \hsize is reduced in double column mode.

```
87    \def\@PageLine{\hbox to \@PageWidth}
88    \def\@PageLineR{\HboxR to \@PageWidth}
```

37.2.7.6   The Definition of Macro \@ProtDSC

\@ProtDSC is a macro to write certain information (see below for which information) to a protocol file. This macro has one parameter, #1, which is some text
identifying where this macro was invoked.

```
 89   \def\@ProtDSC #1{
 90       \WriteProtocol{1}{\string\@ProtDSC: Begin (page \the\pageno)}
 91       \WriteProtocol{2}{"#1",
 92           number of cols: \the\@DSCCurNumberOfColumns}
 93       \WriteProtocol{2}{\string\vsize: \the\vsize,
 94           \string\pagetotal: \the\pagetotal,
 95           \string\pagegoal: \the\pagegoal}
 96       \BoxToProtocol{2}{\@BoxOfPageSoFar}{}
 97       \WriteProtocol{2}{\string\dimen0: \the\dimen0,
 98           \string\outputpenalty: \the\outputpenalty}
 99       \WriteProtocol{1}{\string\@ProtDSC: End}
100   }
```

37.2.7.7   The Definition of Macro `\@CaseReport`

`\@CaseReport` is a macro to report which type of column break is encountered
by our macros. The term "case" refers to the discussion of 37.2.3, p. 168.

```
101   \def\@CaseReport #1{%
102       \WriteProtocol{0}{CASE #1}%
103       \message{CASE #1.}%
104   }
```

37.2.7.8   Macro `\@SetOutputRoutine`

The macro `\@SetOutputRoutine` is called to change in output routines. A macro
is used for switching output routines so that the switch of an output routine can
be written to a protocol file (other than through a call to this macro, output
routines are not switched anywhere else). This macro has the following parameters:

- #1. This parameter should be either `\global` (in which case the redefinition of
  the output routine is done globally) or empty (in which case the redefinition
  of the output routine may be terminated by closing a currently active group).
- #2. The new output routine.
- #3. Some message written to a protocol file.

```
105   \def\@SetOutputRoutine #1#2#3{%
106       #1\output = {#2}%
107       \WriteProtocol{0}{\string\@SetOutputRoutine:}%
108       \WriteProtocol{1}{#3}%
109   }
```

The following *single column output routine* is fairly standard. Whatever is
contained in box register 255 is written out to the `dvi` file. The material in box
register 255 may contain a mixture of double and single column output material.

```
110   \def\@SingleColumnOutputRoutine{%
111       \@EliminateRulesConditional
112       \WriteProtocol{0}{\noexpand\@SingleColumnOutputRoutine called,
113           \noexpand\outputpenalty is \the\outputpenalty}
114       \WriteProtocol{1}{Page number is \the\pageno}
115       \showboxdepth   = 2
116       \showboxbreadth = 1000
117       \@ProtDSC{\string\@SingleColumnOutputRoutine: begin}%
118       \BoxToProtocol{1}{255}{}%
119       \@CheckNumberOfColumns{1}{\@SingleColumnOutputRoutine}%
120       \@ShipAPageBox{%
121           \ifnum\@PageLayoutCodeDSC = 0
122               \vbox to \@PageHeight{\unvbox 255}%
123           \else
124               \vbox to \@PageHeight{\unvbox 255 \vfill}%
125           \fi
126       }%
127   }
```

The macro `\@SetSingleColumnOutput` is called to set up single column
output. The parameter of this macro, #1, is a message that is written out to the
protocol file.

```
128   \def\@SetSingleColumnOutput #1{%
129       \WriteProtocol{0}{\string\@SetSingleColumnOutput: set to single
130                    column output!}%
131       \global\@DSCCurNumberOfColumns = 1
132       \@SetOutputRoutine{\global}{\@SingleColumnOutputRoutine}{#1}%
133       \global\vsize    = \@PageHeight
134       \global\pagegoal = \@PageHeight
135   }
```

`\@SaveCurrentPageOutputRoutine` is an output routine that takes the content of box register 255 and saves it into box register `\@BoxOfPageSoFar`.

```
136   \def\@SaveCurrentPageOutputRoutine{%
137       \global\setbox\@BoxOfPageSoFar = \vbox{\unvbox 255}%
138       \BoxToProtocol{0}{\@BoxOfPageSoFar}%
139           {\string\@SaveCurrentPageOutputRoutine}
140   }
```

`\@ComputeVsizeForDoubleColumns` is a macro which computes the value
for `\vsize` for double column output as twice the available vertical space on the
current page.

```
141   \newcount\@VsizeFactor
142   \def\@ComputeVsizeForDoubleColumns{%
143       \vsize = \@PageHeight
144       \advance\vsize by -\ht\@BoxOfPageSoFar
145       \advance\vsize by -\dp\@BoxOfPageSoFar
146       \multiply\vsize by 2
147       \@VsizeFactor = \vsize
148       \divide\@VsizeFactor by \baselineskip
149       \ifodd\@VsizeFactor
```

```
150        \advance\@VsizeFactor by -1
151      \fi
152      \global\vsize = \@VsizeFactor \baselineskip
153      \@ProtDSC{\string\@ComputeVsizeForDoubleColumns}%
154    }
```

\@ComputeHalfVsize loads dimension register \@HalfVsize with half of the current value of \vsize.

```
155    \newdimen\@HalfVsize
156    \def\@ComputeHalfVsize{%
157      \@HalfVsize = \vsize
158      \divide\@HalfVsize by 2
159    }
```

\BeginDoubleColumns is the macro to switch from single column output to double column output.

```
160    \def\BeginDoubleColumns{%
161      \par
162      \WriteProtocol{0}{\string\BeginDoubleColumns: begin}
163      \@CheckNumberOfColumns{1}{\BeginDoubleOfColumns}%
164      \global\@DSCCurNumberOfColumns = 2
165      \begingroup
166      \@CaseCPageBreakfalse
167      \@EliminateRulesConditional
```

When switching to double column output modus, save what is on the main vertical list in \@BoxOfPageSoFar via a special new output routine, of which the execution is triggered immediately.

```
168      \WriteProtocol{1}{\string\BeginDoubleColumns: put rest of page
169                        into \string\@BoxOfPageSoFar}%
170      \@SetOutputRoutine{}{\@SaveCurrentPageOutputRoutine}%
171        {\string\BeginDoubleColumns: save page built up to now.}%
172      \eject
173      \@ProtDSC{\string\BeginDoubleColumns: 2}%
```

Now switch to double column output.

```
174      \@SetOutputRoutine{}{\@DoubleColumnOutputRoutine}%
175        {Double column output set up by \string\BeginDoubleColumns}%
176      \hsize = \@ColWidth
```

Compute \vsize so it reflects the amount of vertical space for two columns left on the current page.

```
177      \@ComputeVsizeForDoubleColumns
178      \@ProtDSC{\string\BeginDoubleColumns: 3}%
179    }
```

\EndDoubleColumns is the macro to switch back from double column mode to single column mode. Its definition follows now.

```
180    \def\EndDoubleColumns{%
181      \par
182      \WriteProtocol{0}{\string\EndDoubleColumns: begin}
```

```
183        \@CheckNumberOfColumns{2}{\EndDoubleOfColumns}
184        \@ProtDSC{In \string\EndDoubleColumns}
185        \@CaseReport{A}
186        \penalty\@BalancePenalty
```

The following call of \@BuildPageSoFar does *not* generate anything (not even
an empty box) if there is no material in all of the following three box registers: @BoxOfPageSoFar, \@DSCLeftColumnBox and \@DSCRightColumnBox. This
material appears on the main vertical list now.

```
187        \@BuildPageSoFar
```

Back to single column mode.

```
188        \@SetSingleColumnOutput{\string\EndDoubleColumns}
```

Terminate the group which was started by \BeginDoubleColumns.

```
189        \endgroup
190    }
```

\@DoubleColumnOutputRoutine is (you guessed it) the double column output routine. Observe that in addition to the double column material there may
be material remaining in \@BoxOfPageSoFar to be printed on the current page.

```
191    \def\@DoubleColumnOutputRoutine{%
192        \@EliminateRulesConditional
193        \WriteProtocol{0}{\string\@DoubleColumnOutputRoutineput:
194                    begin (penalty: \the\outputpenalty)}
```

Did we have before a case C type of page break?

```
195        \if@CaseCPageBreak
```

Yes, separate columns, right-hand column has arrived!
     Cases D and E are not dealt with.

```
196            \@CaseReport{D/E}%
197            \ifvoid\@DSCLeftColumnBox
198                \errmessage{\string\@DoubleColumnOutputRoutine:
199                    missing left column!}%
200            \else
201                \ifvoid\@DSCRightColumnBox
202                    \global\setbox\@DSCRightColumnBox =
203                        \vbox{\unvbox 255}%
204                    \global\@DSCRightNaturalHeight =
205                        \ht\@DSCRightColumnBox
```

Is the penalty legal? Explicit page break in right-hand column is not allowed.

```
206                    \ifnum\outputpenalty = \@EjectPenalty
207                        \errmessage{\string\@DoubleColumnOutput:
208                            \noexpand\eject in right column illegal.}%
209                        \@CaseReport{D}
210                        \MaxDimen{\dimen0}{\@DSCRightNaturalHeight}%
211                                        {\@DSCLeftNaturalHeight}{}%
212                    \else
```

Case D or case E?

```
213                         \ifnum\outputpenalty = \@BalancePenalty
```

Case E: \EndDoubleColumns was read.

```
214                         \@CaseReport{E}
215                         \MaxDimen{\dimen0}{\@DSCRightNaturalHeight}
216                                    {\@DSCLeftNaturalHeight}{}%
217                       \else
```

Case D.

```
218                         \@CaseReport{D}%
```

That's the remaining space on the page.

```
219                           \dimen0 = \vsize
220                     \fi
221             \fi
222           \@SetColumnBox{\@DSCLeftColumnBox}{}%
223           \@SetColumnBox{\@DSCRightColumnBox}{}%
224           \Vtbox{\@DSCRightColumnBox}{\global}%
225           \ShiftRefPointUpOrDown{\@DSCRightColumnBox}{12pt}%
226           \Vtbox{\@DSCLeftColumnBox}{\global}%
227           \ShiftRefPointUpOrDown{\@DSCLeftColumnBox}{12pt}%
```

Ship left and right column and all the remaining stuff out to the dvi file.

```
228           \ifnum\outputpenalty > -10000
229             \@ShipAPageBox{%
230                 \ifnum\@PageLayoutCodeDSC = 0
231                     \vbox to \@PageHeight{\@BuildPageSoFar}%
232                 \else
233                     \vbox to \@PageHeight{\@BuildPageSoFar
234                                              \vfill}%
235                 \fi
236             }%
237           \else
238             % It's \EndDoubleColumns!
239           \fi
240           \global\@CaseCPageBreakfalse
```

Observe that now the full page is available for double column output.

```
241               \@ComputeVsizeForDoubleColumns
242               \global\pagegoal = \vsize
243             \else
244               \errmessage{\string\@DoubleColumnOutputRoutine:
245                   left / right columns messed up.}
246             \fi
247         \fi
248     \else
```

Case C did not apply *before*. May be it's case C now?

```
249         \ifnum\outputpenalty = \@EjectPenalty
```

Yes, it is case C.

```
250            \@CaseReport{C}
251            \WriteProtocol{1}{\string\@DoubleColumnOutputRoutineput:
252                       penalty \@EjectPenalty call.}
```

There are two separate columns.

```
253            \global\@CaseCPageBreaktrue
254            \ifvoid\@DSCLeftColumnBox
255               \global\setbox\@DSCLeftColumnBox =
256                  \vbox{\unvbox 255}
257               \global\@DSCLeftNaturalHeight = \ht\@DSCLeftColumnBox
```

Compute \vsize on a per column basis.

```
258               \@ComputeHalfVsize
259               \WriteProtocol{2}{*\string\vsize/2 is
260                                 \the\@HalfVsize-%
```

Generate warning if column is too long.

```
261               \dimen1 = \ht\@DSCLeftColumnBox
262               \advance\dimen1 by \dp\@DSCLeftColumnBox
263               \ifdim\dimen1 > \@HalfVsize
264                  \message{WARNING: column is to long }%
265               \fi
```

This is how much is left for the right column.

```
266               \global\pagegoal = \@HalfVsize
267               \global\vsize    = \@HalfVsize
268            \else
269               \errmessage{\string\@DoubleColumnOutputRoutine:
270                  left column box already loaded!}%
271            \fi
272         \else
```

Cases A or B apply. Which one? But first balance, which is needed in either case.

```
273            \@StandardBalanceColumns
274            \ifnum\outputpenalty = \@BalancePenalty
```

Case A: nothing further to do except for reporting it.

```
275               \@CaseReport{A}%
276            \else
```

Case B: nothing further to do except for reporting it.

```
277               \@CaseReport{B}%
```

Ship out the current page.

```
278               \@ShipAPageBox{%
279                  \ifnum\@PageLayoutCodeDSC = 0
280                     \vbox to \@PageHeight{\@BuildPageSoFar}%
281                  \else
282                     \vbox to \@PageHeight{\@BuildPageSoFar
283                                          \vfill}%
284                  \fi
285               }%
```

```
286                    \@ComputeVsizeForDoubleColumns
287                    \global\pagegoal = \vsize
288                \fi
289            \fi
290        \fi
291    }
```

A temporary counter for the following macro is declared now.

```
292    \newcount\@EmptyBoxesBuildPageCount
```

\@BuildPageSoFar builds a box of a page from the partial page collected so
far (potentially empty) and the left (box \@DSCLeftColumnBox) and right (box
\@DSCRightColumnBox) columns.

The following macro produces the current page as it was built so far, but
doesn't produce anything if there is no material at all.

```
293    \def\@BuildPageSoFar{%
294        \@EmptyBoxesBuildPageCount = 0
295        \ifvoid\@BoxOfPageSoFar
296            \advance\@EmptyBoxesBuildPageCount by 1
297        \fi
298        \ifvoid\@DSCLeftColumnBox
299            \advance\@EmptyBoxesBuildPageCount by 1
300        \fi
301        \ifvoid\@DSCRightColumnBox
302            \advance\@EmptyBoxesBuildPageCount by 1
303        \fi
304        \WriteProtocol{1}{\string\@BuildPageSoFar: begin
305                (\noexpand\@EmptyBoxesBuildPageCount is
306                \the\@EmptyBoxesBuildPageCount)}
```

Don't generate anything if all three preceding boxes are empty.

```
307        \ifnum\@EmptyBoxesBuildPageCount < 3
308            \BoxToProtocol{2}{\@BoxOfPageSoFar}{}
309            \BoxToProtocol{2}{\@DSCLeftColumnBox}{}
310            \BoxToProtocol{2}{\@DSCRightColumnBox}{}
311            \unvbox\@BoxOfPageSoFar            % May be empty.
312            \wd\@DSCLeftColumnBox = \@ColWidth      % Left column.
313            \wd\@DSCRightColumnBox = \@ColWidth     % Right column.
314            \@PageLine{%
315                \hskip\@ColIndent
316                \BoxR\@DSCLeftColumnBox
317                \hfil
318                \BoxR\@DSCRightColumnBox
319                \hskip\@ColIndent
320            }
321            \smallskip
322        \fi
323        \WriteProtocol{1}{\string\@BuildPageSoFar: end}
324    }
```

\@SetColumnBox is a macro that is called to adjust the left or right column
box. The dimension register to which the length is being adjusted is \dimen0.

The macro has two parameters:

- #1. \@DSCLeftColumnBox or \@DSCRightColumnBox
- #2. \vfill or empty.

```
325   \def\@SetColumnBox #1#2{%
326       \global\setbox#1 = \vbox to \dimen0{\unvbox#1 #2}%
327       \Vtbox{#1}{\global}%
328       \ShiftRefPointUpOrDown{#1}{12pt}%
329   }
```

Now we will look at macros that do the balanced splitting of the left and right columns. \@StandardBalanceColumns calls \@BalanceColumns (the split amount is computed to be roughly ($\ht255 + \dp255$)/2.

```
330   \def\@StandardBalanceColumns{%
331       \setbox\@DSCTempBox = \vbox{\unvcopy 255}
```

Compute the ideal height of both columns in \dimen0.

```
332       \dimen0 = \ht\@DSCTempBox
333       \advance\dimen0 by \dp\@DSCTempBox
334       \advance\dimen0 by \topskip
335       \divide\dimen0 by 2
336       \WriteProtocol{1}{\string\@StandardBalanceColumns:
337           \noexpand\dimen0 is \the\dimen0, page \the\pageno.}
338       \@BalanceColumns{\dimen0}%
339   }
```

The input to the following macro is in box register 255 and in its only argument, #1, which is a dimension. Box 255 is vertically split and box registers \@DSCLeftColumn and \@DSCRightColumn are loaded from it. The macro balances the two columns so that they are as closely as possible the same in height, but with the left column at least as long as the right column.

```
340   \def\@BalanceColumns #1{%
341       \@ProtDSC{\string\@BalanceColumns: Start}%
342       \@EliminateRulesConditional
343       \ifvoid\@DSCLeftColumnBox\else
344           \errmessage{\string\@BalanceColumns: left column box
345               not empty.}%
346       \fi
347       \ifvoid\@DSCRightColumnBox\else
348           \errmessage{\string\@BalanceColumns: right column box
349               not empty.}%
350       \fi
351       \ifvoid 255
352           \errmessage{\string\@BalanceColumns: box 255 is void.}%
353       \fi
```

Save the complete column (unsplit) box in \@DSCTempBox.

```
354       \setbox\@DSCTempBox = \vbox{\unvbox 255}
```

The ideal height to split off is loaded into \dimen0.

```
355        \dimen0 = #1
356        \splittopskip = \topskip
357        \BoxToProtocol{0}{\@DSCTempBox}{Before \noexpand\vsplit loop}%
358        {%
359            \vbadness = 10000    % Don't report underfull boxes.
360            \loop
361                \global\setbox\@DSCRightColumnBox = \copy\@DSCTempBox
362                \global\setbox\@DSCLeftColumnBox =
363                        \vsplit\@DSCRightColumnBox to \dimen0
364                \WriteProtocol{1}{\string\dimen0: \the\dimen0}%
365                \BoxToProtocol{1}{\@DSCLeftColumnBox}%
366                    {[1] (left column) in \noexpand\vsplit loop}%
367                \BoxToProtocol{1}{\@DSCRightColumnBox}%
368                    {[2] (right column) in \noexpand\vsplit loop}%
```

Do the splitting until the height of the left column is greater or equal to the
height of the right column.

```
369                \NaturalHeight{\dimen3}{\@DSCLeftColumnBox}%
370                \NaturalHeight{\dimen4}{\@DSCRightColumnBox}%
371                \WriteProtocol{3}{\string\dimen3: \the\dimen3}%
372                \WriteProtocol{3}{\string\dimen4: \the\dimen4}%
373                \advance\dimen3 by 1sp
374                \ifdim\dimen4 > \dimen3
375                    \global\advance\dimen0 by 1pt
376            \repeat
377        }
```

Now let the two columns stretch to their natural height.

```
378        \setbox\@DSCLeftColumnBox  = \vbox{\unvbox\@DSCLeftColumnBox}%
379        \setbox\@DSCRightColumnBox = \vbox{\unvbox\@DSCRightColumnBox}%
```

Set \dimen0, used in the following \@SetColumnBox calls.

```
380        \MaxDimen{\dimen0}{\ht\@DSCLeftColumnBox}%
381                        {\ht\@DSCRightColumnBox}{}%
382        \ifcase\@PageLayoutCodeDSC
383           \@SetColumnBox{\@DSCLeftColumnBox}{}%
384           \@SetColumnBox{\@DSCRightColumnBox}{}%
385        \or
386           \@SetColumnBox{\@DSCLeftColumnBox}{\vfill}%
387           \@SetColumnBox{\@DSCRightColumnBox}{\vfill}%
388        \fi
389        \WriteProtocol{1}{\string\@BalanceColumns:
390           balancing done.}%
391    }
```

The macros \DSCHeader and \DSCFooter print headers and footers on a
page. They may be replaced by macros implementing the page layout of your
choice. These macros have no parameters.

```
392    \def\DSCHeader{%
393        \@PageLineR{%
394            \bf Header
```

```
395            \hfil
396            \tt \the\pageno
397        }%
398    }
399    \def\DSCFooter{%
400    %    \@PageLineR{%
401    %        \vrule width \@PageWidth height 1pt depth 2pt
402    %    }%
403        \@PageLineR{%
404            \bf FOOTER
405            \hfil
406            \tt \the\pageno
407        }%
408    }
```

`\@ShipAPageBox` ships a box (#1) to the `dvi` file adding header and footer beforehand. The single parameter, #1, of this macro, is the box register being shipped out. The box's glue must be set properly. Height, depth or width of the box are not adjusted in any way.

```
409    \def\@ShipAPageBox #1{%
410        \WriteProtocol{0}{\string\@ShipAPageBox:
411            called (page \the\pageno)}%
412        \shipout\vbox{%
413            \@EliminateRulesConditional
414            \DSCHeader
415            \vskip 12pt
416            #1
417            \vskip 12pt
418            \DSCFooter
419        }
420        \WriteProtocol{0}%
421            {\string\@ShipAPageBox: done (page \the\pageno)}%
422        \advancepageno
423    }
```

The macro `\@CheckNumberOfColumns` checks the number of columns and generates an error if the number of columns is incorrect. This macro has the following parameters:

- #1. The number of columns currently to be processed.
- #2. The name of the macro that calls this macro.

```
424    \def\@CheckNumberOfColumns #1#2{%
425        \ifnum \@DSCCurNumberOfColumns = #1\relax
426        \else
427            \errmessage{\string\@CheckNumberOfColumns: [\string#2]:
428                currently \the\@DSCCurNumberOfColumns\space columns,
429                should be #1.}
430        \fi
431    }
```

Define a new \bye macro. It must occur in single column mode. Note that you first need to define \bye to be empty so that \bye is "outer free" (otherwise you cannot write \string\bye in the second definition of \bye below).

Note that the following definition can also be used from the plain format output routine.

```
432    \def\bye{}
433    \def\bye{%
434       \@CheckNumberOfColumns{1}{\string\bye: Still in double
435          column mode, forgotten a \string\EndDoubleColumns?}%
436       \vfill\supereject
437       \end
438    }
439    \catcode'\@ = 12
```

• End of <code>out-ds.tip</code> •

37.2.8   The Example Source Code

This section lists the example source code used for testing our macros and for the generation of the output in Figs. 37.3–37.6 on pp. 189–192. An interesting aspect of the example is that the first double column output in this example (first figure) shows a rather strange result of balancing the two columns. The second double column output corrects this balancing problem, but for the price of having a hyphenated line at the bottom of the column break. The first double column output leads to such a poor result, because by accident the text contains two hyphenated lines in a row. You can't have it both ways, with balanced output and no hyphenated lines (the real solution is, of course, to rewrite the text)

Finally, here is the TeX source of the examples! We begin with some initial definitions by loading all needed macros to generate sample paragraphs.

37.2.8.1   Initializations

The example starts with some initializations. Paragraphs will be typeset ragged right, making life easier with narrow columns.

• <code>ex-out-ds.tip</code> •

```
1    \input inputd.tip
2    \InputD{evhvbox.tip}         % 6.13, p. I-206.
3    \InputD{out-ds.tip}          % 37.2.7.1, p. 170.
4    \InputD{samplepa.tip}        % 27.1.3.3, p. III-402.
```

Set up the page layout.

```
5    % \SetUpDSC{28pc}{2.1in}{0.2in}{7.0in}{0}{0}
6    % \SetUpDSC{28pc}{2.1in}{0.2in}{7.0in}{0}{1}
7    \SetUpDSC{28pc}{2.1in}{0.2in}{7.0in}{1}{1}
```

```
 8    \raggedright
```

Increase ragged right stretchability a little more, which is derived from the text generated by \SamplePar.

```
 9    \rightskip = 0pt plus 30pt
```

The following \Title macro (with one argument, #1, the text of the title of some section) generates a section title for the examples below.

```
10    \newcount\TitleCount
11
12    \def\Title #1{%
13        \global\advance\TitleCount by 1
14        \bigskip
15        \centerline{\bf\the\TitleCount.\ \ #1}
16        \medskip
17        \noindent
18        \ignorespaces
19    }
```

37.2.8.2   Case-Related Test Samples

Below you find the collection of the same code. \CaseA generates some example relating to page breaks related to case A.

```
20    \def\CaseA{%
21        \bigskip
22        \centerline{\bf Case A}
23        \bigskip
24        \BeginDoubleColumns
25            \SamplePar{Case A}{3}
26        \EndDoubleColumns
27    }
```

The following macro uses the above macro to actually generate some text.

```
28    \def\CaseATest{%
29        \brokenpenalty = 100 % Default in plain format.
30        \CaseA
31        \brokenpenalty = 10000
32        \CaseA
33        \vfill\eject
34        \brokenpenalty = 100 % Default in plain format.
35    }
36    \CaseATest
```

The next example delivers case B page breaks.

```
37    \def\CaseB{%
38        \bigskip
39        \centerline{Case B}
40        \bigskip
```

```
41        \BeginDoubleColumns
42            \SamplePar{Case B}{4}
43            \SamplePar{Case B}{4}
44            \SamplePar{Case B}{4}
45            \SamplePar{Case B}{4}
46            \SamplePar{Case B}{4}
47            \SamplePar{Case B}{4}
48            \SamplePar{Case B}{4}
49            \SamplePar{Case B}{4}
50            \SamplePar{Case B}{4}
51            \SamplePar{Case B}{4}
52            \SamplePar{Case B}{4}
53            \SamplePar{Case B}{4}
54        \EndDoubleColumns
55        \vfill\eject
56    }
57    \CaseB
```

The `\CaseG` macro does page breaks related to case G.

```
58    \def\CaseG #1#2{%
59        \bigskip
60        \centerline{Case C/G #1/#2}
61        \bigskip
62        \BeginDoubleColumns
63            \SamplePar{Case C/G}{#1}
64            \vfill\eject
65            \SamplePar{Case C/G}{#2}
66            \vfill
67        \EndDoubleColumns
68        End of Case C/G business.
69    }
70    \def\CaseGTest{
71        \CaseG{1}{2}
72        \CaseG{2}{1}
73    }
74    \CaseGTest
```

37.2.8.3   Generation of the Example Text, Page 1

```
75    \Title{First Double Column Output in Our Example}
76    \BeginDoubleColumns
77        \SamplePar{XXX-1}{3}
78        \vfill
79    \EndDoubleColumns
80
81    \BeginDoubleColumns
82        \SamplePar{XXX-2}{2}
83        \vfill
84    \EndDoubleColumns
```

```
85
86  \Title{Forced End of Left and Right Column}
87      Now we show how we can enforce a specific end of the
88  left {\it and\/} the right column.
89  \BeginDoubleColumns
90      \SamplePar{AAA-1}{1}
91      \vfill\eject
92      \SamplePar{AAA-2}{2}
93      And here is some more double column output on top of
94      the next page. It will be balanced by our macros.
95      \SamplePar{AAA-3}{1}
96  \EndDoubleColumns
97  \vfill\eject
```

37.2.8.4   Generation of the Example Text, Page 2

```
98   \Title{Double Column Output Continuing over Two Pages}
99       Now we present some double column output continuing over
100  multiple pages. At the end we force a page break (in single
101  column modus).
102  \BeginDoubleColumns
103      \SamplePar{XXX-1}{4}
104      \SamplePar{XXX-2}{4}
105      \SamplePar{XXX-3}{6}
106      \SamplePar{XXX-4}{4}
107  \EndDoubleColumns
108  \vfill\eject
```

37.2.8.5   Generation of the Example Text, Page 4

```
109  \Title{Manually Balanced Columns}
110  \BeginDoubleColumns
111      Now show how we can balance columns ``manually''. We do so
112  even in the middle of some paragraph! So let us write some text
113  and when it is time for a column break then we simply enter a
114  {\tt \string\vadjust} which contains a {\tt \string\vfill
115  \string\eject}. We did so at the place of the $\bullet$%
116      \vadjust{\vfill\eject}.
117  By the way, this now starts on top of the right column. We will
118  generate a right-hand column with some displayed equations in
119  it. The choice of our column break is kind of unusual because
120  normally the left column is longer than the right column.
121  $$
122      \sum_{i=1}^{100} x_{i} = 0
123  $$
124  $$
```

```
125        \sum_{i=1}^{100} y_{i} = 0
126  $$
127  $$
128        \sum_{i=1}^{100} z_{i} = 0
129  $$
130  \EndDoubleColumns
131
132  \Title{That's it}
133  Back to one column mode; it's time to say bye-bye.
134  \bye
```

• End of **ex-out-ds.tip** •

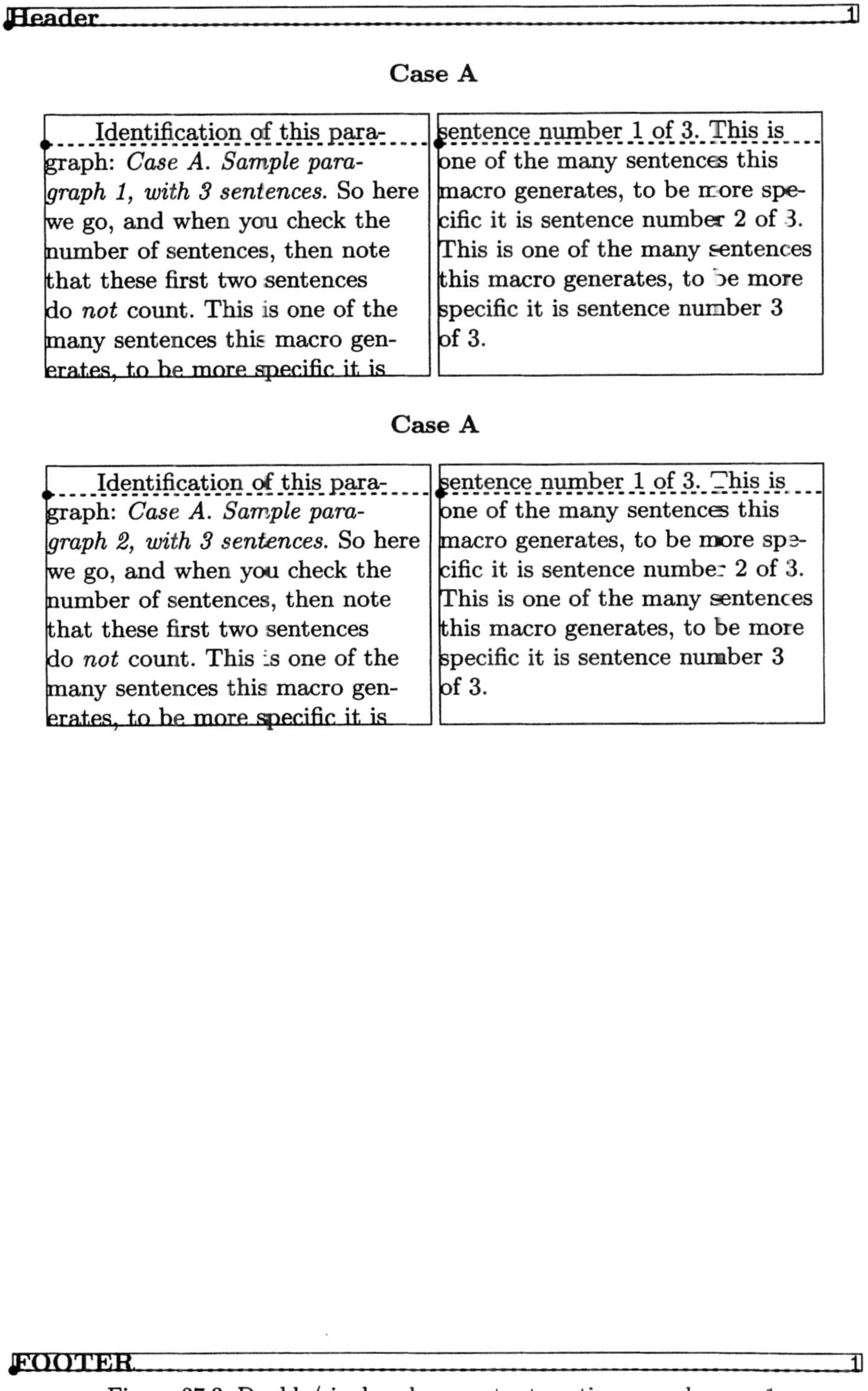

Case A

Identification of this para-graph: *Case A. Sample paragraph 1, with 3 sentences.* So here we go, and when you check the number of sentences, then note that these first two sentences do *not* count. This is one of the many sentences this macro generates, to be more specific it is	sentence number 1 of 3. This is one of the many sentences this macro generates, to be more specific it is sentence number 2 of 3. This is one of the many sentences this macro generates, to be more specific it is sentence number 3 of 3.

Case A

Identification of this para-graph: *Case A. Sample paragraph 2, with 3 sentences.* So here we go, and when you check the number of sentences, then note that these first two sentences do *not* count. This is one of the many sentences this macro generates, to be more specific it is	sentence number 1 of 3. This is one of the many sentences this macro generates, to be more specific it is sentence number 2 of 3. This is one of the many sentences this macro generates, to be more specific it is sentence number 3 of 3.

Figure 37.3. Double/single column output routine, sample page 1.

Case B

Identification of this paragraph: *Case B. Sample paragraph 3, with 4 sentences.* So here we go, and when you check the number of sentences, then note that these first two sentences do *not* count. This is one of the many sentences this macro generates, to be more specific it is sentence number 1 of 4. This is one of the many sentences this macro generates, to be more specific it is sentence number 2 of 4. This is one of the many sentences this macro generates, to be more specific it is sentence number 3 of 4. This is one of the many sentences this macro generates, to be more specific it is sentence number 4 of 4.

Identification of this paragraph: *Case B. Sample paragraph 4, with 4 sentences.* So here we go, and when you check the number of sentences, then note that these first two sentences do *not* count. This is one of the many sentences this macro generates, to be more specific it is sentence number 1 of 4. This is one of the many sentences this macro generates, to be more specific it is sentence number 2 of 4. This is one of the many sentences this macro generates, to be more specific it is sentence number 3 of 4. This is one of the many sentences this macro generates, to be more specific it is sentence number 4 of 4.

Identification of this paragraph: *Case B. Sample paragraph 5, with 4 sentences.* So here we go, and when you check the number of sentences, then note that these first two sentences do *not* count. This is one of the many sentences this macro generates, to be more specific it is sentence number 1 of 4. This is one of the many sentences this macro generates, to be more specific it is sentence number 2 of 4. This is one of the many sentences this macro generates, to be more specific it is sentence number 3 of 4. This is one of the many sentences this macro generates, to be more specific it is sentence number 4 of 4.

Identification of this paragraph: *Case B. Sample paragraph 6, with 4 sentences.* So here we go, and when you check the number of sentences, then note that these first two sentences do *not* count. This is one of the many sentences this macro generates, to be more specific it is sentence number 1 of 4. This is one of the many sentences this macro generates, to be more specific it is sentence number 2 of 4. This is one of the many sentences this macro generates, to be more specific it is sentence number 3 of 4. This is one of the many sentences this macro generates, to be more specific it is sentence number 4 of 4.

Figure 37.4. Double/single column output routine, sample page 2.

Identification of this paragraph: *Case B. Sample paragraph 7, with 4 sentences.* So here we go, and when you check the number of sentences, then note that these first two sentences do *not* count. This is one of the many sentences this macro generates, to be more specific it is sentence number 1 of 4. This is one of the many sentences this macro generates, to be more specific it is sentence number 2 of 4. This is one of the many sentences this macro generates, to be more specific it is sentence number 3 of 4. This is one of the many sentences this macro generates, to be more specific it is sentence number 4 of 4.

Identification of this paragraph: *Case B. Sample paragraph 8, with 4 sentences.* So here we go, and when you check the number of sentences, then note that these first two sentences do *not* count. This is one of the many sentences this macro generates, to be more specific it is sentence number 1 of 4. This is one of the many sentences this macro generates, to be more specific it is sentence number 2 of 4. This is one of the many sentences this macro generates, to be more specific it is sentence number 3 of 4. This is one of the many sentences this macro generates, to be more specific it is sentence number 4 of 4.

Identification of this paragraph: *Case B. Sample paragraph 9, with 4 sentences* So here we go, and when you check the number of sentences, then note that these first two sentences do *not* count. This is one of the many sentences this macro generates, to be more specific it is sentence number 1 of 4. This is one of the many sentences this macro generates, to be more specific it is sentence number 2 of 4. This is one of the many sentences this macro generates, to be more specific it is sentence number 3 of 4. This is one of the many sentences this macro generates, to be more specific it is sentence number 4 of 4.

Identification of this paragraph: *Case B. Sample paragraph 10, with 4 sentences.* So here we go, and when you check the number of sentences, then note that these first two sentences do *not* count. This is one of the many sentences this macro generates, to be more specific it is sentence number 1 of 4. This is one of the many sentences this macro generates, to be more specific it is sentence number 2 of 4. This is one of the many sentences this macro generates, to be more specific it is sentence number 3 of 4. This is one of the many sentences this macro generates, to be more specific it is sentence number 4 of 4.

Figure 37.5. Double/single column output routine, sample page 3.

Identification of this paragraph: *Case B. Sample paragraph 11, with 4 sentences.* So
here we go, and when you check
the number of sentences, then
note that these first two sentences
do *not* count. This is one of the
many sentences this macro generates, to be more specific it is
sentence number 1 of 4. This is
one of the many sentences this
macro generates, to be more specific it is sentence number 2 of 4.
This is one of the many sentences
this macro generates, to be more
specific it is sentence number 3
of 4. This is one of the many sentences this macro generates, to
be more specific it is sentence
number 4 of 4.

Identification of this paragraph: *Case B. Sample paragraph 12, with 4 sentences.* So
here we go, and when you check
the number of sentences, then
note that these first two sentences
do *not* count. This is one of the
many sentences this macro generates, to be more specific it is
sentence number 1 of 4. This is
one of the many sentences this
macro generates, to be more specific it is sentence number 2 of 4.
This is one of the many sentences
this macro generates, to be more
specific it is sentence number 3
of 4. This is one of the many sentences this macro generates, to
be more specific it is sentence
number 4 of 4.

Identification of this paragraph: *Case B. Sample paragraph 13, with 4 sentences.* So
here we go, and when you check
the number of sentences, then
note that these first two sentences
do *not* count. This is one of the
many sentences this macro generates, to be more specific it is
sentence number 1 of 4. This is
one of the many sentences this
macro generates, to be more specific it is sentence number 2 of 4.
This is one of the many sentences
this macro generates, to be more
specific it is sentence number 3
of 4. This is one of the many sentences this macro generates, to
be more specific it is sentence
number 4 of 4.

Identification of this paragraph: *Case B. Sample paragraph 14, with 4 sentences.* So
here we go, and when you check
the number of sentences, then
note that these first two sentences
do *not* count. This is one of the
many sentences this macro generates, to be more specific it is
sentence number 1 of 4. This is
one of the many sentences this
macro generates, to be more specific it is sentence number 2 of 4.
This is one of the many sentences
this macro generates, to be more
specific it is sentence number 3
of 4. This is one of the many sentences this macro generates, to
be more specific it is sentence
number 4 of 4.

Figure 37.6. Double/single column output routine, sample page 4.

Case C/G 1/2

<table>
<tr><td>

Identification of this para-
graph: *Case C/G. Sample para-
graph 15, with 1 sentences.* So
here we go, and when you check
the number of sentences, then
note that these first two sentences
do *not* count. This is one of the
many sentences this macro gen-
erates, to be more specific it is
sentence number 1 of 1.

</td><td>

Identification of this para-
graph: *Case C/G. Sample para-
graph 16, with 2 sentences.* So
here we go, and when you check
the number of sentences, then
note that these first two sentences
do *not* count. This is one of the
many sentences this macro gen-
erates, to be more specific it is
sentence number 1 of 2. This is
one of the many sentences this
macro generates, to be more spe-
cific it is sentence number 2 of 2.

</td></tr>
</table>

End of Case C/G business.

Case C/G 2/1

<table>
<tr><td>

Identification of this para-
graph: *Case C/G. Sample para-
graph 17, with 2 sentences.* So
here we go, and when you check
the number of sentences, then
note that these first two sentences
do *not* count. This is one of the
many sentences this macro gen-
erates, to be more specific it is
sentence number 1 of 2. This is
one of the many sentences this
macro generates, to be more spe-
cific it is sentence number 2 of 2.

</td><td>

Identification of this para-
graph: *Case C/G. Sample para-
graph 18, with 1 sentences.* So
here we go, and when you check
the number of sentences, then
note that these first two sentences
do *not* count. This is one of the
many sentences this macro gen-
erates, to be more specific it is
sentence number 1 of 1.

</td></tr>
</table>

End of Case C/G business.

1. First Double Column Output in Our Example

<table>
<tr><td>

Identification of this para-
graph: *XXX-1. Sample para-
graph 19, with 3 sentences.* So
here we go, and when you check
the number of sentences, then
note that these first two sentences

</td><td>

do *not* count. This is one of the
many sentences this macro gen-
erates, to be more specific it is
sentence number 1 of 3. This is
one of the many sentences this

</td></tr>
</table>

Figure 37.7. Double/single column output routine, sample page 5.

macro generates, to be more specific it is sentence number 2 of 3. This is one of the many sentences	this macro generates, to be more specific it is sentence number 3 of 3.
Identification of this paragraph: *XXX-2. Sample paragraph 20, with 2 sentences.* So here we go, and when you check the number of sentences, then note that these first two sentences do *not* count. This is one of the	many sentences this macro generates, to be more specific it is sentence number 1 of 2. This is one of the many sentences this macro generates, to be more specific it is sentence number 2 of 2.

2. Forced End of Left and Right Column

Now we show how we can enforce a specific end of the left *and* the right column.

Identification of this paragraph: *AAA-1. Sample paragraph 21, with 1 sentences.* So here we go, and when you check the number of sentences, then note that these first two sentences do *not* count. This is one of the many sentences this macro generates, to be more specific it is sentence number 1 of 1.	Identification of this paragraph: *AAA-2. Sample paragraph 22, with 2 sentences.* So here we go, and when you check the number of sentences, then note that these first two sentences do *not* count. This is one of the many sentences this macro generates, to be more specific it is sentence number 1 of 2. This is one of the many sentences this macro generates, to be more specific it is sentence number 2 of 2. And here is some more double column output on top of the next page. It will be balanced by our macros. Identification of this paragraph: *AAA-3. Sample paragraph 23, with 1 sentences.* So here we go, and when you check the number of sentences, then note that these first two sentences do *not* count. This is one of the many sentences this macro generates, to be more specific it is sentence number 1 of 1.

Figure 37.8. Double/single column output routine, sample page 6.

3. Double Column Output Continuing over Two Pages

Now we present some double column output continuing over multiple pages.
At the end we force a page break (in single column modus).

Identification of this paragraph: *XXX-1. Sample paragraph 24, with 4 sentences.* So
here we go, and when you check
the number of sentences, then
note that these first two sentences
do *not* count. This is one of the
many sentences this macro generates, to be more specific it is
sentence number 1 of 4. This is
one of the many sentences this
macro generates, to be more specific it is sentence number 2 of 4.
This is one of the many sentences
this macro generates, to be more
specific it is sentence number 3
of 4. This is one of the many sentences this macro generates, to
be more specific it is sentence
number 4 of 4.

Identification of this paragraph: *XXX-2. Sample paragraph 25, with 4 sentences.* So
here we go, and when you check
the number of sentences, then
note that these first two sentences
do *not* count. This is one of the
many sentences this macro generates, to be more specific it is
sentence number 1 of 4. This is
one of the many sentences this
macro generates, to be more specific it is sentence number 2 of 4.
This is one of the many sentences
this macro generates, to be more
specific it is sentence number 3
of 4. This is one of the many sentences this macro generates, to

be more specific it is sentence
number 4 of 4.

Identification of this paragraph: *XXX-3. Sample paragraph 26, with 6 sentences.* So
here we go, and when you check
the number of sentences, then
note that these first two sentences
do *not* count. This is one of the
many sentences this macro generates, to be more specific it is
sentence number 1 of 6. This is
one of the many sentences this
macro generates, to be more specific it is sentence number 2 of 6.
This is one of the many sentences
this macro generates, to be more
specific it is sentence number 3
of 6. This is one of the many sentences this macro generates, to
be more specific it is sentence
number 4 of 6. This is one of the
many sentences this macro generates, to be more specific it is
sentence number 5 of 6. This is
one of the many sentences this
macro generates, to be more specific it is sentence number 6 of 6.

Identification of this paragraph: *XXX-4. Sample paragraph 27, with 4 sentences.* So
here we go, and when you check
the number of sentences, then
note that these first two sentences
do *not* count. This is one of the
many sentences this macro generates, to be more specific it is
sentence number 1 of 4. This is

Figure 37.9. Double/single column output routine, sample page 7.

Header   8

| one of the many sentences this macro generates, to be more specific it is sentence number 2 of 4. This is one of the many sentences this macro generates, to be more | specific it is sentence number 3 of 4. This is one of the many sentences this macro generates, to be more specific it is sentence number 4 of 4. |

Figure 37.10. Double/single column output routine, sample page 8.

4. Manually Balanced Columns

Now show how we can balance columns "manually". We do so even in the middle of some paragraph! So let us write some text and when it is time for a column break then we simply enter a \vadjust which contains a \vfill\eject. We did so at the place of the •. By the way, this now starts on top of the right column. We will generate a right-hand column with some displayed equations in it. The choice of our column break is kind of unusual because normally the left column is longer than the right column.

$$\sum_{i=1}^{100} x_i = 0$$

$$\sum_{i=1}^{100} y_i = 0$$

$$\sum_{i=1}^{100} z_i = 0$$

5. That's it

Back to one column mode; it's time to say bye-bye.

Figure 37.11. Double/single column output routine, sample page 9.

37.3 Summary

In this chapter we learned:

1. Double column output can be handled in two ways: either left- and right-hand column are looked at as separate pages and then glued together or they are regarded as one double long page which is split into two halves using \vsplit.
2. This chapter presented an output routine that allows for the automatic balancing of columns and switching back and forth between single and double column output mode. The setup just described actually consists of a set of cooperating output routines which are at various times assigned to \output.

38

Tables in TeX Using \halign

We have previously discussed building tables by using hboxes or by using \set-tabs, which in turn uses hboxes. The disadvantage of those approaches (see 6.8, p. I-191, or 6.9, p. I-198) was that the user had to specify the width of every column in advance, either by specifying the width explicitly or implicitly (by specifying the widest entry of each column).

Using \halign, the subject of this and the following three chapters, TeX determines the width of each column by computing the width of the widest entry of each column automatically. Thus, \halign is a lot more flexible than using hboxes to build tables.

In this and the following three chapters when a statement like "a table is done in such a way" is found, then this statement is always to be interpreted as a "table done using \halign," unless stated otherwise.

38.1   The Page Layout of the Chapters on Tables

Note that usually in a document containing tables these tables are allowed to *float*. If a table, because it is too large, cannot be printed at the current location, it is printed on top of the next page (or tables are always printed at the top of pages).

However, in this series the chapters on tables are not allowed to float. The reason for this is that tables are an integral part of the text. In general you will find the following patterns repeated many times in the table chapters: first explain what the subject and motivation of a particular table is, then present its source code, and finally print the output of the source code, the table itself. A floating table would be unnatural as far as the flow of the reading is concerned.

The result of this is that because tables cannot be broken in the middle, some awful looking pages are an unavoidable side effect. The table chapters are set ragged-bottom.

Choosing between the two alternatives, floating tables which interrupt the flow of reading of the table chapters and not-so-great-looking pages we opted for the not-so-great-looking pages.

38.2   Basics

The following Section shows you the theory behind tables built using \halign. First is a discussion of templates and preamble followed by an explanation of the basic primitives in TeX to build tables using \halign.

38.2.1   Templates and Preamble

Typesetting a table using \halign requires you to specify a *preamble* before you specify the table elements. The preamble consists of a list of *templates*.

There is precisely *one template for each column.*

Each template is a *macro-like* construct with precisely *one parameter*, which is applied to every entry in a column. The argument, when this template macro is expanded, is the table entry for the current row. Because "template macros" have precisely one parameter, the place where the table entry will be substituted in the template is *not* marked by "#1" (as would be the case with a macro with one parameter) but by "#" only.

This may all sound very theoretical, so here are some examples for typical templates. The most important and most frequently occurring function of a template is to specify *glue* to determine the alignment of entries within a column.

1.   Left-justified table entries are generated with the template "#\hfil."
2.   Right-justified entries are generated with the template "\hfil#."
3.   Centered entries are generated with template "\hfil#\hfil."

Collectively, the templates form the *preamble*. Remember there is one template per column and one preamble per table. The next Subsection deals with some of the special characters and control sequences you have to know when you build a table using \halign.

We call the *table body* that part of the table that causes the actual printing to occur (the table body is everything except for the preamble).

38.2.2   Special Symbols and Control Sequences in Tables

The following primitives and special characters play a special role in building tables with \halign:

1. \halign is TeX's table-building instruction. This instruction is followed by an opening curly brace, the preamble, the table entries, and a closing curly brace.

   Instead of the curly braces, \bgroup and \egroup can be used, as we will discuss later.

2. There is also the possibility of modifying the natural width of tables using \halign to ⟨dimen⟩ and \halign spread ⟨dimen⟩. This will also be discussed later.

3. The "#" in templates specifies where the *actual table entry* is substituted into the template. There must be exactly one "#" per template.

   The "#" is not used in table entries.

4. The "&" is the *tab character* in \halign-based tables. It specifies advancing to the next column. More precisely, "&" is used in two slightly different instances:

   (a) In the *preamble*, "&" indicates the end of one template and the beginning of the next template.

   (b) In the *table body*, "&" indicates the end of an entry in one column and the beginning of an entry in the next column in the same row.

   The "&" acts as a tab character because its category code is set to 4; see 18.1.6, p. III-5.

5. \cr indicates the end of a row. Like "&" it has a dual function. The first \cr indicates the *end of the preamble* and the beginning of the table body. All the following \crs indicate the *end of the current table row*.

6. \crcr is a special form of a \cr; see 40.6.7, p. 332, for details.

7. \everycr is a token register evaluated at every \cr and every nonredundant \crcr; see 41.8, p. 373.

8. \omit eliminates the template of the current column for the current entry; see 38.3.12, p. 220.

9. \span has two applications:

   (a) In *templates*, \span forces the expansion of the subsequent token when the template is read-in (templates are normally read-in without expansions and are only expanded when the table entries are actually substituted); see 40.6.5, p. 330;

   (b) In the *table body*, \span is used instead of "&", to specify that the previous and the next entry should be regarded as one combined column; see 39.3.2, p. 264, for details.

10. \tabskip glue determines the spacing between the columns of a table; see 38.4, p. 230.

38.2.3   The General Structure of a Table Built with \halign

Instead of lengthy explanations, here is a short table with three columns and four rows to illustrate a table built with \halign. First the source code.

Sample Table **38.1** (Source code).

The following `\tabskip` specification sets the distance between all columns to 10 pt.

```
1    \tabskip = 10pt
```

A table begins with `\halign` followed by an opening curly brace.

```
2    \halign{
```

Now three templates are specified for a table of three columns: the first column is left-justified, the second column is centered, and the third column is right-justified. The percent sign at the end of the third template prevents a space from being inserted on the right-hand side of a table entry of the third column.

```
3        #\hfil&
4        \hfil#\hfil&
5        \hfil#%
```

The following `\cr` terminates the preamble and indicates the beginning of the table body.

```
6    \cr
```

Now a *table body* with four rows is specified.

```
7        ABC-ABC& DEF&      X\cr
8        XXX&     123.00& XX\cr
9        1&       ?&      XXX\cr
10       **&      ???&    XXXX\cr
```

The following closing curly brace terminates this table.

```
11   }
```

Sample Table **38.1** (Output).

ABC-ABC	DEF	X
XXX	123.00	XX
1	?	XXX
**	???	XXXX

As you can see, the table does *not* appear centered on the page the way you probably would like it to appear. To correct this, the whole table is enclosed in a vbox and this vbox is centered using display math mode. This method of centering tables is by the by far most frequently one. Centering tables (including other methods) is discussed in more detail in 40.1, p. 295.

Let me repeat the previous table, now centered using the technique just explained.

Sample Table **38.2** (Source code).

```
 1   $$
 2   \vbox{
 3                                               \tabskip = 10pt
 4       \halign{
 5           #\hfil&
 6           \hfil#\hfil&
 7           \hfil#%
 8       \cr
 9           ABC-ABC& DEF&      X\cr
10           XXX&     123.00& XX\cr
11           1&       ?&       XXX\cr
12           **&      ???&     XXXX\cr
13       }
14   }
15   $$
```

Sample Table **38.2** (Output).

ABC-ABC	DEF	X
XXX	123.00	XX
1	?	XXX
**	???	XXXX

38.2.4   How TeX Evaluates a Tabled Coded by \halign

TeX computes a table built with \halign as follows. You may want to go back to
6.8, p. I-191, to look into how a table is built using hboxes. The parallels should
be obvious. In the hbox-based approach, we first need to determine the width
of the widest entry in each column; then we will do the typesetting. Something
very similar happens in \halign.

The first step in TeX's processing of a table is to read-in the preamble. Then
the following computations are done for each column.

1. TeX collects all table entries of one column (from each of the table rows).
    It substitutes each entry into the template and expands the template. It
    inserts the result into an \hbox. For example, if the template is "\hfil#"
    (right-justified column) and an entry ABC is given, TeX forms the following
    \hbox: \hbox{\hfil ABC}.

    The width of \hbox{\hfil ABC} is now recorded by TeX. Observe that
    at this point the stretchability or shrinkability of any horizontal glue inserted
    by the template (\hfil in the example) is irrelevant; the natural width of
    \hbox{\hfil ABC} and \hbox{ABC} are the same, because the *natural width*
    of \hfil is zero; see 5.4.3, p. I-136.

2. The previous step is repeated for *each* entry of this column through all rows to determine the width w of the *widest* entry of this column.
3. After w is determined, each of the \hboxes formed in the first step is changed from \hbox{...} to \hbox to w {...}.

    Assume the widest entry is 1 in wide ($w = 1$ in); then the resulting \hbox to typeset the above entry reads as follows: \hbox to 1in{\hfil ABC}. This entry will be right-justified within this \hbox, because the width of "ABC" is less than 1 in, and the \hfil glue is inserted to the left of ABC.
4. In the end we have each entry of one column inserted into an hbox of exactly the same width, and within these \hboxes the tables are aligned by horizontal glue inserted by the expansion of templates.

    When you compare this with the approach in 6.8, p. I-191, then the parallels should be obvious.

The above computations are repeated for each column. All the hboxes computed this way from one row are combined into another hbox and \tabskip glue is inserted between the hboxes to determine the spacing *between* those columns.

Figure 38.1 on the next page summarizes the ideas just discussed.

In review, the advantages of using \halign over an hbox-based approach are:

1. \halign automatically determines the width of the widest entry.
2. \halign's templates provide an automatic macro mechanism for each column entry.
3. \halign-based tables have more concise input.

38.3   Simple Tables

38.3.1   Standard Templates

The following three templates have already been introduced. They are the by far most frequently used templates, and will be used for the presentation of simple tables.

1. The template "#\hfil" generates a left-justified column. Compare this with the definition of \leftline in 6.6, p. I-185.
2. The template "\hfil#" generates a right-justified column. Compare this with the definition of \rightline in 6.6, p. I-185.
3. The template "\hfil#\hfil" generates a centered column. Compare this with the definition of \centerline in 6.6, p. I-185.

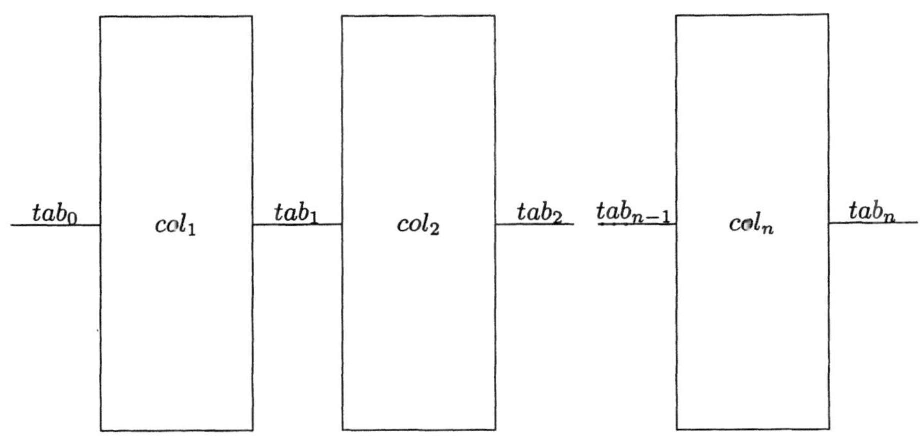

Abbreviations

n: number of columns in the table
col_i: ith column of the table, $(i = 1, \ldots, n)$
$temp_i$:   template for col_i
tab_0:   \tabskip before first column (col_1)
tab_i: \tabskip between col_i and col_{i+1} $(i = 1, \ldots, n-1)$
tab_n: \tabskip after last column (col_n)
           If tab_k is unspecified, use the value of tab_{k-1},
           and continue recursively; tab_0 usually defaults to zero.

Sample Table

```
                                     tab_0      % tab_0 must be specified
                                                % before \halign begins.
\halign{   temp_1                    tab_1 &
           temp_2                    tab_2 &
           ...
           temp_n                    tab_n
\cr                                             % End of preamble.
           entry& entry ...\cr
           entry& entry ...\cr
           ...
}
```

Figure 38.1. Illustration of templates and \tabskip glue in \halign-based tables.

Horizontal glue used to justify entries in templates should be a first-order infinite glue (\hfil), and *not* second-order infinite glue (like \hfill). There are cases where one would like to overwrite a glue inserted through the template with a glue of a higher infinite order. To be successful, a glue of a lower infinite order must be inserted by the template; see 38.3.14, p. 226, for an example.

Numerical alignment, another form of alignment, is discussed in 38.6, p. 242.

38.3.2 A Very Simple Table

The following tables use all three different templates from the previous Subsection. The last column of the following table shows that material other than glue can be inserted by a template.

Sample Table 38.3 (Source code).

```
1   $$\vbox{
2                                   \tabskip = 15pt
3   \halign{                        % Preamble begins here.
4       \hfil#\hfil&                % 1: centered column
5       \hfil#&                     % 2: right-justified column.
6       #\hfil&                     % 3: left-justified column.
7       -#-\hfil                    % 4: entries are between '-'s,
8                                   %    left-justified.
9   \cr                             % Preamble ends here.
10      XX&     YY&     Z&          1\cr        % First row
11      YY&     XX&     ZZ&         2\cr        % Second row
12      AA&     BB&     CCC&        3\cr        % ...
13      TT&     AA&     TTTT&       4\cr
14      BIGG&   BIGGGGGER&  VERY BIG&   300\cr
15  } % \halign
16  }$$
```

Sample Table 38.3 (Output).

XX		YY	Z	-1-
YY		XX	ZZ	-2-
AA		BB	CCC	-3-
TT		AA	TTTT	-4-
BIGG	BIGGGGGER	VERY BIG	-300-

Note that in a "real-world" table, the comments listed after each template should explain the column contents, not the column alignments as is done here.

38.3.3 Leaders in Templates

Instead of using infinite glue (\hfil...) for centering, and left- or right-justifying any columns, *leaders* can be used. Leaders can be interpreted as a form of "visible glue" (see 5.6.1, p. I-147), and they are applied in this way here. The space between columns (generated by the \tabskip glue) is now also clearly visible. Compare the following table with the previous table: the layouts of the two tables are identical.

Sample Table 38.4 (Source code).

```
1   $$\vbox{
2                                          \tabskip = 15pt
3   \halign{
4       \leaders\hrule\hfil #\leaders\hrule\hfil&
5       \leaders\hrule\hfil#&
6       #\leaders\hrule\hfil&
7       -#-\leaders\hrule\hfil
8   \cr
9       XX&        YY&        Z&            1\cr
10      YY&        XX&        ZZ&           2\cr
11      AA&        BB&        CCC&          3\cr
12      TT&        AA&        TTTT&         4\cr
13      BIGG&      BIGGGGGER&  VERY BIG&     300\cr
14  }
15  }$$
```

Sample Table 38.4 (Output).

_XX_	________YY	Z________	-1-__
_YY_	________XX	ZZ_______	-2-__
_AA_	________BB	CCC______	-3-__
_TT_	________AA	TTTT_____	-4-__
BIGG	BIGGGGGER	VERY BIG	-300-

The plain format defines leader-generating macros (5.6.5, p. I-150). They are used in the following example. Note that the leaders \rightarrowfill, \dotfill and \hrulefill correspond to a glue of second-order infinity (\hfill), and *not* first-order infinity (\hfil).

Sample Table 38.5 (Source code).

```
1   $$\vbox{
2                                          \tabskip = 15pt
3   \halign{
4       #\rightarrowfill&
5       \dotfill#&
6       #\hrulefill&
7       -#-\leaders\hrule\hfil
8   \cr
9       XX&        YY&        Z&            1\cr
10      YY&        XX&        ZZ&           2\cr
```

```
11      AA&       BB&       CCC&       3\cr
12      TT&       AA&       TTTT&      4\cr
13      BIGG&     BIGGGGGER& VERY BIG& 300\cr
14    }
15    }$$
```

Sample Table **38.5** (Output).

XX⟶	. . . . YY	Z‗‗‗‗‗‗‗	-1-‗
YY⟶	 XX	ZZ‗‗‗‗‗	-2-‗
AA⟶	. . . . BB	CCC‗‗‗‗	-3-‗
TT⟶	. . . . AA	TTTT‗‗‗	-4-‗
BIGG→	BIGGGGGER	VERY BIG	-300-

38.3.4   Font Changes and Inline Math Mode in Templates

Another common application of templates shown in the following table is to
specify the font to be used for a column in the template. The following example
also contains a column typeset in inline math mode. The inline math mode is
initiated through the template. The table entries do not need to (and must not)
be enclosed in math delimiters.

Sample Table **38.6** (Source code).

```
1   $$\vbox{
2                                           \tabskip = 15pt
3   \halign{
4       \it#\rightarrowfill&         % 1
5       \dotfill\tt#&                % 2
6       \bf#\hfil&                   % 3
7       ${\cal A}_{#}$\hfil         % 4
8   \cr
9       XX&       YY&       Z&        1\cr
10      YY&       XX&       ZZ&       2\cr
11      AA&       BB&       CCC&      3\cr
12      TT&       AA&       TTTT&     4\cr
13      BIGG&     BIGGGGGER& VERY BIG& 300\cr
14    }
15    }$$
```

Sample Table **38.6** (Output).

XX⟶	. . . . YY	**Z**	$\mathcal{A}_1$
YY⟶	 XX	**ZZ**	$\mathcal{A}_2$
AA⟶	. . . . BB	**CCC**	$\mathcal{A}_3$
TT⟶	. . . . AA	**TTTT**	$\mathcal{A}_4$
$BIGG$→	BIGGGGGER	**VERY BIG**	$\mathcal{A}_{300}$

Display math mode cannot be used directly inside a column, and it would not make sense either, because the resulting column would be as wide as the page, leaving no space for any other columns. If the output is supposed to come out as if typeset in display math mode, `\displaystyle` can be used as indicated by the following example. See 14.4.1, p. II-195, for details on this command. `\baselineskip` was increased to insure proper vertical spacing within the table output.

Sample Table 38.7 (Source code).

```
1    $$\vbox{
2        \baselineskip = 36pt
3                                        \tabskip = 15pt
4    \halign{
5        \it #\rightarrowfill&          % 1
6        \dotfill\tt#&                   % 2
7        \bf#\hfil&                      % 3
8        $\displaystyle                  % 4
9            \int_{#}^\infty f(x)dx$\hfil
10   \cr
11       XX&      YY&       Z&           1\cr
12       YY&      XX&       ZZ&          2\cr
13       AA&      BB&       CCC&         3\cr
14       TT&      AA&       TTTT&        4\cr
15       BIGG&    BIGGGGGER&  VERY BIG&    300\cr
16   }
17   }$$
```

Sample Table 38.7 (Output).

$$XX \longrightarrow \quad \dots\dots\text{YY} \quad \mathbf{Z} \qquad \int_{1}^{\infty} f(x)dx$$

$$YY \longrightarrow \quad \dots\dots\text{XX} \quad \mathbf{ZZ} \qquad \int_{2}^{\infty} f(x)dx$$

$$AA \longrightarrow \quad \dots\dots\text{BB} \quad \mathbf{CCC} \qquad \int_{3}^{\infty} f(x)dx$$

$$TT \longrightarrow \quad \dots\dots\text{AA} \quad \mathbf{TTTT} \qquad \int_{4}^{\infty} f(x)dx$$

$$BIGG \rightarrow \quad \text{BIGGGGGER} \quad \mathbf{VERY\ BIG} \qquad \int_{300}^{\infty} f(x)dx$$

Display math mode can be used directly in tables if the table entry is enclosed inside a `\vbox` and `\hsize` is set properly; see 40.3.1, p. 304, for details.

38.3.5   Columns Form Implicit Groups

Columns form *implicit groups*, that is, grouping is done automatically by TeX. Grouping can be bypassed by using \global; see 19.4.10, p. III-106.

Here is an example illustrating implicit grouping of columns: entries in the fourth column of the following table appear in the regular Roman font despite the font change in the third column, a font change not explicitly enclosed in a group. Another example showing implicit grouping in the following sample table appears in the first and second column. A value assigned to counter register \AA in the first column has no effect on the value printed in the second column unless the assignment is done on a global basis. Note that the assignments to \AA do *not* cause any output to be generated in the table itself.

Sample Table 38.8 (Source code).

```
 1  $$\vbox{%
 2  \newcount\AA
 3  \AA = 0
 4                                      \tabskip = 0pt
 5  \halign{
 6      \it#\rightarrowfill           \tabskip = 15pt&    % 1
 7      \hfil#                                           % 2
 8      {\tt\string\AA}: \the\AA&
 9      \bf#\hfil&                                        % 3
10      #\hfil&                                           % 4
11      ${\cal A}_{#}$\hfil           \tabskip = 0pt      % 5
12  \cr
13      \AA=10 XX&     YY&      Z&        This column is&  1\cr
14      \global\AA=4 YY&     XX& ZZ& not in boldface.&2\cr
15      \AA=10 AA&     BB&      CCC&      We have implicit&3\cr
16      TT&           AA&      TTTT&      grouping!&        4\cr
17      \global\AA=2 BIGG&   BIGGGGGER&\it VERY BIG&& 300\cr
18  }
19  }$$
```

Sample Table 38.8 (Output).

$XX\longrightarrow$	YY \AA: 0	**Z**	This column is	$\mathcal{A}_1$
$YY\longrightarrow$	XX \AA: 4	**ZZ**	not in boldface.	$\mathcal{A}_2$
$AA\longrightarrow$	BB \AA: 4	**CCC**	We have implicit	$\mathcal{A}_3$
$TT\longrightarrow$	AA \AA: 4	**TTTT**	grouping!	$\mathcal{A}_4$
$BIGG\rightarrow$	BIGGGGGER \AA: 2	*VERY BIG*		$\mathcal{A}_{300}$

38.3.6   Eliminating a Column

Occasionally you may want to print a table without a specific column, for instance, because this column contains confidential data (which you do not want to show somebody) whereas other columns don't[1]. Here is such a table in which I simply changed the template of the column to be ignored (the second column as far as TeX is concerned) to simply not print the table entry. This is very easy, and certainly easier than changing the table itself. Observe that this eliminated column still exists in TeX's mind although the width is zero.

Note also that the \tabskip glue before and after this column must be set properly. Here is an example where the second column is ignored and where \tabskip is handled correctly.

Sample Table **38.9** (Source code).

```
1    \InputD{gobble.tip}                        % 21.8.6, p. III-186.
2    $$\vbox{
3                                       \tabskip = 15pt
4    \halign{
5       \hfil#&                                        % 1
6       \GobbleOne{#}\relax      \tabskip = 0pt&       % 2
7       #\hfil                   \tabskip = 15pt&      % 3
8       -#-\hfil                                       % 4
9    \cr
10      XX&       YY&       Z&        1\cr
11      YY&       XX&       ZZ&       2\cr
12      AA&       BB&       CCC&      3\cr
13      TT&       AA&       TTTT&     4\cr
14      BIGG&     BIGGGGGER&   VERY BIG&300\cr
15   } % \halign
16   }$$
```

Sample Table **38.9** (Output).

XX	Z	-1-
YY	ZZ	-2-
AA	CCC	-3-
TT	TTTT	-4-
BIGG	VERY BIG	-300-

---

[1] This section was especially written for my students of the Los Alamos National Laboratory, Los Alamos, New Mexico, and of the Lawrence Livermore National Laboratory, Livermore, California, who had to constantly watch their *alien and uncleared* teacher and escort him everywhere.

38.3.7   Determining the Width of a Column

You can determine the width of the widest entry within a table using the method described in this Subsection. Later I present an application where it is actually necessary to determine the width of a column.

Let us first define the macro \WidthSavingBox with two parameters:

- #1. A *table entry*. This entry will be stored on an intermediate basis in a box register, so its width can be determined, and then it will be extracted from this box register so it can be printed.
- #2. A *dimension register*. Each time this macro is called, the value of the dimension register is replaced by the maximum of its previous value and of the width of the box containing the current table entry.

$\mathcal{P}'$ • tabswb.tip •

```
15   \input inputd.tip
16   \InputD{maxmindi.tip}                    % 25.1.14, p. III-332.
17   \catcode'\@ = 11
```

A temporary box is used to save the current table entry.

```
18   \newbox\@WidthSavingBox
```

The macro starts at this point. First the entry is assigned to a box register, then the width of the box, saved in this box register, is compared with the maximum width of the column computed so far. This maximum is replaced, if necessary. This arithmetic must be done on a global basis because columns form implicit groups.

```
19   \def\WidthSavingBox #1#2{%
20       \setbox\@WidthSavingBox = \hbox{#1}%
21       \MaxDimen{#2}{#2}{\wd\@WidthSavingBox}{\global}%
22       \box\@WidthSavingBox
23   }
24   \catcode'\@ = 12
```

• End of tabswb.tip •

We will apply the preceding macro in the source code of this sample table:

Sample Table **38.10** (Source code).

```
1   \InputD{tabswb.tip}                       % 38.3.7, p. 212.
```

Allocate five dimension registers, each containing the width of the widest entry within a column.

```
2   \newdimen\WidthColOne    \WidthColOne = 0pt
3   \newdimen\WidthColTwo    \WidthColTwo = 0pt
4   \newdimen\WidthColThree  \WidthColThree = 0pt
5   \newdimen\WidthColFour   \WidthColFour = 0pt
6   \newdimen\WidthColFive   \WidthColFive = 0pt
```

Now the typesetting of the table begins.

```
 7  $$\vbox{
 8                                  \tabskip = 15pt
 9  \halign{
10      \WidthSavingBox{\it#}{\WidthColOne}\hfil&        % 1
11      \dotfill\WidthSavingBox{\tt#}{\WidthColTwo}&     % 2
12      \WidthSavingBox{\bf#}{\WidthColThree}\hfil& % 3
13      \WidthSavingBox{#}{\WidthColFour}\hfil&        % 4
14      \WidthSavingBox{${\cal A}_{#}$}{\WidthColFive}\hfil % 5
15  \cr
16      XX&     YY&     Z&         This column is&  1\cr
17      YY&     XX&     ZZ& not in boldface.&2\cr
18      AA&     BB&     CCC&       We have implicit&3\cr
19      TT&             AA&      TTTT&    grouping!&            4\cr
20      BIGG& BIGGGGGER&\it VERY BIG&& 300\cr
21  }}$$
```

After the preceding table is printed, another one with the widths of the columns of the previous table is printed.

```
22  The columns in the previous table have the following widths
23  (aren't we lucky that the width of all columns is a floating
24  point number starting with two digits, which makes those
25  entries appear aligned along their decimal point!).
26  $$\vbox{
27                  \tabskip = 10pt
28  \halign{
29      \hfil#\hfil&     % 1. Column number.
30      \the#\hfil       % 2. Width of column.
31  \cr
32      \bf Column&\omit\hfil\bf Width\hfil\cr
33      1& \WidthColOne\cr
34      2& \WidthColTwo\cr
35      3& \WidthColThree\cr
36      4& \WidthColFour\cr
37      5& \WidthColFive\cr
38  }}$$
```

Sample Table **38.10** (Output).

XX	$\dotfill$YY	**Z**	This column is	$\mathcal{A}_1$
YY	$\dotfill$XX	**ZZ**	not in boldface.	$\mathcal{A}_2$
AA	$\dotfill$BB	**CCC**	We have implicit	$\mathcal{A}_3$
TT	$\dotfill$AA	**TTTT**	grouping!	$\mathcal{A}_4$
$BIGG$	BIGGGGGER	$VERY\ BIG$		$\mathcal{A}_{300}$

The columns in the previous table have the following widths (aren't we lucky that the width of all columns is a floating point number starting with two digits, which makes those entries appear aligned along their decimal point!).

Column	Width
1	26.36652pt
2	47.24959pt
3	50.3858pt
4	73.61124pt
5	20.4431pt

38.3.8 Headings and Table Entries

The macro presented in the previous Subsection allows us to solve the following quite common problem: assume you have a table with a heading for a column. This heading is *wider* than any other entry in this column. You now would like to typeset this column in such a way that the *nonheading* entries are left-justified (or right-justified), but centered with respect to the heading of the column. To be more precise: the widest element or elements of the nonheading entries should be centered with respect to the (even wider) heading.

There is no simple trickery with glue to achieve this effect and it is indeed necessary to *measure the width* of the *widest nonheading entry* to determine the glue to be inserted before (in the case of left-justified) or after (in the case of right-justified) nonheading entries. This is how we proceed:

1. The table needs to be typeset twice. The very first typesetting's only purpose is to determine the width of the widest entry among the nonheading entries. Note in particular that the heading at this point is ignored.
2. The second time the table is typeset with the heading and with a different template than was used the first time the table was typeset.
3. To be able to do step 1 without generating an actual output, the table is enclosed inside a vbox, and this vbox is assigned to a box register, but the box register's content is discarded.

Here is an example:

Sample Table **38.11** (Source code).

```
 1     \nonstopmode
 2     \input inputd.tip
 3     \InputD{tabswb.tip}                        % 38.3.7, p. 212.
```

First define a macro that contains the table body (excluding heading), so
that the table body can be replicated easily.

```
 4     \def\TableBody{%
 5              1&   X\cr
 6             10&   XX\cr
 7            100&   XXX\cr
 8           1000&   XXXX\cr
 9          10000&   XXXXX\cr
10     }
```

Two dimension registers for the width of the widest non-heading entries of
the two columns are allocated next.

```
11     \newdimen\WidthColOne     \WidthColOne = 0pt
12     \newdimen\WidthColTwo     \WidthColTwo = 0pt
```

Typeset the table the very first time to determine the widest entry of the nonheading entries. No glue in the templates is used (because it is irrelevant). The
setting of \tabskip is immaterial too.

```
13     \setbox0 = \vbox{
14         \halign{%
15             \WidthSavingBox{\bf#}{\WidthColOne}&     % 1
16             \WidthSavingBox{#}{\WidthColTwo}%         % 2
17         \cr
18             \TableBody
19         }
20     }
```

Next we need to determine the width of each of the headings, subtract the width
of the widest non-heading entry and then divide this dimension by two. The
resulting dimension is the indentation to be inserted into the new templates in
the form of horizontal glue.

Define two macros which contain the text of the headings.

```
21     \def\HeadingOne{\bf Heading One}
22     \def\HeadingTwo{\bf Another Wide Heading}
```

Have two registers for the glue to be inserted to the right of the two columns.

```
23     \newdimen\ColumnOneGlue
24     \newdimen\ColumnTwoGlue
```

Compute the difference between the width of the heading and the width of
the widest entry (excluding the heading) and then divide it by two, for the first
column.

```
25     \setbox 0 = \hbox{\HeadingOne}
26     \ColumnOneGlue = \wd0
27     \advance\ColumnOneGlue by -\WidthColOne
```

```
28    \divide\ColumnOneGlue by 2
```

Do the same for the second column.

```
29    \setbox 0 = \hbox{\HeadingTwo}
30    \ColumnTwoGlue = \wd0
31    \advance\ColumnTwoGlue by -\WidthColTwo
32    \divide\ColumnTwoGlue by 2
```

Now do the table again, although this time it is printed and the table body entries (excluding headings) will be printed properly.

```
33    $$\vbox{
34                                        \tabskip = 10pt
35    \halign{
36        \hskip\the\ColumnOneGlue plus 1fil
37            #%
38        \hskip\ColumnOneGlue&
39        \hskip\the\ColumnTwoGlue plus 1fil
40            #%
41        \hskip\ColumnTwoGlue
42    \cr
43        \omit\HeadingOne& \omit\HeadingTwo\cr
44        \TableBody
45    }}$$
```

Sample Table **38.11** (Output).

Heading One	Another Wide Heading
1	X
10	XX
100	XXX
1000	XXXX
10000	XXXXX

38.3.9 Repeated Use of Table Entries

There are cases where you need to use a table entry more than once in one template (the following Subsection gives a more practical example than we give here). Because you cannot use more than one # per template, you cannot use the table entry more than once unless you apply the following trick: use the one instance you have to access the table entry to define a macro where the replacement text of this macro is the table entry. The macro just defined can subsequently be expanded as often as necessary.

Here is an example:

Sample Table **38.12** (Source code).

```
1   $$
2       \vbox{
3                                       \tabskip = 10pt
4           \halign{
```

Save the first column entry in macro \ColumnOne.

```
5               \def\ColumnOne{#}%                    % 1
6               \hfil\ColumnOne--\ColumnOne\hfil&
```

Save the second column entry in macro \ColumnTwo. Do so on a global basis, because you also want to use this entry in the third column (columns form implicit groups).

```
7               \gdef\ColumnTwo{#}%                   % 2
8               \hfil\ColumnTwo/\ColumnTwo/\ColumnTwo&
9               \hfil\ColumnTwo--#%                   % 3
10          \cr
11          A& B& C\cr
12          AA& BB& CC\cr
13          AAAAA& X& QQQQQ\cr
14          }
15      }
16  $$
```

Sample Table **38.12** (Output).

A–A	B/B/B	B–C
AA–AA	BB/BB/BB	BB–CC
AAAAA–AAAAA	X/X/X	X–QQQQQ

38.3.10   Repeated Entries in Columns

The next table is presented under the assumption that an entry is repeated in the same column a couple of times. Let's assume the first column (any other column or columns could be chosen, of course) in a row is frequently identical to the entry of the first column of the previous row. If this is true, an asterisk should be printed instead of repeating the same entry itself. This can be done easily in TeX as follows.

Sample Table **38.13** (Source code).

```
 1    \InputD{compst.tip}                          % 25.1.17.1, p. III-334.
 2    $$\vbox{
 3                                          \tabskip = 15pt
```

The macro \OldEntryColumnOne saves the entry of the previous row. Because no row preceeds the first row, this macro is initialized to expand to "nothing."

```
 4    \def\OldEntryColumnOne{}
```

The table begins here.

```
 5    \halign{%
```

Save the new table entry.

```
 6        \def\NewEntryColumnOne{#}%        % 1
```

Compare the old and new entry. If they are the same, print a centered asterisk.

```
 7        \if\StringsEqualConditional{\OldEntryColumnOne}%
 8                              {\NewEntryColumnOne}%
 9           \hfil*\hfil
10        \else
```

If the entries are different, print the new entry, left-justified, and redefine the old entry to be the new entry. Because columns form implicit groups, this redefinition must be done globally. Also note that \xdef, rather than \gdef, must be used to define \OldEntryColumnOne, because \NewEntryColumn is redefined when the next expansion of this template is started, and therefore \OldEntryColumnOne *must* expand to a complete copy of the previous entry. The preceding Subsection also explained the necessity of using a macro to save a table entry if that table entry must be used more than once.

```
11           \NewEntryColumnOne\hfil
12           \xdef\OldEntryColumnOne{\NewEntryColumnOne}%
13        \fi&
```

Here are the templates of the second and third column followed by the table body.

```
14           \hfil#&                       % 2
15           \bf#\hfil                      % 3
16        \cr
17           XXX&      YY&      Z\cr
18           XXX&      XX&      ZZ\cr
19           XXX&      BB&      CCC\cr
20           YYYY&     AA&      TTTT\cr
21           YYYY&     BIGGGGGER& X\cr
22           ZZ&       AA&      TTTT\cr
23           YYYYY&    BIGGGGGER& X\cr
24           YYYYY&    AA&      TTTT\cr
25           YYYYY&    BIGGGGGER& X\cr
26        }
27    }$$
```

Sample Table **38.13** (Output).

<pre>
XXX   YY   Z
        *   XX   ZZ
        *   BB   CCC
YYYY   AA   TTTT
        *   BIGGGGGER   X
ZZ   AA   TTTT
YYYYY   BIGGGGGER   X
        *   AA   TTTT
        *   BIGGGGGER   X
</pre>

38.3.11   Templates Without Glue

Occasionally you may want to use a template without glue to control the alignment, although this case is an exception. No glue is needed if all entries within one column have the same width. I show an example here where the second column is set in the typewriter font and all table entries of this column are four characters long. Thus, the condition that all table entries have the same width is fulfilled.

Templates without glue are also used for templates that only contain struts (see 39.4, p. 271) or that only contain vertical rules (in both of those cases the entries have the same widths in every row).

Sample Table **38.14** (Source code).

```
1   $$\vbox{
2                                   \tabskip = 15pt
3   \halign{
4       \it#\rightarrowfill&        % 1
5       \tt#&                       % 2
6       \bf#\hfil&                  % 3
7       ${\cal A}_{#}$\hfil         % 4
8   \cr
9       XX&      YY-Y&    Z&        1\cr
10      YY&      XX/L&    ZZ&       2\cr
11      AA&      1020&    CCC&      3\cr
12      TT&      AATT&    TTTT&     4\cr
13      BIGG&    (80)&    VERY BIG& 300\cr
14  }}$$
```

Sample Table **38.14** (Output).

<pre>
XX⟶   YY-Y   Z   𝒜₁
YY⟶   XX/L   ZZ   𝒜₂
AA⟶   1020   CCC   𝒜₃
TT⟶   AATT   TTTT   𝒜₄
BIGG→   (80)   VERY BIG   𝒜₃₀₀
</pre>

Now let me discuss the case where the user *should* have specified some glue to control the alignment, but has not done so. Columns without template glue print left-justified, without any warning from TeX. (It is poor programming style not to specify glue in left-justified columns and in general I discourage making use of TeX's default behavior).

Sample Table **38.15** (Source code).

```
1    $$
2    \vbox{
3                                        \tabskip = 10pt
4    \halign{
5        #&
6        #&
7        #%
8    \cr
9        This&      is&     a poorly\cr
10       programmed& table.& Don't\cr
11       do&        it&      this\cr
12       way!!!!\cr
13   }}$$
```

Sample Table **38.15** (Output).

This	is	a poorly
programmed	table.	Don't
do	it	this
way!!!!		

38.3.12   Omitting Templates, \omit

Templates are defined with the majority of the entries of a column in mind. But not all entries in a column may fit the template. Therefore TeX provides \omit to eliminate a template. The rules for applying \omit are:

1. \omit must be specified as the very first token of a table entry. If not, TeX will report a misplaced \omit. It can be preceded by space tokens (see below for details).
2. \omit applies only to the current entry of the current column. If the template for another row of the current column has to be eliminated, \omit must be repeated.
3. An application of \omit can be looked at in two different ways:
   (a)   No template is provided by TeX and the entry is used "as is,"
   (b)   The trivial template # is provided by TeX instead of the template usually applicable to that column.

4.   \omit is an essential ingredient of the definition of \multispan. In particular,
     \multispan{1} is equivalent to \omit; see 39.3.2, p. 264.

Here is an example using \omit. Note in particular that \omit eliminates
any glue controlling the alignment of entries, because this glue is inserted through
the template (at least that is what is done normally). Therefore, in general, when
\omit is used, alignment glue must be provided as part of the table entry itself.

Sample Table **38.16** (Source code).

```
1    $$\vbox{
2                                          \tabskip = 15pt
3    \halign{
4        #\rightarrowfill&
5        \dotfill#&
6        \bf#\hfil&
7        ${\cal A}_{#}$\hfil
8    \cr
9        XX&        YY&        Z&              1\cr
10       YY&        XX&        ZZ&             2\cr
11       AA&        \omit LEFT\hfil&
12                  CCC&              \omit$\infty$\hfil\cr
13       TT&        AA&        TTTT&           4\cr
14       BIGG&    BIGGGGGER&   VERY BIG&    300\cr
15   }}$$
```

Sample Table **38.16** (Output).

XX$\longrightarrow$	. . . . YY	**Z**	$\mathcal{A}_1$
YY$\longrightarrow$	 XX	**ZZ**	$\mathcal{A}_2$
AA$\longrightarrow$	LEFT	**CCC**	∞
TT$\longrightarrow$	. . . . AA	**TTTT**	$\mathcal{A}_4$
BIGG$\rightarrow$	BIGGGGGER	**VERY BIG**	$\mathcal{A}_{300}$

One important detail ought to be considered, about how TeX determines
whether \omit applies to the current column. TeX actually *expands* the first
token of the column (if that should be necessary). It then *skips space tokens* and
keeps expanding until it finds a non-space token. If this first non-space token is
\omit, this \omit is applied. Here is a table that shows this detail. (Later, we
see a table where this aspect of the expansion of table entries is important).

Sample Table **38.17** (Source code).

```
1    \input inputd.tip
2    \InputD{mspaces.tip}                     % 21.4.7, p. III-165.
3    $$\vbox{
```

Macro \FunnyOmit expands to four spaces and \omit.

```
4    \def\FunnyOmit{%
5        \FourSpaces
6        \omit
7    }
```

The table begins here.

```
 8                                                  \tabskip = 15pt
 9   \halign{
10       #\rightarrowfill&
11       \dotfill#&
12       \bf#\hfil&
13       ${\cal A}_{#}$\hfil
14   \cr
15       XX&      YY&      Z&           1\cr
16       YY&      XX&      ZZ&          2\cr
```

Here follows the call of \FunnyOmit.

```
17       AA&      \FunnyOmit LEFT\hfil&
18                CCC&          \omit$\infty$\hfil\cr
```

The rest is as before.

```
19       TT&      AA&      TTTT&        4\cr
20       BIGG&    BIGGGGGER&  VERY BIG&  300\cr
21   }}$$
```

Sample Table 38.17 (Output).

XX⟶	...........YY	**Z**	$\mathcal{A}_1$
YY⟶	XX	**ZZ**	$\mathcal{A}_2$
AA⟶	LEFT	**CCC**	∞
TT⟶	...........AA	**TTTT**	$\mathcal{A}_4$
BIGG→	BIGGGGGER	**VERY BIG**	$\mathcal{A}_{300}$

38.3.13 Omitting and Skipping Columns, Empty Rows

It is not necessary to have an entry for each column in each row. Here is a table where some rows do not extend throughout all columns. Just type \cr wherever you want to end a row and proceed to the next row.

Sample Table 38.18 (Source code).

```
 1   $$\vbox{
 2                                        \tabskip = 15pt
 3   \halign{
 4       #\rightarrowfill&
 5       \dotfill#&
 6       #\hfil&
 7       -#-\hfil
 8   \cr
 9       XX\cr
10       YY&      XX\cr
11       AA&      BB&       CCC\cr
12       TT&      AA&      TTTT&        4\cr
```

```
13      BIGG&    BIGGGGGER&   VERY BIG&    300\cr
14   }}$$
```

Sample Table **38.18** (Output).

```
XX⟶
YY⟶        ...........XX
AA⟶        ...........BB   CCC
TT⟶        ...........AA   TTTT       -4-
BIGG→      BIGGGGGER   VERY BIG   -300-
```

When nothing is entered for a particular table column entry, then note that as a practical matter that *usually* means no entry for the particular column will be printed. But *actually* what happens is that the template for this column is still expanded only with nothing substituted in place of the #. This can lead to undesirable results if the template itself generates text or leaders, even though the table entry itself, #, is empty. If this happens, use \omit to eliminate the template (another solution is to change the template, so the template contains TeX code which tests whether # is empty or not). Here is an example (look at column 1, rows 1 and 2).

Sample Table **38.19** (Source code).

```
1    $$\vbox{
2                                       \tabskip = 15pt
3    \halign{
4       \it#\rightarrowfill&         % 1
5       \dotfill\tt#&                % 2
6       \bf#\hfil&                   % 3
7       ${\cal A}_{#}$\hfil          % 4
8    \cr
9       &        YY&       Z&        1\cr
10      \omit&   XX&       ZZ&       2\cr
11      AA&      BB&       CCC&      3\cr
12      TT&      AA&       TTTT&     4\cr
13      BIGG&    BIGGGGGER&   VERY BIG&    300\cr
14   }}$$
```

Sample Table **38.19** (Output).

$$
\begin{array}{llll}
\longrightarrow & \ldots\ldots\text{YY} & \mathbf{Z} & \mathcal{A}_1 \\
 & \ldots\ldots\text{XX} & \mathbf{ZZ} & \mathcal{A}_2 \\
AA\longrightarrow & \ldots\ldots\text{BB} & \mathbf{CCC} & \mathcal{A}_3 \\
TT\longrightarrow & \ldots\ldots\text{AA} & \mathbf{TTTT} & \mathcal{A}_4 \\
BIGG\rightarrow & \text{BIGGGGGER} & \mathbf{VERY\ BIG} & \mathcal{A}_{300}
\end{array}
$$

To generate an *empty row* in a table, the first thing you might consider is entering two consecutive \crs. But as far as TeX is concerned, there is still a *first column*, although the entry in the first column is empty. This is relevant when the template of the first column generates printing material, as it does in the following table. Then it is *insufficient* to write \cr to generate an empty row. Write \omit\cr instead (see the explanation of the preceding table for details). Here is the example table:

Sample Table **38.20** (Source code).

```
 1  $$\vbox{
 2                                      \tabskip = 15pt
 3  \halign{
 4      \it#\rightarrowfill&            % 1
 5      \dotfill\tt##&                  % 2
 6      \bf#\hfil&                      % 3
 7      ${\cal A}_{#}$\hfil             % 4
 8  \cr
 9      XX&      YY&      Z&            1\cr
10          \cr
11      YY&      XX&      ZZ&           2\cr
12      AA&      BB&      CCC&          3\cr
13          \cr
14          \cr
15      TT&      AA&      TTTT&         4\cr
16      BIGG&    BIGGGGGER&   VERY BIG&    300\cr
17          \omit\cr
18      MORE&    OF&      THIS&         10\cr
19          \omit\cr
20          \omit\cr
21      NONE&    OF&      THIS&         1000\cr
22  }}$$
```

Sample Table **38.20** (Output).

$XX\longrightarrow$	$\dots\dots$YY	**Z**	$\mathcal{A}_1$
$\longrightarrow$			
$YY\longrightarrow$	$\dots\dots$XX	**ZZ**	$\mathcal{A}_2$
$AA\longrightarrow$	$\dots\dots$BB	**CCC**	$\mathcal{A}_3$
$\longrightarrow$			
$\longrightarrow$			
$TT\longrightarrow$	$\dots\dots$AA	**TTTT**	$\mathcal{A}_4$
$BIGG\longrightarrow$	BIGGGGGER	**VERY BIG**	$\mathcal{A}_{300}$
$MORE\longrightarrow$	$\dots\dots$OF	**THIS**	$\mathcal{A}_{10}$
$NONE\longrightarrow$	$\dots\dots$OF	**THIS**	$\mathcal{A}_{1000}$

There is one last trick I would like to show. Instead of writing \omit\cr
for each empty row in the previous table, you can redefine \par to expand to
\omit\cr. Leaving an empty line in the input (suggesting an empty row in the
table) will translate into an empty row in the table (the reason for this is that
empty lines in the input are converted into \pars). The redefinition of \par is
local to the table because it is enclosed inside the \vbox enclosing the table.
Here is the complete example (the output is identical to the previous table):

Sample Table 38.21 (Source code).

```
1   $$\vbox{
2   \def\par{\omit\cr}
3                                               \tabskip = 15pt
4   \halign{
5       \it#\rightarrowfill&        % 1
6       \dotfill\tt#&               % 2
7       \bf#\hfil&                  % 3
8       ${\cal A}_{#}$\hfil         % 4
9   \cr
10      XX&      YY&      Z&            1\cr
11         \cr
12      YY&      XX&      ZZ&           2\cr
13      AA&      BB&      CCC&          3\cr
14         \cr
15         \cr
16      TT&      AA&      TTTT&         4\cr
17      BIGG&    BIGGGGGER&   VERY BIG&    300\cr
18
19      MORE&    OF&      THIS&         10\cr
20
21
22      NONE&    OF&      THIS&         1000\cr
23   }}$$
```

Sample Table 38.21 (Output).

$XX \longrightarrow$	$\dots\dots$YY	Z	$\mathcal{A}_1$
$\longrightarrow$			
$YY \longrightarrow$	$\dots\dots$XX	ZZ	$\mathcal{A}_2$
$AA \longrightarrow$	$\dots\dots$BB	CCC	$\mathcal{A}_3$
$\longrightarrow$			
$\longrightarrow$			
$TT \longrightarrow$	$\dots\dots$AA	**TTTT**	$\mathcal{A}_4$
$BIGG \longrightarrow$	BIGGGGGER	**VERY BIG**	$\mathcal{A}_{300}$
$MORE \longrightarrow$	$\dots\dots$OF	**THIS**	$\mathcal{A}_{10}$
$NONE \longrightarrow$	$\dots\dots$OF	**THIS**	$\mathcal{A}_{1000}$

Also, \par could have been redefined to \noalign{\vskip\baselineskip};
see 39.2, p. 256, for an explanation of \noalign.

38.3.14   Using Glue of a Higher Order to Change Justification

In 38.3.12, p. 220, \omit was used to eliminate a template for the specific purpose
of eliminating the glue to change the alignment for an entry. Now a different
approach for solving the same problem is discussed: infinite glue of a higher
order is used to "overpower" a glue of a lower order to change the alignment of
a table entry; see 5.4.4, p. I-137, for an explanation of this technique.

In the following table in the fourth row of the third column, the \hfill
glue is part of the entry. Therefore, the expanded template reads "\bf\hfill
TTTT\hfil." This, in turn, is equivalent to the following: "\bf \hfill TTTT." If
\hfil instead of \hfill had been used, the expanded template would read "\bf
\hfil TTTT\hfil," generating a centered entry because the same glue appears
on both sides of TTTT.

When you compare this approach with the \omit-based approach, observe
that \omit eliminates the whole template, whereas in the solution here, the glue
provided through the template is simply overpowered through glue from the table
entry. Any other instruction in the template (such as font change instructions)
remains in effect.

Sample Table 38.22 (Source code).

```
1   $$\vbox{
2                                       \tabskip = 15pt
3   \halign{
4       #\rightarrowfill&
5       \dotfill#&
6       \bf#\hfil&
7       ${\cal A}_{#}$\hfil
8   \cr
9       XX&      YY&      Z&            1\cr
10      YY&      XX&      ZZ&           2\cr
11      AA&      \omit LEFT\hfil&
12                        CCC&          \omit$\infty$\cr
13      TT&      AA&      \hfill    TTTT&   4\cr
14      BIGG&    BIGGGGGER&   VERY BIG&    300\cr
15  }}$$
```

Sample Table 38.22 (Output).

XX$\longrightarrow$	$\ldots\ldots\ldots$YY	Z	$\mathcal{A}_1$
YY$\longrightarrow$	$\ldots\ldots\ldots$XX	ZZ	$\mathcal{A}_2$
AA$\longrightarrow$	LEFT	CCC	∞
TT$\longrightarrow$	$\ldots\ldots\ldots$AA	TTTT	$\mathcal{A}_4$
BIGG$\rightarrow$	BIGGGGGER	VERY BIG	$\mathcal{A}_{300}$

38.3.15   Static Columns

A "static" column is defined as a column which contains the same text through-out the whole table. It is natural to put this text, which is repeated in each row, inside the template instead of repeating it each time as a table entry. The template of this column *must* contain a #, although nothing will ever be substituted, because all table entries will be empty.

Another instance of static columns occurs in the usage of struts (see 39.4, p. 271) and vertical rules (see 39.5, p. 275). Struts and vertical rules are normally allocated their own columns. The table entries of those "columns" are usually empty. The templates of these columns contain the rules or struts to be inserted.

Here is a table with a static column (column 2).

Sample Table **38.23** (Source code).

```
$$\vbox{
                                      \tabskip = 15pt
\halign{%
    #\rightarrowfill&         % 1
    compare with#&            % 2: "#" mandatory here.
    \bf#\hfil&                % 3
    ${\cal A}_{#}$\hfil       % 4
\cr
    XX&      &       Z&       1\cr
    YY&      \omit&  ZZ&      2\cr
    AA&      &       CCC&     \omit$\infty$\cr
    TT&      &       TTTT&    4\cr
    BIGG&    &       VERY BIG& 300\cr
}}$$
```

Sample Table **38.23** (Output).

$$
\begin{array}{llll}
\text{XX}\longrightarrow & \text{compare with} & \textbf{Z} & \mathcal{A}_1 \\
\text{YY}\longrightarrow &   & \textbf{ZZ} & \mathcal{A}_2 \\
\text{AA}\longrightarrow & \text{compare with} & \textbf{CCC} & \infty \\
\text{TT}\longrightarrow & \text{compare with} & \textbf{TTTT} & \mathcal{A}_4 \\
\text{BIGG}\rightarrow & \text{compare with} & \textbf{VERY BIG} & \mathcal{A}_{300}
\end{array}
$$

38.3.16   Repeated Use of the Same Template(s)

Sometimes you may want to use a certain template or a certain set of templates for more than one column. TeX has a way to do exactly that: if you put an *extra* "&" before one of the templates, TeX will repeat the following template or templates in a circular fashion, and this repetition will continue as often as necessary. Because of the "as often as necessary repetition," it is no longer possible to determine how many columns a table has by simply looking at the preamble.

In the following table the templates of the third and fourth column are also used for the fifth and sixth column. The same templates would also be used for the seventh and eight columns (and so forth) if we specified actual entries for those columns.

Sample Table **38.24** (Source code).

```
 1  $$\vbox{
 2                                      \tabskip = 10pt
 3  \halign{
 4      #\rightarrowfill&                       % 1
 5      \dotfill#&&          % Observe '&&' here!    % 2
 6      \bf#\hfil&                               % 3, 5, 7, ...
 7      ${\cal A}_{#}$\hfil                      % 4, 6, 8, ...
 8  \cr
 9      XX&      YY&      Z&              1&      XX&  -1\cr
10      YY&      XX&      ZZ&             2&      YY&  -2\cr
11      AA&      \omit LEFT\hfil&
12                      CCC&             \omit$\infty$&
13                                       LLL LLL\cr
14      TT&      AA&      TTTT&          4&      LL&  -4\cr
15      BIGG&    BIGGGGGER&  VERY BIG&   300&    L&   -300\cr
16  }}$$
```

Sample Table **38.24** (Output).

XX⟶	...........YY	**Z**	$\mathcal{A}_1$	**XX**	$\mathcal{A}_{-1}$
YY⟶	XX	**ZZ**	$\mathcal{A}_2$	**YY**	$\mathcal{A}_{-2}$
AA⟶	LEFT	**CCC**	∞	**LLL LLL**	
TT⟶	...........AA	**TTTT**	$\mathcal{A}_4$	**LL**	$\mathcal{A}_{-4}$
BIGG→	BIGGGGGER	**VERY BIG**	$\mathcal{A}_{300}$	**L**	$\mathcal{A}_{-300}$

Let me point out a special case where the repetition starts with the very first column. In this case, the extra "&" has to precede the first template and immediately follows the opening curly brace of \halign.

Sample Table **38.25** (Source code).

```
 1  $$\vbox{
 2                                      \tabskip = 15pt
 3  \halign{
 4      &                               % 'Repetition &'
 5      \hfil-#-\hfil&                   % 1, 3, 5, ...
 6      X:#\hfil                         % 2, 4, 6, ...
 7  \cr
 8      AAA&     BBB&     CCC&     DDD&     EEE\cr
 9      A&       B&       C&       D&       E\cr
10      AA&      BB&      CC&      DD&      EE\cr
11  }}$$
```

Sample Table **38.25** (Output).

-AAA-	X:BBB	-CCC-	X:DDD	-EEE-
-A-	X:B	-C-	X:D	-E-
-AA-	X:BB	-CC-	X:DD	-EE-

I personally believe that it is better programming style to give a template for each column (even if this involves repeating certain template specifications), because the table is ultimately easier to understand—especially when comments are added to each template identifying the content of the column. But there are cases where the above approach has its merits, especially when a table-building macro is defined and the final number of columns is unknown.

I would also like to point out that this repetition construct is *not* very general. Once can only specify that *starting* with a specific template, *all* remaining templates are to be repeated as often as necessary. For example, it is not possible to specify a preamble with a special template for the first and last column, and a different template for all columns in between.

38.3.17 Spaces After the Tab Character "&" Are Ignored

Spaces *after* the tab character "&" are ignored. Note that spaces *before* "&" are *not* ignored. In the following table, I print each column entry in double quotation marks (introduced by the template) to make any spaces clearly visible.

Sample Table **38.26** (Source code).

```
 1   $$\vbox{
 2                                       \tabskip = 15pt
 3   \halign{
 4        #\hfil&          % 1
 5        ''#''\hfil&       % 2
 6        ''#''\hfil        % 3
 7   \cr
 8        1.&           A&B\cr
 9        2.&           A& B\cr
10        3.&           A&  B\cr
11        4.&          A &B\cr
12        5.&        A  &B\cr
13        6.&      A     &B\cr
14        7.&          A & B\cr
15        8.&          A & B \cr
16        9.&          A & B          \cr
17   }}$$
```

Sample Table **38.26** (Output).

1.   "A"   "B"
2.   "A"   "B"
3.   "A"   "B"
4.   "A "   "B"
5.   "A "   "B"
6.   "A "   "B"
7.   "A "   "B"
8.   "A "   "B "
9.   "A "   "B "

38.4   Spacing Between Table Columns, \tabskip

\tabskip is the glue responsible for the spacing between columns. TeX essentially treats the columns of a table independently of each other. (This is not quite true for entries that span more than one column, but let's ignore that for now). After the columns have been computed, \tabskip glue is inserted between each of the column's hboxes (this process is repeated for each row).

More precisely, \tabskip glue is inserted in the following places in a table with n columns (compare this with Fig. 38.1, p. 205):

1.   *Before* the first column. This glue is identified as tab_0.
2.   *Between* the columns. Glue tab_i is inserted between columns i and $i + 1$, where $i = 1, \ldots, n$.
3.   *After* the last column. This glue is identifies as tab_n.

Note that all \tabskip specifications you have seen so far are specifications where the glue has no stretchability and shrinkability (this is true for most applications). Also note that in all preceding tables only tab_0 was specified. If this is the case, tab_i will be set to the same value as tab_0. This is discussed in more detail shortly.

\tabskip is also used in \valign-based tables (the vertical counter part of \halign); see 41.5, p. 355.

38.4.1   Identical \tabskip for All Columns

If you want to set all tab_i to the same value, it is sufficient to set tab_0 to the desired value. In general, the rule is that unless tab_i is specified, it inherits the value of tab_{i-1}. Therefore, if only tab_0 is specified, all other tab_i $(i > 0)$ are set to tab_0. Observe that the tab_0 specification precedes \halign; see Fig. 38.1, p. 205.

Here is a table with all tab_i set to zero: now all columns touch each other.

Sample Table **38.27** (Source code).

```
1   $$\vbox{
2                                       \tabskip = 0pt
3   \halign{
4       #\rightarrowfill&
5       \dotfill#&
6       \bf#\hfil&
7       ${\cal A}_{#}$\hfil
8   \cr
9       XX&      YY&      Z&            1\cr
10      YY&      XX&      ZZ&           2\cr
11      AA&      \omit LEFT\hfil&CCC&3\cr
12      TT&      AA&      TTTT&         4\cr
13      BIGG&    BIGGGGGER&   VERY BIG&300\cr
14  }}$$
```

Sample Table **38.27** (Output).

$$
\begin{array}{llll}
\text{XX} \longrightarrow \dots\dots\text{YY}\mathbf{Z} & & & \mathcal{A}_1 \\
\text{YY} \longrightarrow \dots\dots\text{XX}\mathbf{ZZ} & & & \mathcal{A}_2 \\
\text{AA} \longrightarrow \text{LEFT} & \mathbf{CCC} & & \mathcal{A}_3 \\
\text{TT} \longrightarrow \dots\dots\text{AA}\mathbf{TTTT} & & & \mathcal{A}_4 \\
\text{BIGG} \rightarrow \text{BIGGGGGER}\mathbf{VERY\ BIG} & & & \mathcal{A}_{300}
\end{array}
$$

\tabskip can be set to a negative value, which makes columns overlap. Here is an example of a table where this is done.

Sample Table **38.28** (Source code).

```
1   $$\vbox{
2                                       \tabskip = -10pt
3   \halign{
4       #\rightarrowfill&
5       \dotfill#&
6       \bf#\hfil&
7       ${\cal A}_{#}$\hfil
8   \cr
9       XX&      YY&      Z&            1\cr
10      YY&      XX&      ZZ&           2\cr
11      AA&      \omit LEFT\hfil&CCC&3\cr
12      TT&      AA&      TTTT&         4\cr
13      BIGG&    BIGGGGGER&   VERY BIG&300\cr
14  }}$$
```

Sample Table **38.28** (Output).

```
XX——→.........YZY          𝒜₁
YY——→.........XZZ          𝒜₂
AA——LEFT        CCC        𝒜₃
TT——→.........AATTT        𝒜₄
BIGGBIGGGGGEVERY BIG₃₀₀
```

38.4.2   Tables with Different \tabskips for Each Column

Now let us look at a table where all the individual tab_is are specified. Assume that we want to leave a distance of 45 pt between the first and second column (tab_1) in a four column table ($n = 4$), 15 pt between the second and the third column (tab_2), and 30 pt between the third and the fourth column (tab_3). Tab_0 and tab_4 (which is tab_n) will be set to zero, which is a good idea in general, as we will see later.

Refer to Fig. 38.1, p. 205. This figure shows that tab_0 is specified by a \tabskip specification *before* \halign and that tab_i is specified at the end of the ith template. Note that although templates and \tabskip specifications are really two separate items, they are specified together. When TeX saves a template for later expansion, it saves those templates by themselves, so all \tabskip specifications are removed.

Observe below how the input is structured. All \tabskip instructions have been moved to the right and align to separate the \tabskip specifications from the template specifications themselves. This is not necessary, but highly recommended.

Sample Table **38.29** (Source code).

```
1   $$\vbox{
2                                   \tabskip = 0pt       % 0 (tab_0)
3   \halign{
4       #\rightarrowfill           \tabskip = 45pt&      % 1 (tab_1)
5       \dotfill#\relax            \tabskip = 15pt&      % 2 (tab_2)
6       \bf#\hfil                  \tabskip = 30pt&      % 3 (tab_3)
7       ${\cal A}_{#}$\hfil        \tabskip = 0pt        % 4 (tab_4)
8   \cr
9       XX&     YY&     Z&         1\cr
10      YY&     XX&     ZZ&        2\cr
11      AA&     \omit LEFT\hfil&CCC&  -1\cr
12      TT&     AA&     TTTT&      4\cr
13      BIGG&   BIGGER& VERY BIG&  300\cr
14  }}$$
```

Sample Table **38.29** (Output).

XX $\longrightarrow$	.....YY	Z	$\mathcal{A}_1$
YY $\longrightarrow$	XX	ZZ	$\mathcal{A}_2$
AA $\longrightarrow$	LEFT	CCC	$\mathcal{A}_{-1}$
TT $\longrightarrow$	.....AA	TTTT	$\mathcal{A}_4$
BIGG $\rightarrow$	BIGGER	VERY BIG	$\mathcal{A}_{300}$

38.4.3   Spaces and Templates

The following discussion is important because it deals with the issue of "unintentional spaces" in table templates. The *second template* of the *previous table* (the \tabskip specification is not part of the template itself and was therefore removed), when, written as a token list, reads \dotfill • # • \relax. Note the \relax. If \relax had been omitted, the template would read (again written as a list of tokens), as follows: \dotfill • # • ␣. In the latter case, when a table element is substituted in place of the "#," each table entry is always followed by a space. This is not the case in the first form of the template.

There is another way to specify the second template. This time the space introduced by the end-of-line character is removed by the "%," and spaces at the beginning of a line, before \tabskip, are ignored. Therefore, the template generated now is identical in its effect to the first version of the template presented previously. Here is the second form of this template:

```
1   \dotfill#%
2                    \tabskip = 15pt&
```

From a practical matter I should not that frequently it is not visible whether an entry is "padded" by a space on the right-hand side, because all entries of a column are padded this way. But there are cases where this space is clearly visible. For instance, this is true if two columns are supposed to touch each other. If all entries in the left column of the two columns are padded by a space on the right-hand side, the output will show nontouching columns; see 38.6, p. 242, for an example.

Here is another example where one must be careful to ensure the correct handling of spaces in templates. The correct template to center an entry reads \hfil#\hfil, and *not* \hfil# \hfil! In the second case, the text that is centered is the table entry followed by a space (which distorts the centering slightly).

The template \hfil #\hfil is correct, of course, because spaces after control words (here the first \hfil) are ignored.

38.4.4   Incomplete \tabskip Specifications

It is *not necessary* to specify all tab_i values. We already saw that if only tab_0 is set all remaining tab_i will be set to the same value. The rules according to which a missing tab_i is computed are as follows:

1. If tab_0 is *specified*, it is set to the specified value. If it is *not specified*, assume it is zero (see 38.4.6.1, p. 238, for details). My recommendation is to *always* set tab_0 to the desired value, because it may be set to a value different than the default and different than what you expect it to be.
2. The value for tab_i ($i > 0$) is determined as follows: if tab_i is explicitly *specified*, it is obviously set to the specified value. If it is *not specified*, it inherits the value of tab_{i-1}.

Here is a more complex example of incomplete \tabskip specifications. Look at the following preamble.

```
 1   \halign{
 2       #\hfil                          \tabskip = 15pt &        % 1
 3       #\hfil &                                                 % 2
 4   '   #\hfil                          \tabskip = 30pt &        % 3
 5       #\hfil                          \tabskip = 20pt &        % 4
 6       #\hfil &                                                 % 5
 7       #\hfil                                                   % 6
 8   \cr
 9           . . . . .
10   }
```

Here is the "expanded version" of the same table with each tab_i specified. See 38.4.6, p. 237, for details on tab_0.

```
 1                                       \tabskip = 0pt
 2   \halign{
 3       #\hfil                          \tabskip = 15pt &        % 1
 4       #\hfil                          \tabskip = 15pt &        % 2
 5       #\hfil                          \tabskip = 30pt &        % 3
 6       #\hfil                          \tabskip = 20pt &        % 4
 7       #\hfil                          \tabskip = 20pt &        % 5
 8       #\hfil                          \tabskip = 20pt          % 6
 9   \cr
10           . . . . .
11   }
```

38.4.5   The Special Importance of First and Last \tabskip

I would now like to draw your attention to tab_0 and tab_n. In the following example, the table is included inside a \VboxR (a visible vbox). The table is not centered, because the vbox is not enclosed inside display math mode.

Sample Table **38.30** (Source code).

```
1   \InputD{box-mac.tip}                    % 9.3.14, p. I-343.
2   \bigskip
3   \VboxR{
4                            \tabskip = 0pt
5   \halign{
6      #\rightarrowfill    \tabskip = 45pt&      % 1
7      \dotfill#\relax     \tabskip = 5pt&       % 2
8      \bf#\hfil           \tabskip = 15pt&      % 3
9      ${\cal A}_{#}$\hfil \tabskip = 0pt        % 4
10  \cr
11     XX&        YY&       Z&            1\cr
12     YY&        XX&       ZZ&           2\cr
13     AA&        \omit LEFT\hfil&CCC&3\cr
14     TT&        AA&       TTTT&         4\cr
15     BIGG&      BIGGER&   VERY BIG&     300\cr
16  }}
17  \bigskip
```

Sample Table **38.30** (Output).

XX⟶	..... YY Z	$\mathcal{A}_1$
YY⟶	..... XX ZZ	$\mathcal{A}_2$
AA⟶	LEFT CCC	$\mathcal{A}_3$
TT⟶	..... AA TTTT	$\mathcal{A}_4$
BIGG⟶	BIGGER VERY BIG	$\mathcal{A}_{300}$

Because the table is not included inside display math mode, no extra vertical glue before and after the table is inserted by TeX to offset the table. A \bigskip was added before and after the table to improve the vertical spacing.

In the following table, tab_0 is set to 40 pt, which gives you the *impression* that the table was moved 40 pt to the right. This is not really correct as far as TeX is concerned: the table was made 40 pt wider on the left side and therefore it seemed to have moved to the right by this amount.

Sample Table **38.31** (Source code).

```
1   \InputD{box-mac.tip}                    % 9.3.14, p. I-343.
2   \bigskip
3   \VboxR{
4                            \tabskip = 40pt
5   \halign{
6      #\rightarrowfill    \tabskip = 45pt&      % 1
7      \dotfill#\relax     \tabskip = 5pt&       % 2
8      \bf#\hfil           \tabskip = 15pt&      % 3
9      ${\cal A}_{#}$\hfil \tabskip = 0pt        % 4
10  \cr
11     XX&        YY&       Z&            1\cr
12     YY&        XX&       ZZ&           2\cr
13     AA&        \omit LEFT\hfil&CCC\cr
14     TT&        AA&       TTTT&         4\cr
```

```
15      BIGG&    BIGGER& VERY BIG&    300\cr
16   }}
```

Sample Table 38.31 (Output).

```
┌──────────────────────────────────────────────────────┐
│    XX⟶              ..... YY  Z            A₁         │
│    YY⟶              ..... XX  ZZ           A₂         │
│    AA⟶              LEFT     CCC                       │
│    TT⟶              ..... AA  TTTT         A₄         │
│....BIGG⟶...........BIGGER.VERY.BIG....A₃₀₀            │
└──────────────────────────────────────────────────────┘
```

Here is an example where the desired distance of the table from the *right margin* of the text is 20 pt. We cannot use the previous approach because in order to set tab_0 to the proper value we would have to know the width of the table (excluding tab_0 or tab_n). Because we have no access to this information, we take the following different approach:

1. Construct a table of width \hsize (\halign to \hsize). This construct is similar to the \hbox to ⟨dimen⟩ construct, which builds an \hbox of width ⟨dimen⟩; see 38.7.1, p. 248, for details.
2. Let tab_0 stretch as far as necessary (0pt plus 1fil).
3. Set tab_n to the desired distance of the last column and the right edge (20 pt).

Again a visible vbox using \VboxR is used to enclose the table. This clearly indicates that for TEX the table is as wide as the whole page.

Sample Table 38.32 (Source code).

```
 1   \bigskip
 2   \VboxR{
 3                            \tabskip = 0pt plus 1fil
 4   \halign to \hsize{
 5       #\rightarrowfill      \tabskip = 45pt&       % 1
 6       \dotfill#\relax       \tabskip = 10pt&       % 2
 7       \bf#\hfil             \tabskip = 15pt&       % 3
 8       ${\cal A}_{#}$\hfil \tabskip = 20pt         % 4
 9   \cr
10       XX&     YY&     Z&              1\cr
11       YY&     XX&     ZZ&             2\cr
12       AA&     \omit LEFT\hfil&CCC\cr
13       TT&     AA&     TTTT&           4\cr
14       BIGG&   BIGGER& VERY BIG&       300\cr
15   }}
16   \bigskip
```

Sample Table 38.32 (Output).

```
┌──────────────────────────────────────────────────────────────────┐
│        XX⟶              ..... YY  Z               A₁              │
│        YY⟶              ..... XX  ZZ              A₂              │
│        AA⟶              LEFT     CCC                               │
│        TT⟶              ..... AA  TTTT            A₄              │
│....BIGG⟶...............BIGGER..VERY.BIG....A₃₀₀....                │
└──────────────────────────────────────────────────────────────────┘
```

Here is an example of a table pushed all the way to the right. The table is very similar to the previous table, but now $tab_n = 0$.

Sample Table 38.33 (Source code).

```
 1   \bigskip
 2   \vbox{
 3                              \tabskip = 0pt plus 1fil
 4   \halign to \hsize{
 5       #\rightarrowfill      \tabskip = 45pt&          % 1
 6       \dotfill#\relax       \tabskip = 10pt&          % 2
 7       \bf#\hfil             \tabskip = 15pt&          % 3
 8       ${\cal A}_{#}$\hfil \tabskip = 0pt            % 4
 9   \cr
10       XX&       YY&      Z&          1\cr
11       YY&       XX&      ZZ&         2\cr
12       AA&       \omit LEFT\hfil&CCC\cr
13       TT&       AA&      TTTT&       4\cr
14       BIGG&     BIGGER&  VERY BIG&   300\cr
15   }}
16   \bigskip
```

Sample Table 38.33 (Output).

XX$\longrightarrow$	$\dots\dots$ YY	**Z**	$\mathcal{A}_1$
YY$\longrightarrow$	$\dots\dots$ XX	**ZZ**	$\mathcal{A}_2$
AA$\longrightarrow$	LEFT	**CCC**	
TT$\longrightarrow$	$\dots\dots$ AA	**TTTT**	$\mathcal{A}_4$
BIGG$\rightarrow$	BIGGER	**VERY BIG**	$\mathcal{A}_{300}$

What has been discussed in this Subsection is used only infrequently. Normally, a table is enclosed inside a vbox, which is centered using display math mode. The examples' main purpose was to demonstrate the special role of tab_0 and tab_n. In general, it is a good practice to set tab_0 and tab_n to zero. If you don't do that then what *you* perceive as the width of the table (measured from the leftmost point of the first column to the rightmost point of the last column) is different from how TEX computes the width of the table, because in TEX's measurements (unlike yours), tab_0 and tab_n are included in the width of the table.

38.4.6   \tabskip Default, tab_0

The default for tab_0 is *normally* zero." Here is a brief explanation of how tab_0 is set (what does *normally* mean after all in the preceding sentence?):

1. When TEX starts-up, the value for \tabskip is zero. Therefore, unless tab_0 is set, tab_0 for a table is zero. The format you use could set \tabskip to a different value which would determine the default for tab_0, of course.

2. Now observe that as a practical matter most tables are enclosed inside vboxes, and tab_0 is normally set inside the vbox, as you have seen in the preceding examples. Because a vbox forms an implicit group, setting tab_0 inside the box enclosing the table means tab_0 for the next table will still be zero (unless specifically set).

Again note that this is *not necessarily true*: tab_0 can be changed on a global basis (`\global\tabskip = ...`), or a table need not be enclosed inside a `\vbox`.

Here is an example where the table is not enclosed inside a `\vbox` and therefore tab_0 is 10 pt for *both* tables.

```
1                            \tabskip = 10pt
2    \halign{%
3         ...
4    \cr
5         ...\cr
6    }
7
8    \halign{
9         ...
10   }
```

To be on the safe side you should consider always setting tab_0 or enclosing all tables inside vboxes, and changing tab_0 only inside these vboxes. Also, look at the definition of `\ialign` in the following Subsubsection.

38.4.6.1   Initialized \halign, \ialign

The plain format defines a macro `\ialign` (initialized <u>align</u>), which performs the following three functions:

- Set `\tabskip` (tab_0) to zero.
- Clear the token parameter `\everycr`. `\everycr` is a token register evaluated at each `\cr` and nonredundant `\crcr`; see 41.8, p. 373.
- Start `\halign`.

Here is the definition of `\ialign`.

```
1    \def\ialign{%
2        \everycr = {}%
3        \tabskip = 0pt
4        \halign
5    }
```

And here is an application of `\ialign` (the `\tabskip` setting of line 1 is irrelevant, because `\ialign` sets `\tabskip` to zero):

Sample Table **38.34** (Source code).

```
1                                          \tabskip = 3in % Irrelevant.
2   $$\vbox{
3   \ialign{
4       #\rightarrowfill              \tabskip = 15pt&
5       \dotfill#&
6       #\hfil&
7       -#-\hfil                      \tabskip = 0pt
8   \cr
9       XX\cr
10      YY&       XX\cr
11      AA&       BB&       CCC\cr
12      TT&       AA&       TTTT&         4\cr
13      BIGG&     BIGGGGGER&  VERY BIG&   300\cr
14  }}$$
```

Sample Table **38.34** (Output).

$$
\begin{array}{llll}
\text{XX} \longrightarrow & & & \\
\text{YY} \longrightarrow & \dotfill\text{XX} & & \\
\text{AA} \longrightarrow & \dotfill\text{BB} & \text{CCC} & \\
\text{TT} \longrightarrow & \dotfill\text{AA} & \text{TTTT} & \text{-4-} \\
\text{BIGG} \rightarrow & \text{BIGGGGGER} & \text{VERY BIG} & \text{-300-}
\end{array}
$$

38.4.6.2   A Table-Related Macro for Math Modes, \@lign

The following macro is used inside displayed equation alignments. It does not start a table, but resets \tabskip and \everycr.

```
1   \def\@lign{%
2       \tabskip = 0pt
3       \everycr = {}%
4   }
```

38.4.7   Another Example

Assume you have to typeset a table with three columns. The distance from the left text margin to the first table column should be 20 pt; the distance between the last table column and the right text margin should be 40 pt; the distance *between* the columns should be *equal*. Here is how TeX can do this:

1. The width of the table is set to \hsize (\halign to \hsize). Note $n = 3$.
2. Tab_0 and tab_n are used to control the white space on the left- and right-hand side of the table body.

3. Tab_i for $0 < i < n$ (these are the \tabskips between the columns) is set to
   0pt plus 1fil so an equal amount of space will be generated between the
   columns.

Sample Table 38.35 (Source code).

```
 1   \medskip
 2   \vbox{
 3                                  \tabskip = 20pt
 4   \halign to \hsize{%
 5       #\hfil                     \tabskip = 0pt plus 1fil&
 6       #\hfil                     \tabskip = 0pt plus 1fil&
 7       #\hfil                     \tabskip = 40pt
 8   \cr
 9       ABC ABC&          DEF DEF DEF&      CCC\cr
10       ABC&              DEF XXX&          XXX XXX\cr
11       ABC ABC&          DEF XXX DEF&      CCC XXX\cr
12   }}
13   \medskip
```

Sample Table 38.35 (Output).

ABC ABC	DEF DEF DEF	CCC
ABC	DEF XXX	XXX XXX
ABC ABC	DEF XXX DEF	CCC XXX

38.5   Controlling the Spacing Between Columns by Glue in Templates

Instead of using \tabskip glue to control the spacing between columns, horizontal glue can be added to templates. Such glue would need to have a nonzero natural width, since this glue does *not* only control the alignment of entries within a column, but in effect, makes entries artificially wider by padding them with white space.

Before I discuss this approach, I would like to point out that the \tabskip-based approach of controlling the spacing between columns is in my eyes conceptually cleaner and easier. I discourage the approach I will discuss now (but discuss it anyway because it is common practice).

In the example below \tabskip is set to zero. Therefore the columns (as far as TeX's notion is concerned) will touch each other. The source code of the following table is:

Sample Table **38.36** (Source code).

```
1   $$\vbox{
2                                                \tabskip = 0pt
3   \halign{
4       #\rightarrowfill&
5       \dotfill#&
6       \bf#\hfil&
7       ${\cal A}_{#}$\hfil
8   \cr
9       XX&        YY&      Z&            1\cr
10      YY&        XX&      ZZ&           2\cr
11      AA&        \omit LEFT\hfil\qquad&& 3\cr
12      TT&        AA&      TTTT&         4\cr
13      BIGG&      BIGGER&  VERY BIG&     300\cr
14  }
15  }$$
```

Sample Table **38.36** (Output).

$$
\begin{aligned}
&\text{XX} \longrightarrow \ldots\ldots\text{YYZ} \qquad\qquad \mathcal{A}_1\\
&\text{YY} \longrightarrow \ldots\ldots\text{XXZZ} \qquad\qquad \mathcal{A}_2\\
&\text{AA} \longrightarrow \text{LEFT} \qquad\qquad\qquad\quad \mathcal{A}_3\\
&\text{TT} \longrightarrow \ldots\ldots\text{AATTTT} \qquad\quad \mathcal{A}_4\\
&\text{BIGG} \rightarrow .\,\text{BIGGER}\textbf{VERY BIG}\mathcal{A}_{300}
\end{aligned}
$$

In the next table the distance between the first and second column is the
length of one quad (10p pt in the current font), the distance between the second
and the third column is two quads, and the distance between the third and the
fourth column is 0.5 in. This statement was not technically accurate: the first
column was extended by one quad of horizontal space on the right-hand side
because each entry of the first column is padded by one quad of white space on
the right-hand side. This makes the first and the second column *appear* to be
separated by one quad of white space. Similar statements apply to the description
of the spacing between the other columns.

Sample Table **38.37** (Source code).

```
1   $$\vbox{
2                                                \tabskip = 0pt
3   \halign{
4       #\rightarrowfill\quad&
5       \dotfill#\qquad&
6       \bf#\hskip 0.5in&
7       ${\cal A}_{#}$\hfil
8   \cr
9       XX&        YY&      Z&            1\cr
10      YY&        XX&      ZZ&           2\cr
11      AA&        \omit LEFT\hfil\qquad&& 3\cr
12      TT&        AA&      TTTT&         4\cr
13      BIGG&      BIGGER&  VERY BIG&     300\cr
14  }}$$
```

Sample Table **38.37** (Output).

XX⟶	..... YY	**Z**	$\mathcal{A}_1$
YY⟶	 XX	**ZZ**	$\mathcal{A}_2$
AA⟶	LEFT		$\mathcal{A}_3$
TT⟶	..... AA	**TTTT**	$\mathcal{A}_4$
BIGG→	BIGGER	**VERY BIG**	$\mathcal{A}_{300}$

I discourage this approach of controlling the spacing between columns for the following reason: using horizontal spacing generated from templates the distance between the columns, when you look at the *typeset* table, does not coincide with TeX's way of determining the width of the columns (the distance between the first and the second column is zero, as far as TeX is concerned, because \tabskip is zero). Things are even more confusing if \tabskip is nonzero and the spacing between columns is a mixture of padding glue in templates and \tabskip glue.

38.6   Numerical Alignment

Numerical alignment of real numbers ("numbers with a decimal point") is discussed in this Section. Table entries are aligned at their decimal point. For a number such as 12.34 the "12" is called *integer part* and "34" is called *fractional part*.

There are two different methods to achieve numerical alignment:

1. Treat the integer part as a right-justified column and the fractional part as a separate left-justified column, and make the distance between the two columns zero. This is discussed in 38.6.1 on this page.
2. Print the numbers right-justified and pad the numbers with "too few digits in their fractional part" with spaces. This approach is discussed in 38.6.3, p. 247.

38.6.1   The Right-Justified / Left-Justified Solution

As mentioned before, the idea in this approach for numerically aligning numbers is to look at the integer part of numbers as a *right-justified column*, and to look at their fractional parts as a *left-justified column*. The distance between the two columns is set to zero. In other words, a numerically aligned column is effectively split up into two columns with zero distance between the columns. The main disadvantage of this approach is that what *you* perceive as *one column*, is treated as *two columns by TeX*. Each time you make a statement about column i you have to specify whether you mean what you perceive as the ith column or what TeX thinks is the ith column.

The columns of the table below which contain the integer parts of the numbers are set in math mode, so a minus sign ($-$) instead of a hyphen (-) is printed for negative numbers. The left-justified columns containing the fractional parts do not need to be typeset in math mode because no minus sign will ever be printed in those columns.

The use of `\relax` to prevent a space from sneaking in at the very end of a template was discussed in 38.4.2, p. 232. This is necessary here because otherwise the entry in the first row and second column would read "100. 0" instead of "100.0."

Sample Table 38.38 (Source code).

```
1    $$\vbox{
2                                 \tabskip = 0pt
3    \halign{
4         #\rightarrowfill       \tabskip = 45pt&      % 1
5         \hfil$#.$\relax         \tabskip = 0pt&       % 2 and 3
6                 $#$\hfil         \tabskip = 15pt&
7         \hfil$#.$\relax         \tabskip = 0pt&       % 4 and 5
8                 $#$\hfil         \tabskip = 15pt&
9         #\hfil                  \tabskip = 0pt        % 6
10   \cr
11        X&              100&0&          200&90&        X\cr
12        XXX&        1,000&23&        30,000&1004&      XX\cr
13        XXXX&     -20,100&0004&        -200&1&         XXX\cr
14        XXXXX&    -80,100&00&     -200,000&19&         XXXX\cr
15        XXXXXX& -20,100&00&        -2,000&190&         XXXXX\cr
16        XX&             100&0&          200&90&        X\cr
17   }}$$
```

Sample Table 38.38 (Output).

X$\longrightarrow$	100.0	200.90	X
XXX$\longrightarrow$	1,000.23	30,000.1004	XX
XXXX$\longrightarrow$	$-20,100.0004$	-200.1	XXX
XXXXX$\longrightarrow$	$-80,100.00$	$-200,000.19$	XXXX
XXXXXX$\rightarrow$	$-20,100.00$	$-2,000.190$	XXXXX
XX$\longrightarrow$	100.0	200.90	X

38.6.2   The Decimal Point as an Alignment Character

I continue discussing the previous approach and try to make the input a little more convenient. This is done by making the decimal point of numerically aligned tables act as a tab character, so that the numbers can be entered using decimal points. There is no need to write "&"s to separate the integer part from the decimal part.

To make the decimal point act as a tab character, set its category code to 4, which is of course the same category code assigned to the ampersand by the plain format; see 18.1.6, p. III-5.

The category code change of the period is made on a "table global basis," inside the \vbox enclosing the table. Macro \PeriodText is used to print the period. Note in the following example that a \tabskip specification must *not* contain a period (for a decimal point), because by the time the template is read in, the period's category code was already changed.

Sample Table 38.39 (Source code).

```
 1   \InputD{verb-bas.tip}                            % 18.3.1, p. III-27.
 2   $$\vbox{
 3       \catcode'\. = 4
 4                                    \tabskip = 0pt
 5   \halign{
 6       #\rightarrowfill             \tabskip = 45pt&      % 1
 7       \hfil$#$\relax               \tabskip = 0pt&       % 2 and 3
 8           \PeriodText$#$\hfil \tabskip = 15pt&
 9       \hfil$#$\relax               \tabskip = 0pt&       % 4 and 5
10           \PeriodText$#$\hfil \tabskip = 15pt&
11       #\hfil                       \tabskip = 0pt       % 6
12   \cr
13       X&          100.0&           200.90&       X\cr
14       XXX&        1,000.23&        30,000.1004&  X\PeriodText X X\cr
15       XXXX&   -20,100.0004&        -200.1&       X X X X\cr
16       XXXXX&  -80,100.00&    -200,000.19&        X X X X\cr
17       XXXXXX& -20,100.00&         -2,000.190&    X X X\cr
18       XX&         100.0&           200.90&       X X\cr
19       X&          100&\omit&       200&\omit&    X\cr
20   }}$$
```

Sample Table 38.39 (Output).

X⟶	100.0	200.90	X
XXX⟶	1,000.23	30,000.1004	X.X X
XXXX⟶	−20,100.0004	−200.1	X X X X
XXXXX⟶	−80,100.00	−200,000.19	X X X X
XXXXXX→	−20,100.00	−2,000.190	X X X
XX⟶	100.0	200.90	X X
X⟶	100	200	X

The previous approach has the problem that \PeriodText must be written to print a period. Can we do better than that? Let's try an approach where the category code change of the period is initiated through templates. Because templates form implicit groups, there is no need to undo the category code change of the period. That also eliminates the problem of \tabskip settings containing a period, as in \tabskip = 12.3pt.

The decimal point in the table for numerically aligned entries is generated by the second column, the one that contains the part of the number after the decimal period. There is no need to use \PeriodText in the template itself to generate the printed decimal period because the templates are read-in without these category code changes in effect.

Here is a table identical to the previous table (identical except for changes in the templates and the last column):

Sample Table **38.40** (Source code).

```
 1   $$\vbox{
 2                                      \tabskip = 0pt
 3   \halign{
 4       #\rightarrowfill          \tabskip = 45pt&      % 1
 5       \catcode'.=4 \hfil$#$\relax \tabskip = 0pt&     % 2 and 3
 6              .$#$\hfil           \tabskip = 15pt&
 7       \catcode'.=4 \hfil$#$\relax \tabskip = 0pt&     % 4 and 5
 8              .$#$\hfil           \tabskip = 15pt&
 9       #\hfil                    \tabskip = 0pt        % 6
10   \cr
11       X&            100.0&          200.90&      X\cr
12       XXX&        1,000.23&      30,000.1004&    X.X.X\cr
13       XXXX&    -20,100.0004&        -200.1&      X.X.X.X\cr
14       XXXXX&   -80,100.00&    -200,000.19&       X.X.X.X\cr
15       XXXXXX& -20,100.00&        -2,000.190&     X.X.X\cr
16       XX&           100.0&          200.90&      X.X\cr
17       X&            100&\omit&      200&\omit&    X\cr
18   }}$$
```

Sample Table **38.40** (Output).

X⟶	100.0	200.90	X
XXX⟶	1,000.23	30,000.1004	X.X.X
XXXX⟶	−20,100.0004	−200.1	X.X.X.X
XXXXX⟶	−80,100.00	−200,000.19	X.X.X.X
XXXXXX→	−20,100.00	−2,000.190	X.X.X
XX⟶	100.0	200.90	X.X
X⟶	100	200	X

It is important to write all numbers with a decimal point. In case there is a number with no decimal point, you must use & to delimit the number, and you must use \omit for the following column (which contains the decimal point, not to be printed in this case).

There is one problem with the above solution. To demonstrate the problem I will enter a number like .23, with nothing preceding the decimal point (it was entered as .23, and not as 0.23). In table35.41, the second row shows you how you probably *would like* to enter some numbers, and the third row shows you how you *must do* it.

Sample Table **38.41** (Source code).

```
 1  $$\vbox{
 2                                      \tabskip = 0pt
 3  \halign{
 4      #\rightarrowfill           \tabskip = 45pt&    % 1
 5      \catcode'.=4 \hfil$#$\relax \tabskip = 0pt&     % 2 and 3
 6             .$#$\hfil            \tabskip = 15pt&
 7      \catcode'.=4 \hfil$#$\relax \tabskip = 0pt&     % 4 and 5
 8             .$#$\hfil            \tabskip = 15pt&
 9      #\hfil                     \tabskip = 0pt    % 6
10  \cr
11      X&          100.0&          200.90&        X\cr
12      X&             .23&             .87&        X.y.z\cr    % Row 2.
13      X&      \relax .23&      \relax .87&        X.y.z\cr    % Row 3.
14  }}$$
```

Sample Table **38.41** (Output).

$$
\begin{array}{lllll}
X\rightarrow & 100.0 & 200.90 & X \\
X\rightarrow & .23..87 & X.y.z \\
X\rightarrow & .23 & .87 & X.y.z
\end{array}
$$

Here is what happens in the case of the second column, second row: TEX skips spaces after a & to determine whether an \omit is coming or not. When TEX finally sees the period of .23, it still has *not* invoked the template of the first column. This is because TEX must first determine whether the first token of a table entry is \omit. TEX has already converted the period into a token (with category code 12), it then determines that there is *no* \omit and now the template for the second column is inserted. However, it is too late: the period of .23, which is supposed to have category code 4, has already been assigned category code 12, and the following category code changes are no longer necessary. This is where everything goes wrong. The second column is actually terminated by the & in 0.23&, but this & was supposed to terminate the third column! From now on everything is off by one column, and you can see the results yourself. The same problem repeats itself when TEX comes to .87.

In the third row, the \relax of \relax .23 causes TEX to expand the second template in a timely fashion, and the problem is corrected (TEX uses \relax to determine the absence of \omit).

38.6.3   Aligning Numbers Using a Dummy Character

We will now discuss the second approach for achieving numerically aligned entries. Each numerically aligned table column is now *one* right-justified column (not two columns as in the previous solution). A prerequisite for the following solution to work is that all digits have the same width, which is true for most fonts, including TEX's Roman font. Each entry will be padded on the right-hand side by the appropriate amount of white space so that although TEX thinks all entries are right-justified, all entries are actually aligned along their decimal point. If the entry (or entries) with the most digits after the decimal point has k entries after the decimal point, an entry must be padded by $k - l$ times the width of a digit on the right-hand side if the entry has l digits after the decimal point.

To do this conveniently, when entering the table we use a one-character-long instruction to insert the padding spaces: the "?" is made an active character, which means it acts like a macro. "?" will expand to horizontal white space which is as wide as one digit of the font used.

Because the "?" is now active, macro \QuestionMarkText must be used to print a question mark. This macro is defined in 18.3.1, p. III-26.

Finally here is the example.

Sample Table **38.42** (Source code).

```
1    \InputD{verb-bas.tip}                    % 18.3.1, p. III-27.
```

Allocate a dimension register to save the width of a digit.

```
2    \newdimen\DigitWidth
```

Store the width of a digit in \DigitWidth. Digit "0" is used, but any digit could be used.

```
3    \setbox0 = \hbox{\rm 0}
4    \DigitWidth = \wd0
```

Now start display math mode for the table.

```
5    $$
```

Now define "macro ?." This macro definition occurs inside the implicit group of the display math mode just started and therefore disappears at the very end.

```
6    \catcode'\? = \active
7    \def?{\kern\DigitWidth}
```

Now the table itself begins enclosed inside a \vbox.

```
8    \vbox{
9                            \tabskip = 30pt
10   \halign{
11       #\rightarrowfill&                    % 1
12       \hfil$#$&                            % 2
13       \hfil$#$&                            % 3
14       #\hfil                               % 4
15   \cr
```

```
16      X&              100.0???&      200.90??&          X\QuestionMarkText\cr
17      XXX&          1000.23??&    30000.1004&          XX\cr
18      XXXX&      20100.0004&        200.1???&          XXX\cr
19      XXXXX&    80100.00??&  200000.19??&          XXX\cr
20      XXXXXX& 20100.00??&      2000.190?&          XX\cr
21      XX&            100.0???&        200.0???&          X\cr
22   }}$$
```

Sample Table 38.42 (Output).

X———————→	100.0	200.90	X?
XXX———————→	1000.23	30000.1004	XX
XXXX———————→	20100.0004	200.1	XXX
XXXXX———————→	80100.00	200000.19	XXX
XXXXXX→	20100.00	2000.190	XX
XX———————→	100.0	200.0	X

38.7   The Width of a Table

The discussion in this section deals with the width of tables and how this width can be changed in TeX.

38.7.1   \halign spread and \halign to

The following discussion on TeX's way of determining the overall width of a table is analogous to the way the width of an \hbox is determined. There are three ways to specify an \hbox: \hbox{...}, \hbox spread ⟨dimen⟩ {...}, and \hbox to ⟨dimen⟩ {...}. The same three possibilities exist for \halign:

1. \halign {...}. The width of the table is the *natural width* of the table. The natural width is the sum of the widths of all columns and the sum of all \tabskip values ($tab_0 \ldots tab_n$).

    Observe that tab_0 and tab_n are included in this computation! I stated before that it is a good idea to set both of these values to zero. The reason should be obvious: if any of the values is not zero, your perception of the width of the table is different from what TeX computes as its natural width.

    Stretchabilities and shrinkabilities of any of the \tabskip glues are irrelevant here. Actually, \tabskip is normally set to a glue that cannot stretch or shrink. This is similar to the shrinkabilities and stretchabilities of glues that are ignored when it comes to the computation of the *natural width* of an \hbox.

Most of the tables built with \halign use this form of the instruction. Also note that in that case \tabskip is a *glue* and not a dimension, although in that particular form of a table \tabskip is usually set to a glue without stretchability and shrinkability, because specifying non-zero stretchabilities and shrinkabilities does not make any sense.

2. \halign to ⟨dimen⟩: TEX generates a table that is ⟨dimen⟩ wide. If a horizontal box is built using \hbox to ⟨dimen⟩, the \hbox must contain glue that can shrink or stretch. Likewise a table built using the \halign to ⟨dimen⟩ approach must contain \tabskip glue that can shrink or shrink. Observe that none of the glues *contained in the templates* will shrink or stretch. The columns themselves are totally unaffected by this shrinking and stretching of the table as a whole.

3. \halign spread ⟨dimen⟩ instructs TEX to increase the natural width of the table by ⟨dimen⟩. Similar to \halign to ⟨dimen⟩, stretching or shrinking is done using the \tabskip glues only. \halign spread directly relates to \hbox spread the way \halign to related to \hbox to.

Here we present a \halign spread ⟨dimen⟩ example. Later a \halign to ⟨dimen⟩ example is shown; see 40.1.2, p. 296. First print a table with an unmodified width:

Sample Table **38.43** (Source code).

```
1   $$\vbox{
2                            \tabskip = 0pt
3   \halign{
4       #\hrulefill          \tabskip = 45pt plus 60pt&
5       \dotfill#\qquad      \tabskip = 0pt  plus 30pt&
6       \bf#\hfil            \tabskip = 15pt&
7       ${\cal A}_{#}$\hfil  \tabskip = 0pt
8   \cr
9       XX&     YY&     Z&          1\cr
10      YY&     XX&     ZZ&         2\cr
11      AA&     \omit LEFT\hfil\qquad&   CCC\cr
12      TT&     AA&     TTTT&       4\cr
13      BIGG&   BIGGER& VERY BIG&   300\cr
14  }}$$
```

Sample Table **38.43** (Output).

XX__	..... YY	**Z**	$\mathcal{A}_1$
YY__	 XX	**ZZ**	$\mathcal{A}_2$
AA__	LEFT	**CCC**	
TT__	..... AA	**TTTT**	$\mathcal{A}_4$
BIGG	BIGGER	**VERY BIG**	$\mathcal{A}_{300}$

Now spread the preceding table out by 60 pt. The stretchability of tab_1 is 60 pt and the stretchability of tab_2 is 30 pt, whereas tab_0, tab_3 and tab_4 do not stretch. In 5.3.8, p. I-133, we discussed how the stretching of glue is proportionally divided among those glues that can stretch. Therefore 2/3 of the necessary 60 pt to stretch is taken from tab_1 and 1/3 from tab_2. Here is the new table:

Sample Table 38.44 (Source code).

```
1    $$\vbox{
2                                   \tabskip = 0pt
3    \halign spread 60pt{
4        #\hrulefill              \tabskip = 45pt plus 60pt&
5        \dotfill#\qquad          \tabskip = 0pt  plus 30pt&
6        \bf#\hfil                \tabskip = 15pt&
7        ${\cal A}_{#}$\hfil      \tabskip = 0pt
8    \cr
9        XX&      YY&      Z&            1\cr
10       YY&      XX&      ZZ&           2\cr
11       AA&      \omit LEFT\hfil\qquad&  CCC\cr
12       TT&      AA&      TTTT&         4\cr
13       BIGG&    BIGGER&  VERY BIG&     300\cr
14   }}$$
```

Sample Table 38.44 (Output).

XX___	..... YY	Z	$\mathcal{A}_1$
YY___	 XX	ZZ	$\mathcal{A}_2$
AA___	LEFT	CCC	
TT___	..... AA	**TTTT**	$\mathcal{A}_4$
BIGG	BIGGER	**VERY BIG**	$\mathcal{A}_{300}$

The preceding table is therefore equivalent to the following table:

```
1    ...
2                                   \tabskip = 0pt
3    \halign{
4        #\hrulefill              \tabskip = 85pt&      % 45 + (2/3)*60
5        \dotfill#\qquad          \tabskip = 20pt&      %  0 + (1/3)*60
6        \bf#\hfil                \tabskip = 15pt&
7        ${\cal A}_{#}$\hfil      \tabskip = 0pt
8    \cr
9    ...
```

38.7.2   Presetting the Width of Each Column

There are cases where you may want to specify the width of each column and bypass TeX's mechanism that sets the width of a column to the widest entry of the column. Obviously, you can typeset such a table using \hboxes of predetermined width (\hbox to ...), but because of \halign's convenient notation, let me discuss how you can use \halign.

First specify a row with "no real table entries," meaning that this row will not generate any visible output. Each entry of this row is an \hskip specification of the desired column width. The assumption is that no column entry is wider than the \hskip of its column.

Here is an example of such a table. The first row, which determines the width of each column, contributes an empty row in the output. This is, in general, undesirable; it can be prevented by using a strut for controlling the vertical spacing of the table, and then eliminating the strut for this row using \omit. In effect, a zero height and depth row is generated this way; see 7.4, p. I-235, and 38.3.12, p. 220.

Sample Table **38.45** (Source code).

```
 1   $$\vbox{
 2                               \tabskip = 15pt
 3   \halign {
 4       #\hrulefill&            % 1
 5       \dotfill#&              % 2
 6       \bf#\hfil&              % 3
 7       ${\cal A}_{#}$\hfil     % 4
 8   \cr
 9       \omit\hskip 60pt&
10       \omit\hskip 60pt&
11       \omit\hskip 90pt&
12       \omit\hskip 45pt\cr
13       KK      YY&     Z&              1\cr
14       YY&     XX&     ZZ&             2\cr
15       AA&     \omit LEFT\hfil\qquad&CCC\cr
16       TT&     AA&     TTTT&           4\cr
17       BIGG&   BIGGER& VERY BIG&       300\cr
18   }}$$
```

Sample Table **38.45** (Output).

KK YY_______	.............Z	1	
YY_______	XX	**ZZ**	$\mathcal{A}_2$
AA_______	LEFT	**CCC**	
TT_______	..........AA	**TTTT**	$\mathcal{A}_4$
BIGG_______	....BIGGER	**VERY BIG**	$\mathcal{A}_{300}$

Here is a different approach using \hboxes of specified width in the template. In the second and the fourth column, I used a visible horizontal box. There is no need to specify a row of \hskips because the width of columns is encoded into the templates.

Also note that a table entry that is wider than its column's width as specified in the template will not cause any warning to be generated: its column will simply be made wider.

Sample Table 38.46 (Source code).

```
1    \InputD{box-mac.tip}                          % 9.3.14, p. I-343.
2    $$\vbox{
3                                    \tabskip = 15pt
4    \halign {
5        \hbox  to 60pt{#\hrulefill}&             % 1
6        \HboxR to 60pt{\dotfill#}&               % 2
7        \hbox  to 90pt{\bf#\hfil}&               % 3
8        \HboxR to 45pt{${\cal A}_{#}$\hfil}%%  4
9    \cr
10       KK      YY&      Z&              1\cr
11       YY&     XX&      ZZ&             2\cr
12       AA&     \omit LEFT\hfil\qquad&CCC\cr
13       TT&     AA&      TTTT&           4\cr
14       BIGG&   BIGGER&  VERY BIG&       300\cr
15   }}$$
```

Sample Table 38.46 (Output).

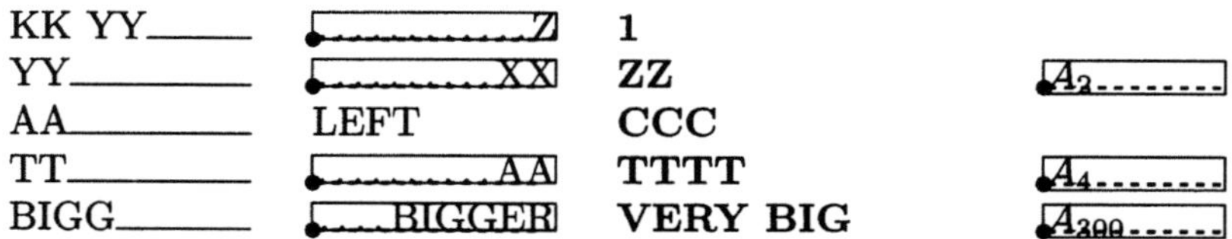

38.8 Summary

In this chapter we learned:

- Tables can be coded very easily in TEX using \halign. TEX computes the width of each column automatically.
- A table begins with the specification of a preamble. The preamble contains one template for each column. Templates are a macro-like mechanism with a separate macro for each column.

- The main purpose of the templates is to hold glue specifications controlling the alignment of columns. The table entry is substituted in place of the # in the template. Templates such as `#\hfil`, `\hfil#` and `\hfil#\hfil` are used to generate right-justified, left-justified or centered entries. Leaders instead of glue can be used to make the justification visible.
- & is used to proceed from one column to the next and `\cr` is used to terminate a row (the very first `\cr` in a table terminates the preamble). Spaces after the & are ignored in TeX.
- Templates can also be used to change fonts and to typeset a column in display math mode.
- To center a heading over a sequence of right or left-justified entries where the heading is wider than any of the entries, requires the determination of the width of the widest entry to allow for proper centering.
- It is possible to replace automatically repeated entries in a column by some special symbol.
- `\omit` can be used to eliminate an unsuitable templates on a one time basis.
- `\hfill` glue can be used to overrite `\hfil`-based glue inserted by templates if a change of alignment in a table should become necessary.
- The distance between columns is determined by `\tabskip`. This glue is inserted before the first column of a table, between the columns of a table, and after the last column of a table. The very first and the very last `\tabskip` glue should be set to zero, particularly when horizontal rules are used in tables. If a value for `\tabskip` is *not* specified, the value of the preceding `\tabskip` is used. `\ialign` initializes the very first `\tabskip` to zero.
- The horizontal spacing between columns in a table can also be controlled by extra glue to the left or right of column entries.
- It is possible to have columns align on their decimal point. There are two ways to achieve this. One is to treat the integer and fractional part of a number as two separate columns. In the other approach the fractional part of a number is padded with white space on the right-hand side.
- With the `\halign to` ⟨dimen⟩ and the `halign spread` ⟨dimen⟩ constructs the width of a table can be adjusted. Only the `\tabskip` glues will stretch and shrink if the width of a table is adjusted this way (glues in the templates are not affected).

39
More Tables in TeX Using \halign

More tables in TeX mean more wonderful things we can do. For instance, we will learn about vertical spacing in tables, \noalign, entries spanning multiple columns, struts, and vertical and horizontal rules in tables.

As explained in 38.1, p. 199, there may be some pages with rather poor page breaks. We apologize.

39.1 Vertical Spacing of Tables

You can control the vertical spacing in tables, the distance between rows in a table. There are two different ways to do so:

1. Set \baselineskip to the desired value; this will be discussed in the remainder of this Section.
2. Use struts; see 39.4, p. 271.

\baselineskip determines the spacing of lines in a table, the same way it determines the line spacing of text in paragraphs; see 7.3.4, p. I-220, for details. Here is an example of a table where \baselineskip was set to 17 pt.

Since \baselineskip cannot be changed in the middle of a table, the value of \baselineskip that was valid *before* \halign was entered is used throughout the whole table.

Sample Table **39.1** (Source code).

```
1   $$\vbox{
2   \baselineskip = 17pt
3                                   \tabskip = 15pt
4   \halign{
5       #\hrulefill&
6       \dotfill##&
7       \bf#\hfil&
8       ${\cal A}_{#}$\hfil
9   \cr
10      XX&     YY&     Z&          1\cr
11      YY&     XX&     ZZ&         2\cr
12      AA&     \omit LEFT\hfil\qquad&CCC\cr
13      TT&     AA&     TTTT&       4\cr
14      BIGG&   BIGGER& VERY BIG&   300\cr
15  }}$$
```

Sample Table **39.1** (Output).

XX__	.......YY	**Z**	$\mathcal{A}_1$
YY__	XX	**ZZ**	$\mathcal{A}_2$
AA__	LEFT	**CCC**	
TT__	.......AA	**TTTT**	$\mathcal{A}_4$
BIGG	.BIGGER	**VERY BIG**	$\mathcal{A}_{300}$

\openup increases \baselineskip (and also other values) by the specified value. Because \baselineskip's default is 12 pt, the previous table could
also have been typeset by specifying \openup 5pt (instead of \baselineskip =
17pt).

39.2   Inserting Material Not Subject to Alignment, \noalign

\noalign allows one to insert vertical material between table rows. The easiest
way to imagine the workings of \noalign is to think of TeX first processing
the table rows independently, putting all \noaligns aside. After the table rows
are computed, the material of \noaligns is inserted between the box of the
current row and the box of the next row. To be precise: the \noalign material is
inserted *before* the interline glue, which is normally inserted between table rows
(this interline glue is based on either \baselineskip or \lineskip).

The material of \noalign is not subject to alignment in the table but is
independent of the rest of the table. The \noalign material is inserted in curly
braces following the \noalign as in \noalign{\vskip 3pt}.

The \noalign material forms an implicit group; see 19.4.10, p. III-106. In
other words, changes inside \noalign are local unless prefixed by \global.

\noaligns can only occur after a \cr. This includes the \cr ending the preamble, thus \noalign material can precede the very first row. If you think about it, this is the only sensible way to do it: \noalign material is inserted between rows so it must be specified between rows, after a \cr terminating one row and before the material of the first column of the next row is entered.

Note also that one \noalign can be followed by another \noalign. This will be shown shortly.

Here are some applications of \noalign:

1. To increase the vertical spacing between two specific rows. This is useful, for instance, for increasing the distance between a row containing headers and the following first "information row" of a table. \noalign{\vskip 5pt} would do just that; see 39.2.1 on this page for an example.
2. To insert horizontal rules (\noalign{\hrule}) between table rows. For examples see 39.2.2 on the next page.
3. To insert a penalty to control page breaks within a table. For page breaks within a table to be possible, the table must *not* be included inside a \vbox. A \noalign-inserted penalty may also control a \vsplit; see 41.3.3, p. 349.
4. To insert paragraphs, see 40.4.1, p. 312, for an application where paragraphs are inserted between table rows.
5. \noalign can be also used together with \valign; see 41.5, p. 355, for details.

39.2.1 Changing the Vertical Spacing with \noalign

The following demonstration uses \noalign to increase the vertical spacing between table rows. The distance between the second and third row of the tables is increased by 10 pt. A negative \vskip inside \noalign can be used to move rows closer, as is also shown in the table below (third and fourth row).

Sample Table **39.2** (Source code).

```
 1   $$\vbox{
 2                                        \tabskip = 0pt
 3   \halign{
 4       #\hrulefill                      \tabskip = 15pt&
 5       \dotfill#&
 6       \bf#\hfil&
 7       ${\cal A}_{#}$\hfil              \tabskip = 0pt
 8   \cr
 9       XX&      YY&      Z&             1\cr
10       YY&      XX&      ZZ&            2\cr\noalign{\vskip 10pt}
11       AA&      \omit LEFT\hfil\qquad&CCC&\omit$\infty$\hfil\cr
12                                        \noalign{\vskip -6pt}
13       TT&      AA&      TTTT&          4\cr
14       BIGG&    BIGGER&  VERY BIG&      300\cr
15   }}$$
```

Sample Table **39.2** (Output).

XX__	.......YY	**Z**	$\mathcal{A}_1$
YY__	XX	**ZZ**	$\mathcal{A}_2$
AA__	LEFT	CCC	∞
TT__	.......AA	**TTTT**	$\mathcal{A}_4$
BIGG	.BIGGER	**VERY BIG**	$\mathcal{A}_{300}$

39.2.2   Inserting Horizontal Rules with \noalign

To insert a horizontal line into a table is another application of \noalign. Note that the rule extends to the full width of the table. Because neither tab_0 nor tab_n are zero, the rules extend to the left of the first column and to the right of the last column.

Sample Table **39.3** (Source code).

```
1   $$\vbox{
2                               \tabskip = 15pt
3   \halign{
4       #\hrulefill&
5       \dotfill#&
6       \bf#\hfil&
7       ${\cal A}_{#}$\hfil
8   \cr
9                                       \noalign{\hrule}
10      XX&      YY&      Z&           1\cr
11      YY&      XX&      ZZ&          2\cr
12      AA&      \omit LEFT\hfil\qquad&CCC&\omit$\infty$\hfil\cr
13      TT&      AA&      TTTT&        4\cr\noalign{%
14                                      \vskip 5pt
15                                      \hrule height 2pt
16                                      \vskip 5pt
17                                      }
18      BIGG&    BIGGER&  VERY BIG&    300\cr
19                                      \noalign{\hrule}
20  }}$$
```

Sample Table **39.3** (Output).

XX__	.......YY	**Z**	$\mathcal{A}_1$
YY__	XX	**ZZ**	$\mathcal{A}_2$
AA__	LEFT	**CCC**	∞
TT__	.......AA	**TTTT**	$\mathcal{A}_4$
BIGG	.BIGGER	**VERY BIG**	$\mathcal{A}_{300}$

To emphasize the importance of tab_0 and tab_n once more, here is a table where these two parameters are set to different values. This causes an asymmetrical horizontal rule to be printed.

Sample Table **39.4** (Source code).

```
 1   $$\vbox{
 2                                       \tabskip = 0pt
 3   \halign{
 4       #\hrulefill                      \tabskip = 15pt&
 5       \dotfill#&
 6       \bf#\hfil&
 7       ${\cal A}_{#}$\hfil              \tabskip = 40pt
 8   \cr
 9                                        \noalign{\hrule}
10       XX&      YY&      Z&            1\cr
11       YY&      XX&      ZZ&           2\cr\noalign{\vskip 10pt}
12       AA&      \omit LEFT\hfil\qquad&CCC&\omit$\infty$\hfil\cr
13       TT&      AA&      TTTT&          4\cr\noalign{\vskip 5pt
14                                               \hrule height 2pt
15                                               \vskip 5pt}
16       BIGG&    BIGGER& VERY BIG&      300\cr
17                                        \noalign{\hrule}
18   }}$$
```

Sample Table **39.4** (Output).

XX⎯	.......YY	Z	$\mathcal{A}_1$
YY⎯	XX	ZZ	$\mathcal{A}_2$
AA⎯	LEFT	CCC	∞
TT⎯	.......AA	TTTT	$\mathcal{A}_4$
BIGG	.BIGGER	VERY BIG	$\mathcal{A}_{300}$

39.2.3   Multiple \noaligns

It is possible to have consecutive \noaligns. In the source code of the previous table, source code lines 13–15 could have been replaced by:

```
1   ..... TTTT& -4\cr      \noalign{\vskip 5pt}
2                          \noalign{\hrule height 2pt}
3                          \noalign{\vskip 5pt}
```

39.2.4   Double Printed Horizontal Rules

Here is a \noalign example where the same horizontal rule is printed twice. This is done in three steps:

1. The horizontal rule is printed.
2. A negative backspace occurs through a negative \vskip. The amount of backspacing is determined by the thickness (sum of height and depth) of the rule just printed.
3. Step 1 is repeated and the rule is printed again.

Here is the example table (see 41.3.4, p. 351, for an application of this seemingly useless operation).

Sample Table 39.5 (Source code).

```
1   $$\vbox{
2                                            \tabskip = 15pt
3   \halign{
4       #\hrulefill&
5       \dotfill#&
6       \bf#\hfil&
7       ${\cal A}_{#}$\hfil
8   \cr
9       XX&     YY&     Z&              1\cr
10      YY&     XX&     ZZ&             2\cr
11      AA&     \omit LEFT\hfil\qquad&CCC&\omit$\infty$\hfil
12                                      \cr
13      TT&     AA&     TTTT&           4\cr
14              \noalign{\hrule height 2pt  % 1. Print rule first time.
15                              \vskip -2pt  % 2. Back up.
16                              \hrule height 2pt} % 3. Print rule again.
17      BIGG&   BIGGER& VERY BIG&       300\cr
18  }}$$
```

Sample Table 39.5 (Output).

XX___	.......YY	**Z**	$\mathcal{A}_1$
YY___	XX	**ZZ**	$\mathcal{A}_2$
AA___	LEFT	**CCC**	∞
TT___	.......AA	**TTTT**	$\mathcal{A}_4$
BIGG	.BIGGER	**VERY BIG**	$\mathcal{A}_{300}$

39.2.5   Horizontal Rules Made Part of the \vbox Enclosing the Table

A horizontal rule preceding or following a table can be entered in two ways:

1. It can be entered using \noalign as shown in the previous tables.
2. The rule can be made part of the \vbox enclosing the table, not part of the
    table itself.

Here is an example demonstrating the second approach:

Sample Table **39.6** (Source code).

```
 1   $$\vbox{
```

The following rule is *not* part of the table.

```
 2   \hrule
 3                                             \tabskip = 15pt
 4   \halign{
 5       #\hrulefill&
 6       \dotfill#&
 7       \bf#\hfil&
 8       ${\cal A}_{#}$\hfil
 9   \cr
10       XX&      YY&      Z&           1\cr
11       YY&      XX&      ZZ&          2\cr      \noalign{\vskip 10pt}
12       AA&      \omit LEFT\hfil\qquad&CCC&\omit$\infty$\hfil
13                                      \cr
14       TT&      AA&      TTTT&        4\cr      \noalign{%
15                                                   \vskip 5pt
```

Note that the following \hrule again does not really belong to the table because
it is contained in a \noalign.

```
16                                                \hrule height 2pt
17                                                \vskip 5pt
18                                      }%
19       BIGG&    BIGGER&  VERY BIG&    300\cr
20   }
21   \hrule
22   }$$
```

Sample Table **39.6** (Output).

XX___	.......YY	Z	$\mathcal{A}_1$
YY___	XX	**ZZ**	$\mathcal{A}_2$
AA___	LEFT	**CCC**	∞
TT___	.......AA	**TTTT**	$\mathcal{A}_4$
BIGG	. BIGGER	**VERY BIG**	$\mathcal{A}_{300}$

Why does inserting \hrules outside the table itself work? The width of a \vbox is determined by the width of the widest component of the \vbox. The two horizontal rules have no explicit width specification. Therefore as far as the width computation of the \vbox is concerned, only the width of the table is relevant and the \vbox inherits the width of the table. The rules in turn inherit the width from the \vbox. Therefore the horizontal rules, although outside of the \halign, inherit the width of the table.

Selective horizontal rules, which span selected columns only, are discussed in 39.6.6, p. 289.

By the way, the previous example can be extended to include vertical rules to the left and right of the table, as the following table shows:

Sample Table **39.7** (Source code).

```
 1  $$
 2      \vbox{
 3          \hrule
 4          \hbox{%
 5              \vrule
 6                                          \tabskip = 15pt
 7              \vbox{%
 8                  \halign{
 9                      #\hrulefill&
10                      \dotfill#&
11                      \bf#\hfil&
12                      ${\cal A}_{#}$\hfil
13                  \cr
14                      XX&     YY&     Z&          1\cr
15                      YY&     XX&     ZZ&         2\cr
16                                          \noalign{\vskip 10pt}
17                      AA&     \omit LEFT\hfil\qquad&CCC&
18                                          \omit$\infty$\hfil\cr
19                      TT&     AA&     TTTT&       4\cr
20                                          \noalign{%
21                                              \vskip 5pt
22                                              \hrule height 2pt
23                                              \vskip 5pt
24                                          }
25                      BIGG&   BIGGER& VERY BIG&   300\cr
26                  }%
27              }%
28              \vrule
29          }
30          \hrule
31      }
32  $$
```

Sample Table **39.7** (Output).

XX__	.......YY	Z	$\mathcal{A}_1$
YY__	XX	ZZ	$\mathcal{A}_2$
AA__	LEFT	CCC	∞
TT__	.......AA	TTTT	$\mathcal{A}_4$
BIGG	.BIGGER	VERY BIG	$\mathcal{A}_{300}$

39.3   Entries Spanning Multiple Columns

Entries spanning two or more columns are necessary, particularly for headings of table columns. There are two different solutions to the problem:

1.   Use \hidewidth. See the following Subsection for details.
2.   Use \multispan. See 39.3.2 on the next page for details.

39.3.1   Macro \hidewidth

The macro \hidewidth is defined as follows:

```
1   \def\hidewidth{\hskip -1000pt plus 1fill}
```

What does this macro do when applied to an entry? It hides the width of the entry, as the name of the macro suggests. For example, assume \hidewidth is inserted to the right of some entry, as in XX\hidewidth. When TeX computes the natural width of this entry, it adds the width of XX and the natural width of \hidewidth, which is -1000 pt. Therefore the resulting value, which is negative, does *not* contribute to the width of this column because this entry is not the widest entry (in a reasonable table, a table where the XX material is not wider than 1000 pt).

If the entry of this table is changed to \hskip ⟨dimen⟩ XXX\hidewidth, then this entry could be moved right to any desired horizontal position (by changing ⟨dimen⟩), but the width computation of the column in which this entry occurs would still not be affected.

Observe, on the other hand, \hidewidth entries still participate in the alignment game. An entry like XX\hidewidth is left-aligned within the current column. Because \hidewith uses a glue of order fill, it overpowers the glue in properly designed templates (that is, templates using \hfil glue to control the alignment).

Here is an example that shows an application of \hidewidth. It also shows that you have to do your arithmetic right: TeX will not prevent you from running text of one column into another column.

Sample Table **39.8** (Source code).

```
1   $$\vbox{
2                                           \tabskip = 15pt
3   \halign{
4       #\hrulefill                         \tabskip = 45pt&
5       \dotfill#&
6       \bf#\hfil                           \tabskip = 15pt&
7       ${\cal A}_{#}$\hfil
8   \cr
```

The following dimension of 60 pt was found by trial and error and seems to center the heading.

```
9       \omit\hskip 60pt \bf A Heading goes here\hidewidth\cr
10                                          \noalign{\medskip}
11      XX&      YY&      Z&          1\cr
12      YY&      XX&      ZZ&         2\cr
13      AA&      \omit LEFT\hfil\qquad&   CCC&      \omit$\infty$\hfil\cr
14      TT&      AA&      A hidewidth example\hidewidth&4\cr
15      BIGG&    BIGGER&  VERY BIG&   300\cr
16  }}$$
```

Sample Table **39.8** (Output).

A Heading goes here

XX__	.......YY	**Z**	$\mathcal{A}_1$
YY__	XX	**ZZ**	$\mathcal{A}_2$
AA__	LEFT	**CCC**	∞
TT__	.......AA	**A hidewidth ex**$\mathcal{A}$**mple**	
BIGG	. BIGGER	**VERY BIG**	$\mathcal{A}_{300}$

39.3.2   \span and \omit, \multispan

It is more common to use \multispan than \hidewidth to generate entries spanning multiple columns, but before discussing \multispan, we must look at \span.

\span has two totally *unrelated* applications in the typesetting of tables.

1. If \span is used in a *template*, the token following it, and only this token, is expanded when the template is read-in. Normally the tokens of a template are not expanded until the table body is processed and a table entry is substituted in place of the # of the template. This application of \span is discussed in 40.6.5, p. 330.

2. If \span is used otherwise (with table entries, not as part of the preamble),
    it allows entries to span multiple columns.

    In this application of \span it replaces the &.

    The material located before and after \span is processed in the ordinary
    way, but afterwards it is placed into a single box instead of two. The width
    of this combination is the sum of the width of the individual columns and
    the width of the \tabskip glue between them; therefore the spanning box
    will line up with nonspanning boxes in the rows that are being spanned.

\span (as specified in item 2 of the preceding list and as it is discussed in
this Subsection) is rarely applied directly. Instead \multispan is used, which
generates the proper number of \omits and \spans. Here is the definition of
\multispan:

```
1   \newcount\mscount
2
3   \def\multispan #1{%
4       \omit
5       \mscount = #1
6       \loop
7           \ifnum\mscount > -1
8               \@span
9           \repeat
10  }
11  \def\sp@n{%
12      \span
13      \omit
14      \advance\mscount by -1
15  }
```

For instance, \multispan3 expands to \omit\span\omit\span\omit, which
means that the following entry can span from the current column into the two
following columns *and* that the templates of these three columns are ignored.
An important consequence of this is that \multispan-defined entries are without
templates, and therefore glue must be provided by the user to align entries with
the number of spanned columns.

One question you might ask yourself is why an extra macro \sp@n is defined
and why the replacement text of this macro is not incorporated into the \loop
... \repeat construct of \multispan itself. One might think that \multispan
could have been defined as follows (no macro \sp@n would be necessary in this
case):

```
1   \def\multispan #1{%
2       \omit
3       \mscount = #1
4       \loop
5           \ifnum\mscount > -1
6               \span
7               \omit
8               \advance\mscount by -1
9           \repeat
```

```
10    }
```

However, this definition had been used, then upon trying to find the closing token of `\loop` (which is `\repeat`) TₑX would find a `\span` token and that token is illegal in this context.

Note that `\multispan{1}` is equivalent to `\omit`.

Here is an example using `\multispan` to enter a heading into a table. Leaders, instead of ordinary glue, are used to make the glues centering the headers more visible.

Sample Table 39.9 (Source code).

```
1    $$\vbox{
2                                    \tabskip = 0pt
3    \halign{                        \tabskip = 15pt
4        #\hrulefill                 \tabskip = 45pt&
5        \dotfill##&
6        \bf#\hfil                   \tabskip = 15pt&
7        ${\cal A}_{#}$\hfil         \tabskip = 0pt
8    \cr
9        \multispan{2}%
10           \leaders\hrule\hfil\bf
11               Columns 1 and 2%
12           \leaders\hrule\hfil&
13       \multispan{2}%
14           \leaders\hrule\hfil\bf
15               Columns 3 and 4%
16       \leaders\hrule\hfil\cr
17       XX&      YY&      Z&              1\cr
18       YY&      XX&      ZZ&             2\cr    \noalign{\vskip 10pt}
19       AA&      \omit LEFT\hfil\qquad&   CCC\cr
20       TT&      AA&      TTTT&           4\cr
21                                        \noalign{\vskip 5pt
22                                            \hrule height 2pt
23                                            \vskip 5pt}
24       BIGG&    BIGGER&  VERY VERY BIG&  300\cr
25    }}$$
```

Sample Table 39.9 (Output).

___Columns 1 and 2___		___Columns 3 and 4___	
XX__	.......YY	Z	$\mathcal{A}_1$
YY__	XX	ZZ	$\mathcal{A}_2$
AA__	LEFT	CCC	
TT__	.......AA	**TTTT**	$\mathcal{A}_4$
BIGG	. BIGGER	**VERY VERY BIG**	$\mathcal{A}_{300}$

If you use tables with vertical rules and `\multispan` you have to be careful when computing the number of columns in a table. This is discussed in 39.3.4, p. 287.

39.3.3   A Problem with \multispan

Let me demonstrate a problem with \multispan. Here is a three-column table with a \multispan-generated heading above the first two columns.

Sample Table **39.10** (Source code).

```
1   $$\vbox{
2                                          \tabskip = 10pt
3   \halign{
4       \hfil#&
5       #\hfil&
6       \hfil#\hfil
7   \cr
8       \multispan{2}
9           \leaders\hrule\hfil
10          Headerline I
11          \leaders\hrule\hfil&
12                              Header II\cr
13      1& A&                   XX\cr
14      10& AB&                 XXX\cr
15      100& ABC&               XXXXX\cr
16      1000& ABCD&             XX\cr
17      100000000& ABCDEFGHIJ&      XX\cr
18  }}$$
```

Sample Table **39.10** (Output).

Headerline I		Header II
1	A	XX
10	AB	XXX
100	ABC	XXXXX
1000	ABCD	XX
100000000	ABCDEFGHIJ	XX

Everything is OK with the previous table. But now look at the following table. In this table the heading over the first two lines is "too wide." By this I mean the following: define w_1 and w_2 to be the width of the first and second column *ignoring* the header row (the first row) of the table. Then $w_1 + tab_1 + w_2$ is less than the width of the header above columns 1 and 2.

Sample Table **39.11** (Source code).

```
1   $$
2   \vbox{
3                                          \tabskip = 10pt
4   \halign{
5       \hfil#&
6       #\leaders\hrule\hfil&
7       \hfil#\hfil
8   \cr
9   \multispan{2}%
```

```
10              \leaders\hrule\hfil Header I is very, very long.
11              \leaders\hrule\hfil&
12                              Header II\cr
13        1&  A&                XX\cr
14       10&  AB&               XXX\cr
15      100&  ABC&              XXXXX\cr
16     1000&  ABCD&             XX\cr
17   }}$$
```

Sample Table **39.11** (Output).

Header I is very, very long.		Header II
1	A________________	XX
10	AB________________	XXX
100	ABC________________	XXXXX
1000	ABCD________________	XX

Here are two important observations of how TℇX handles this situation:

1. The second column (the *last* column as far as the header that is too wide is concerned) takes "all the heat," since it is the width of this column and only this column that is increased, because of an "oversized \multispar."
2. TℇX does not print any warning message in this case.

The next Subsection discusses a common special case of the problem just outlined.

39.3.4   Headers Over Left- and Right-Justified Columns

For the following, an understanding of the previous Subsection is essential. There is a special case of left- or right-justified columns that occurs rather frequently:

1. The header of a column is wider than any of the other (nonheader) entries in that column.
2. All the nonheader entries should be printed either left- or right-justified.
3. The nonheader entries, looked at as a block, should be centered with respect to the header of this column.

A solution was already discussed in 38.3.8, p. 214. Here we discuss other approaches, which are in general much less successful than the solution offered at the given reference.

One attempt to solve this problem is to use three columns; the middle column will hold all the nonheader entries, and the header will be allowed to expand into the first and last column. Furthermore, the \tabskip glue between the first and second columns, as well as between the second and third columns, will be allowed to stretch as far as necessary. The hope is that the header will stretch these two \tabskips and therefore appear centered above the middle column, exactly as needed.

Well, it does not work, which should be obvious from the table of the last Subsection: the third and last column "takes all the heat" and the part of the header that is wider than the widest entry of the nonheader entries causes the third column to be widened, but does not cause any of the \tabskip glues to stretch.

Sample Table 39.12 (Source code).

```
 1   $$
 2   \vbox{
 3                               \tabskip = 0pt
 4   \halign{
 5       #\hfil                  \tabskip = 0pt plus 1fil&
 6       #\hfil                  \tabskip = 0pt plus 1fil&
 7       #\hfil                  \tabskip = 0pt
 8   \cr
 9       \multispan{3}%
10          \leaders\hrule\hfil
11              This is a header which is quite long%
12          \leaders\hrule\hfil\cr
13      &ABCDEF\cr
14      &ABCDEF XXX\cr
15      &DODO\cr
16      &12\cr
17   }}$$
```

Sample Table 39.12 (Output).

```
This is a header which is quite long
ABCDEF
ABCDEF XXX
DODO
12
```

39.3.4.1   An Approximate Solution

Let me discuss some possible solutions: we can artificially establish nonzero width columns 1 and 3 as we do in the following example. But the question now is how wide the first and third columns should be. In the following example, a width that was larger than necessary was chosen because the leaders to the left and right of the header become visible. This is a trial and error approach, not very satisfactory.

Sample Table 39.13 (Source code).

```
1   $$
2   \vbox{
3                                    \tabskip = 0pt
4   \halign{
5       \hskip 30pt#&                            % 1
6       #\hfil&                                  % 2
7       \hskip 30pt#%                            % 3
8   \cr
9       \multispan{3}%
10          \leaders\hrule\hfil
11              This is a header which is quite long%
12          \leaders\hrule\hfil\cr
13      &ABCDEF\cr
14      &ABCDEF XXX\cr
15      &DODO\cr
16      &12\cr
17  }}$$
```

Sample Table 39.13 (Output).

This is a header which is quite long
ABCDEF
ABCDEF XXX
DODO
12

The above solution can be written differently by moving the \hskip of the first and third column into the \tabskip. The preamble of such a table would look like this:

```
1                                    \tabskip = 0pt
2   \halign{
3       #\relax              \tabskip = 30pt&    % 1
4       #\hfil               \tabskip = 30pt&    % 2
5       #\relax              \tabskip = 0pt      % 3
6   \cr
```

The previous solution can be simplified somewhat by having only one column and making nonheader entries in that column artificially wider. This leads to the following table (the \hskips are irrelevant for the headline because the headline is printed without this template).

Sample Table **39.14** (Source code).

```
 1   $$
 2   \vbox{
 3                                   \tabskip = 0pt
 4   \halign{
 5       \hskip 30pt
 6           #%
 7       \hskip 30pt plus 1fil
 8   \cr
 9       \omit
10           \leaders\hrule\hfil
11               This is a header which is quite long%
12           \leaders\hrule\hfil\cr
13       ABCDEF\cr
14       ABCDEF XXX\cr
15       DODO\cr
16       12\cr
17   }}$$
```

Sample Table **39.14** (Output).

This is a header which is quite long
ABCDEF
ABCDEF XXX
DODO
12

All of this playing around should satisfy you that the solution of 38.3.8, p. 214, is the best one after all.

39.4   Struts and the Vertical Spacing of Tables

I now discuss the use of struts to control the vertical spacing in tables. So far we discussed \baselineskip to control the vertical spacing in tables (39.1, p. 255). We also saw that \noalign can be used to change the spacing between selected rows. 39.5, p. 275, discusses vertical rules in tables and justifies why it is necessary to use struts in the first place.

A strut (see 7.4, p. I-235) is a *vertical rule of zero width, with height and depth adding up to the desired vertical line spacing.* Inserting a strut into any box makes this box of such height and depth that no interline glue is necessary— boxes can simply be stacked on top of each other and they appear to be properly spaced. The preceding reference also discusses the various definitions of struts. See also 7.4.3, p. I-239, which discusses macros to compute struts automatically.

When struts are used, the interline glue must be disabled by calling \off-interlineskip; see 7.3.11, p. I-230, for details. The changes caused by \offinterlineskip are local in the tables below because \offinterlineskip is called inside the \vbox containing the table, and \vboxes form implicit groups.

39.4.1   A Table with Strut-Controlled Spacing

In the following table observe that the strut is allocated its own column (this is a good practice):

Sample Table 39.15 (Source code).

```
1    $$\vbox{
2    \def\MyStrut{\vrule height 8pt depth 4pt width 0pt}
3    \offinterlineskip
4                               \tabskip = 0pt
5    \halign{
6        \MyStrut#&                                        % 1: "strut column"
7        #\hrulefill          \tabskip = 15pt&    % 2
8        \dotfill#&                               % 3
9        \bf#\hfil&                               % 4
10       ${\cal A}_{#}$\hfil \tabskip = 0pt       % 5
11   \cr
12       &   XX&     YY&     Z&          1\cr
13       &   YY&     XX&     ZZ&         2\cr
14       &   AA&     BB&     CCC&        3\cr
15                               \noalign{\hrule}
16       &   TT&     AA&     TTTT&       4\cr
17       &   BIGG&   BIGGER& VERY BIG&   300\cr
18   }}$$
```

Sample Table 39.15 (Output).

XX__	..... YY	**Z**	$\mathcal{A}_1$
YY__	 XX	**ZZ**	$\mathcal{A}_2$
AA__	..... BB	**CCC**	$\mathcal{A}_3$
TT__	..... AA	**TTTT**	$\mathcal{A}_4$
BIGG	BIGGER	**VERY BIG**	$\mathcal{A}_{300}$

A strut definition \MyStrut had to be given and the dimensions of the strut had to be computed. It is much easier to use the macro \ComputeStrut and let TEX do the work. This is done in the next sample table.

Observe the way \tabskip was set in the previous table: both tab_0 and tab_n ($n = 5$) are zero, but tab_1 is also set to zero because the first column (as far as TEX is concerned) is the "strut column" and this column has with of zero (as should the glue surrounding this column).

Compare this with the following table where \tabskip is always 15 pt. Note that the table seems to be asymmetrical within the box containing the table:

Sample Table **39.16** (Source code).

```
 1   \InputD{setstrut.tip}                       % 7.4.3.1, p. I-239.
 2   $$\VboxR{
 3   \ComputeStrut
 4   \offinterlineskip
 5                                   \tabskip = 15pt
 6   \halign{
 7      \MyStrut##&                              % 1
 8      #\hrulefill&                             % 2
 9      \dotfill##&                              % 3
10      \bf#\hfil&                               % 4
11      ${\cal A}_{#}$\hfil                      % 5
12   \cr
13      &   XX&     YY&      Z&          1\cr
14      &   YY&     XX&      ZZ&         2\cr
15      &   AA&     BB&      CCC&        3\cr
16      &   TT&     AA&      TTTT&       4\cr
17      &   BIGG&   BIGGER&  VERY BIG&   300\cr
18   }}$$
```

Sample Table **39.16** (Output).

XX__	.....YY	**Z**	$\mathcal{A}_1$
YY__	XX	**ZZ**	$\mathcal{A}_2$
AA__	.....BB	**CCC**	$\mathcal{A}_3$
TT__	.....AA	**TTTT**	$\mathcal{A}_4$
.....BIGG...BIGGER...VERY.BIG...$\mathcal{A}_{300}$....			

39.4.2   Using \offinterlineskip but No Struts

Building a table without struts but with \offinterlineskip destroys the vertical spacing. Rows now touch each other; see 7.4.2, p. I-237. The next table has this problem.

Sample Table **39.17** (Source code).

```
 1   \InputD{setstrut.tip}                       % 7.4.3.1, p. I-239.
 2   $$\vbox{
 3   \ComputeStrut
 4   \offinterlineskip
 5                                   \tabskip = 0pt
 6   \halign{
 7      #\hrulefill           \tabskip = 15pt&     % 1
 8      \dotfill##&                                % 2
 9      \bf#\hfil&                                 % 3
10      ${\cal A}_{#}$\hfil   \tabskip = 0pt       % 4
11   \cr
12      XX&      YY&      Z&          1\cr
```

```
13      YY&      XX&      ZZ&        2\cr
14      AA&      BB&      CCC&       3\cr
15      TT&      AA&      TTTT&      4\cr
16      BIGG&    BIGGER&  VERY BIG&  300\cr
17  }}$$
```

Sample Table **39.17** (Output).

$$\begin{array}{llll}
\text{XX} & \cdots\cdots \text{YY} & \text{Z} & \mathcal{A}_1 \\
\text{YY} & \cdots\cdots \text{XX} & \text{ZZ} & \mathcal{A}_2 \\
\text{AA} & \cdots\cdots \text{BB} & \text{CCC} & \mathcal{A}_3 \\
\text{TT} & \cdots\cdots \text{AA} & \text{TTTT} & \mathcal{A}_4 \\
\text{BIGG} & \text{BIGGER} & \text{VERY BIG} & \mathcal{A}_{300}
\end{array}$$

If it were not for the subscripts in the last column, the bottoms of the capital letters of one row would touch the tops of the capital letters in the next row. To show this, I print the previous table, but leave out the last column.

Sample Table **39.18** (Source code).

```
1   \InputD{setstrut.tip}                  % 7.4.3.1, p. I-239.
2   $$\vbox{
3   \ComputeStrut
4   \offinterlineskip
5                              \tabskip = 0pt
6   \halign{
7       #\hrulefill            \tabskip = 15pt&      % 1
8       \dotfill#&                                   % 2
9       \bf#\hfil              \tabskip = 0pt        % 3
10  \cr
11      XX&      YY&      Z\cr
12      YY&      XX&      ZZ\cr
13      AA&      BB&      CCC\cr
14      TT&      AA&      TTTT\cr
15      BIGG&    BIGGER&  VERY BIG\cr
16  }}$$
```

Sample Table **39.18** (Output).

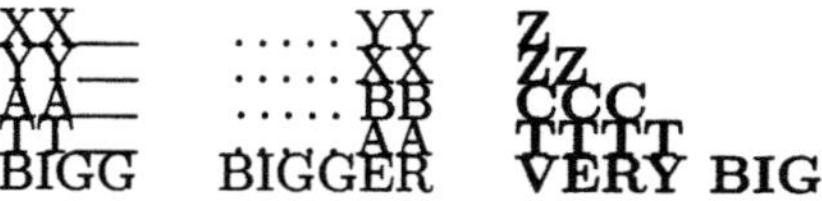

39.4.3 Using a Different Strut

By changing the strut, you can obviously change the vertical spacing: in the following table \baselineskip is changed, then \ComputeStrut is called and generates a strut that is equivalent to vertical spacing of 18 pt.

Sample Table **39.19** (Source code).

```
1    \InputD{setstrut.tip}                          % 7.4.3.1, p. I-239.
2    $$\vbox{
3    \baselineskip = 18pt
4    \ComputeStrut
5    \offinterlineskip
6                                         \tabskip = 0pt
7    \halign{
8        \MyStrut#&                                          % 1
9        #\hrulefill           \tabskip = 15pt&              % 2
10       \dotfill#&                                          % 3
11       \bf#\hfil&                                          % 4
12       ${\cal A}_{#}$\hfil      \tabskip = 0pt            % 5
13   \cr
14       &   XX&    YY&     Z&           1\cr
15       &   YY&    XX&     ZZ&          2\cr
16       &   AA&    BB&     CCC&         3\cr
17       &   TT&    AA&     TTTT&        4\cr
18       &   BIGG&  BIGGER& VERY BIG&    300\cr
19   }}$$
```

Sample Table **39.19** (Output).

XX__	..... YY	**Z**	$\mathcal{A}_1$
YY__	 XX	**ZZ**	$\mathcal{A}_2$
AA__	..... BB	**CCC**	$\mathcal{A}_3$
TT__	..... AA	**TTTT**	$\mathcal{A}_4$
BIGG	BIGGER	**VERY BIG**	$\mathcal{A}_{300}$

39.5   Vertical Rules in Tables

A look at vertical rules in tables explains why struts must be used in tables with vertical rules.

39.5.1   A Simple Example

In the following table I changed the width of the strut \vrule to 2 pt (struts by definition consist of a zero-width rule), and a vertical rule appeared in the table:

Sample Table 39.20 (Source code).

```
1    \InputD{setstrut.tip}                            % 7.4.3.1, p. I-239.
2    $$\vbox{
3    \ComputeStrut
4    \offinterlineskip
5                                      \tabskip = 0pt
6    \halign{
7       \MyStrut width 2pt#&                          % 1: strut
8       #\hrulefill            \tabskip = 15pt&        % 2
9       \dotfill#&                                     % 3
10      \bf#\hfil&                                     % 4
11      ${\cal A}_{#}$\hfil       \tabskip = 0pt       % 5
12   \cr
13      &    XX&     YY&     Z&           1\cr
14      &    YY&     XX&     ZZ&          2\cr
15      &    AA&     BB&     CCC&         3\cr
16      &    TT&     AA&     TTTT&        4\cr
17      &    BIGG&   BIGGER& VERY BIG&    300\cr
18   }}$$
```

Sample Table 39.20 (Output).

XX___	..... YY	**Z**	$\mathcal{A}_1$
YY___	 XX	**ZZ**	$\mathcal{A}_2$
AA___	..... BB	**CCC**	$\mathcal{A}_3$
TT___	..... AA	**TTTT**	$\mathcal{A}_4$
BIGG	BIGGER	**VERY BIG**	$\mathcal{A}_{300}$

Observe that what appears to be one vertical rule is really a vertical rule composed of five rules stacked on top of each other, one per row. The table has five rows, each row contributing one piece of the vertical rule. To illustrate this more clearly, change the width of the vertical rule in every row as was done in the following table:

Sample Table 39.21 (Source code).

```
1    \InputD{setstrut.tip}                            % 7.4.3.1, p. I-239.
2    $$\vbox{
3    \ComputeStrut
4    \offinterlineskip
5                                      \tabskip = 0pt
6    \halign{
7       \MyStrut width #&                             % 1: "strut"
8       #\hrulefill            \tabskip = 15pt&        % 2
9       \dotfill#&                                     % 3
10      \bf#\hfil&                                     % 4
11      ${\cal A}_{#}$\hfil       \tabskip = 0pt       % 5
12   \cr
13      1pt&    XX&     YY&     Z&           1\cr
14      4pt&    YY&     XX&     ZZ&          2\cr
15      1pt&    AA&     BB&     CCC&         3\cr
```

```
16      4pt&    TT&     AA&     TTTT&       4\cr
17      1pt&    BIGG&   BIGGER& VERY BIG&   300\cr
18   }}$$
```

Sample Table 39.21 (Output).

$$
\begin{array}{|llll l}
\mathrm{XX}\underline{} & \ldots\ldots\mathrm{YY} & \mathbf{Z} & \mathcal{A}_1 \\
\mathrm{YY}\underline{} & \ldots\ldots\mathrm{XX} & \mathbf{ZZ} & \mathcal{A}_2 \\
\mathrm{AA}\underline{} & \ldots\ldots\mathrm{BB} & \mathbf{CCC} & \mathcal{A}_3 \\
\mathrm{TT}\underline{} & \ldots\ldots\mathrm{AA} & \mathbf{TTTT} & \mathcal{A}_4 \\
\mathrm{BIGG} & \mathrm{BIGGER} & \mathbf{VERY\ BIG} & \mathcal{A}_{300}
\end{array}
$$

39.5.2   Separating Struts and Vertical Rules

In the preceding sample table, a strut was converted into a vertical rule. It is
much cleaner to put struts and vertical rules in separate columns. It is also good
practice not to include vertical rules with regular text columns.

Let us now assume that a strut is allocated its own column. To generate
a vertical rule anywhere simply write \vrule. There is no need to specify the
height and depth of this rule, because it inherits the height and depth from the
strut, which is the highest and deepest item in each row.

Here is an example:

Sample Table 39.22 (Source code).

```
1    $$\vbox{
2    \ComputeStrut
3    \offinterlineskip
4                                    \tabskip = 0pt
5    \halign{
6        \MyStrut#&                              % 1: strut
7        \vrule#\relax           \tabskip = 15pt&    % 2
8        #\hrulefill&                            % 3
9        \dotfill#&                              % 4
10       \bf#\hfil&                              % 5
11       ${\cal A}_{#}$\hfil     \tabskip = 0pt      % 6
12   \cr
13       &&  XX&     YY&     Z&          1\cr
14       &&  YY&     XX&     ZZ&         2\cr
15       &&  AA&     BB&     CCC&        3\cr
16       &&  TT&     AA&     TTTT&       4\cr
17       &&  BIGG&   BIGGER& VERY BIG&   300\cr
18   }}$$
```

Sample Table **39.22** (Output).

XX__	.....YY	**Z**	$\mathcal{A}_1$
YY__	XX	**ZZ**	$\mathcal{A}_2$
AA__	.....BB	**CCC**	$\mathcal{A}_3$
TT__	.....AA	**TTTT**	$\mathcal{A}_4$
BIGG	BIGGER	**VERY BIG**	$\mathcal{A}_{300}$

39.5.3   Multiple Vertical Rules

The next example is a table with two vertical rules. It is very simple to modify
the previous table: a new seventh "vertical rule only" column is added.

Sample Table **39.23** (Source code).

```
1    \InputD{setstrut.tip}                        % 7.4.3.1, p. I-239.
2    $$\vbox{
3    \ComputeStrut
4    \offinterlineskip
5                                  \tabskip = 0pt
6    \halign{
7        \MyStrut#&                               % 1: strut
8        \vrule#\relax            \tabskip = 15pt& % 2
9        #\hrulefill&                             % 3
10       \dotfill##&                              % 4
11       \bf#\hfil&                               % 5
12       ${\cal A}_{#}$\hfil&                     % 6
13       \vrule#\relax            \tabskip = 0pt  % 7
14   \cr
15       &&  XX&     YY&     Z&          1&\cr
16       &&  YY&     XX&     ZZ&         2&\cr
17       &&  AA&     BB&     CCC&        3\cr        % Error!
18       &&  TT&     AA&     TTTT&       4&\cr
19       &&  BIGG&   BIGGER& VERY BIG&   300&\cr
20   }}$$
```

Sample Table **39.23** (Output).

XX__	.....YY	**Z**	$\mathcal{A}_1$	
YY__	XX	**ZZ**	$\mathcal{A}_2$	
AA__	.....BB	**CCC**	$\mathcal{A}_3$	
TT__	.....AA	**TTTT**	$\mathcal{A}_4$	
BIGG	BIGGER	**VERY BIG**	$\mathcal{A}_{300}$	

The above table contains a deliberate mistake: the rightmost vertical line is
interrupted because the input of the third row ends in 3\cr instead of 3&\cr.
Without the "&" the seventh template, which generates the last vertical rule, is
never invoked.

Here is another table with double vertical rules:

Sample Table **39.24** (Source code).

```
1    \InputD{setstrut.tip}                          % 7.4.3.1, p. I-239.
2    $$\vbox{
3    \ComputeStrut
4    \offinterlineskip
5                                      \tabskip = 15pt
6    \halign{
7        \MyStrut#&                         % 1: strut
8        \vrule \hskip 2pt \vrule#&         % 2
9        #\hrulefill&                       % 3
10       \vrule width 2pt#&                 % 4
11       \dotfill#&                         % 5
12       \bf#\hfil&                         % 6
13       ${\cal A}_{#}$\hfil&               % 7
14       \vrule width 1pt \hskip 1pt \vrule width 2pt #% 8
15   \cr
16       &&  XX&&    YY&     Z&        1&\cr
17       &&  YY&&    XX&     ZZ&       2&\cr
18       &&  AA&&    BB&     CCC&      3&\cr
19       &&  TT&&    AA&     TTTT&     4&\cr
20       &&  BIGG&&  BIGGER& VERY BIG& 300&\cr
21   }}$$
```

Sample Table **39.24** (Output).

XX__	..... YY	Z	$\mathcal{A}_1$
YY__	 XX	ZZ	$\mathcal{A}_2$
AA__	..... BB	CCC	$\mathcal{A}_3$
TT__	..... AA	TTTT	$\mathcal{A}_4$
BIGG	BIGGER	VERY BIG	$\mathcal{A}_{300}$

39.6   Vertical and Horizontal Rules

Here are some examples where horizontal and vertical rules occur at the same
time. \tabskip must be set properly! Note how vertical rules can be interrupted
in selective columns by using \omit.

Sample Table 39.25 (Source code).

```
 1   \InputD{setstrut.tip}                              % 7.4.3.1, p. I-239.
 2   $$\vbox{
 3   \ComputeStrut
 4   \offinterlineskip
 5   \hrule
 6                            \tabskip = 0pt
 7   \halign{
 8       \MyStrut#&                                      % 1: strut
 9       \vrule#\relax         \tabskip = 15pt&          % 2
10       #\hrulefill&                                    % 3
11       \dotfill#&                                      % 4
12       \vrule#&                                        % 5
13       \bf#\hfil&                                      % 6
14       ${\cal A}_{#}$\hfil&                            % 7
15       \vrule#\relax         \tabskip = 0pt            % 8
16   \cr
17       &&  XX&      YY&&     Z&           1&\cr
18       &\omit&YY&   XX&&     ZZ&          2&\cr
19       &&  AA&      BB&\omit&CCC&         3&\cr
20                       \noalign{\hrule}
21       &&  TT&      AA&&     TTTT&        4&\cr
22       &&  BIGG&    BIGGER&&VERY BIG&     300&\cr
23   }          % \halign
24   \hrule
25   }$$
```

Sample Table 39.25 (Output).

XX__	..... YY	Z	$\mathcal{A}_1$
YY__	 XX	ZZ	$\mathcal{A}_2$
AA__	..... BB	CCC	$\mathcal{A}_3$
TT__	..... AA	TTTT	$\mathcal{A}_4$
BIGG	BIGGER	VERY BIG	$\mathcal{A}_{300}$

39.6.1   \tabskips and Struts

The previous tables showed how \tabskip is set properly. Now let me show what can go wrong. If the value of \tabskip preceding the first column (tab_0) or the value of \tabskip after the last column (tab_n) is not zero, horizontal rules extend beyond the vertical rules, a situation that is usually undesirable. Here is an example:

Sample Table **39.26** (Source code).

```
1    \InputD{setstrut.tip}                              % 7.4.3.1, p. I-239.
2    $$\vbox{
3    \ComputeStrut
4    \offinterlineskip
5    \hrule
6                                       \tabskip = 15pt
7    \halign{
8        \MyStrut#&                    % 1
9        \vrule#&                      % 2
10       #\hrulefill&                  % 3
11       \dotfill#&                    % 4
12       \vrule#&                      % 5
13       \bf#\hfil&                    % 6
14       ${\cal A}_{#}$\hfil&          % 7
15       \vrule#\relax                 % 8
16   \cr
17       && XX&     YY&&    Z&              1&\cr
18       && YY&     XX&&    ZZ&             2&\cr
19       && AA&     BB&&    CCC&            3&\cr
20       && TT&     AA&&    TTTT&           4&\cr
21       && BIGG&   BIGGER&&VERY BIG&       300&\cr
22   }       % \halign
23   \hrule
24   }$$
```

Sample Table **39.26** (Output).

XX__	. . . . . YY	Z	$\mathcal{A}_1$
YY__	 XX	ZZ	$\mathcal{A}_2$
AA__	. . . . . BB	CCC	$\mathcal{A}_3$
TT__	. . . . . AA	TTTT	$\mathcal{A}_4$
BIGG	BIGGER	VERY BIG	$\mathcal{A}_{300}$

Sometimes, of course, you *do* want horizontal and vertical lines to "stick out." Here is an example of a table with this property.

Sample Table **39.27** (Source code).

```
1    \InputD{setstrut.tip}                              % 7.4.3.1, p. I-239.
2    $$\vbox{
3    \ComputeStrut
4    \offinterlineskip
5                                       \tabskip = 30pt
6    \halign{
7        \MyStrut#\relax               \tabskip = 0pt&     % 1
8        \vrule#\relax                 \tabskip = 10pt&    % 2
9        #\hrulefill&                                      % 3
10       \dotfill#&                                        % 4
11       \vrule#&                                          % 5
12       \bf#\hfil&                                        % 6
```

```
13      ${\cal A}_{#}$\hfil&                                    % 7
14      \vrule#\relax                      \tabskip = 30pt      % 8
15   \cr
16   height 30pt depth 0pt
17      &&\omit
18            &  \omit&&\omit&        \omit&\cr   \noalign{\hrule}
19      &&  XX&       YY&&    Z&          1&\cr
20      &&  YY&       XX&&    ZZ&         2&\cr
21      &&  AA&       BB&&    CCC&        3&\cr
22      &&  TT&       AA&&    TTTT&       4&\cr
23      &&  BIGG&     BIGGER&&VERY BIG&   300&\cr \noalign{\hrule}
24   height 30pt depth 0pt
25      &&\omit
26            &  \omit&&\omit&        \omit&\cr
27   }}$$
```

Sample Table 39.27 (Output).

XX⎯ ..... YY	Z	$\mathcal{A}_1$	
YY⎯ XX	ZZ	$\mathcal{A}_2$	
AA⎯ ..... BB	**CCC**	$\mathcal{A}_3$	
TT⎯ ..... AA	**TTTT**	$\mathcal{A}_4$	
BIGG BIGGER	**VERY BIG**	$\mathcal{A}_{300}$	

39.6.2   Double Vertical Rules

Here is an example of a table with double vertical rules and horizontal rules. As shown before, double vertical rules are combined into one template.

Sample Table 39.28 (Source code).

```
1    \InputD{setstrut.tip}                  % 7.4.3.1, p. I-239.
2    $$\vbox{
3    \ComputeStrut
4    \offinterlineskip
5    \hrule
6                                           \tabskip = 0pt
7    \halign{
8       \MyStrut#&                                           % 1
9       \vrule \hskip 2pt \vrule#\relax \tabskip = 15pt&     % 2
10      #\hrulefill&                                         % 3
11      \dotfill#&                                           % 4
12      \vrule\hskip 4pt \vrule#&                            % 5
```

```
13        \bf#\hfil&                                               % 6
14        ${\cal A}_{#}$\hfil&                                     % 7
15        \vrule\hskip 2pt\vrule#\relax     \tabskip = 0pt         % 8
16   \cr
17        &&  XX&       YY&&       Z&           1&\cr
18        &&  YY&       XX&&       ZZ&          2&\cr
19        &&  AA&       BB&&       CCC&         3&\cr
20        &&  TT&       AA&&       TTTT&        4&\cr
21        &&  BIGG&     BIGGER&&VERY BIG&       300&\cr
22   }
23   \hrule
24   }$$
```

Sample Table 39.28 (Output).

<table>
<tr><td>XX___</td><td>. . . . . YY</td><td>Z</td><td>$\mathcal{A}_1$</td></tr>
<tr><td>YY___</td><td>. . . . . XX</td><td>ZZ</td><td>$\mathcal{A}_2$</td></tr>
<tr><td>AA___</td><td>. . . . . BB</td><td>CCC</td><td>$\mathcal{A}_3$</td></tr>
<tr><td>TT___</td><td>. . . . . AA</td><td>TTTT</td><td>$\mathcal{A}_4$</td></tr>
<tr><td>BIGG</td><td>BIGGER</td><td>VERY BIG</td><td>$\mathcal{A}_{300}$</td></tr>
</table>

39.6.3 Fine-Tuning Vertical Rules

A \noalign {\vskip ...} type of construct can no longer be used to increase the vertical distance between two rows of a table; this would interrupt vertical rules as indicated by the next table.

Sample Table 39.29 (Source code).

```
1    \InputD{setstrut.tip}                     % 7.4.3.1, p. I-239.
2    $$\vbox{
3    \ComputeStrut
4    \offinterlineskip
5    \hrule
6                                              \tabskip = 0pt
7    \halign{
8        \MyStrut#&
9        \vrule#\relax                         \tabskip = 15pt&
10       #\hrulefill&
11       \dotfill#&
12       \vrule#&
13       \bf#\hfil&
14       ${\cal A}_{#}$\hfil&
15       \vrule#\relax                         \tabskip = 0pt
16   \cr
17       &&  XX&       YY&&       Z&           1&\cr
18       &&  YY&       XX&&       ZZ&          2&\cr       \noalign{\vskip 9pt}
19       &&  AA&       BB&&       CCC&         3&\cr
```

```
20        &&  TT&       AA&&      TTTT&        4&\cr
21        &&  BIGG&     BIGGER&&VERY BIG&      300&\cr
22    }   % \halign
23    \hrule
24    }$$
```

Sample Table 39.29 (Output).

XX__ ..... YY	Z	$\mathcal{A}_1$
YY__ XX	ZZ	$\mathcal{A}_2$
AA__ ..... BB	CCC	$\mathcal{A}_3$
TT__ ..... AA	TTTT	$\mathcal{A}_4$
BIGG BIGGER	VERY BIG	$\mathcal{A}_{300}$

There are various ways to increase the space between the second and the third row:

1. Make the second row artificially deeper. This makes all \vrules deeper.
2. Make the third row artificially higher. This makes all \vrules of the third row higher.
3. Use a combination of the two previous methods.
4. Insert an additional row, in which the height and depth add up to the desired increase in the space between the second and the third row.

Here is an example where the depth of the strut in the second row is set to 10 pt so that the first row becomes artificially deeper (it reaches a depth of 10 pt). If we specified \DeeperStrut{10pt}, instead of depth 10pt, a strut that is 10 pt *deeper* than a regular strut would have been generated. At \baselineskip = 12pt, the depth of a regular strut is computed to be 3.6 pt, so \DeeperStrut{10pt} would generate a row 13.6 pt deep.

Sample Table 39.30 (Source code).

```
1    \InputD{setstrut.tip}              % 7.4.3.1, p. I-239.
2    $$\vbox{
3    \ComputeStrut
4    \offinterlineskip
5    \hrule
6                                       \tabskip = 0pt
7    \halign{
8        \MyStrut#&
9        \vrule#\relax                  \tabskip = 15pt&
10       #\hrulefill&
11       \dotfill#&
12       \vrule#&
13       \bf#\hfil&
14       ${\cal A}_{#}$\hfil&
15       \vrule#\relax                  \tabskip = 0pt
16   \cr
17       &&  XX&       YY&&      Z&           1&\cr
```

```
18      depth 10pt&
19        &    YY&       XX&&     ZZ&           2&\cr
20       &&    AA&       BB&&     CCC&          3&\cr
21       &&    TT&       AA&&     TTTT&         4&\cr
22       &&    BIGG&     BIGGER&&VERY BIG&      300&\cr
23  }
24  \hrule
25  }$$
```

Sample Table 39.30 (Output).

XX__	..... YY	$\mathbf{Z}$	$\mathcal{A}_1$
YY__	 XX	$\mathbf{ZZ}$	$\mathcal{A}_2$
AA__	..... BB	$\mathbf{CCC}$	$\mathcal{A}_3$
TT__	..... AA	$\mathbf{TTTT}$	$\mathcal{A}_4$
BIGG	BIGGER	$\mathbf{VERY\ BIG}$	$\mathcal{A}_{300}$

Here is the other solution. The newly inserted row (between the second and the third text rows) contains nothing in the various text columns. This row is 6 pt high (the sum of **height** 4pt and **depth** 2pt).

Sample Table 39.31 (Source code).

```
1   \InputD{setstrut.tip}              % 7.4.3.1, p. I-239.
2   $$\vbox{
3   \ComputeStrut
4   \offinterlineskip
5   \hrule
6                              \tabskip = 0pt
7   \halign{
8       \MyStrut#&                          % 1
9       \vrule#\relax          \tabskip = 15pt&   % 2
10      #\hrulefill&                        % 3
11      \dotfill##&                         % 4
12      \vrule#&                            % 5
13      \bf#\hfil&                          % 6
14      ${\cal A}_{#}$\hfil&                % 7
15      \vrule#\relax          \tabskip = 0pt    % 8
16  \cr
17       &&    XX&       YY&&     Z&           1&\cr
18       &&    YY&       XX&&     ZZ&          2&\cr
19      height 4pt depth 2pt&&
20        \omit&      \omit&& \omit&        \omit&\cr
21       &&    AA&       BB&&     CCC&         3&\cr
22       &&    TT&       AA&&     TTTT&        4&\cr
23       &&    BIGG&     BIGGER&&VERY BIG&     300&\cr
24  }
25  \hrule
26  }$$
```

Sample Table 39.31 (Output).

XX__	. . . . . YY	Z	$\mathcal{A}_1$
YY__	 XX	ZZ	$\mathcal{A}_2$
AA__	. . . . . BB	CCC	$\mathcal{A}_3$
TT__	. . . . . AA	TTTT	$\mathcal{A}_4$
BIGG	BIGGER	VERY BIG	$\mathcal{A}_{300}$

This last solution can be used nicely for the insertion of two horizontal rules that are separated by a small amount of vertical space (smaller than the height and depth of the strut). Here is an example of such a table:

Sample Table 39.32 (Source code).

```
1   \InputD{setstrut.tip}                        % 7.4.3.1, p. I-239.
2   $$\vbox{
3   \ComputeStrut
4   \offinterlineskip
5   \hrule
6                               \tabskip = 0pt
7   \halign{
8       \MyStrut#&                               % 1
9       \vrule#\relax           \tabskip = 15pt& % 2
10      #\hrulefill&                             % 3
11      \dotfill#&                               % 4
12      \vrule#&                                 % 5
13      \bf#\hfil&                               % 6
14      ${\cal A}_{#}$\hfil&                     % 7
15      \vrule#\relax           \tabskip = 0pt   % 8
16  \cr
17      && XX&       YY&&     Z&          1&\cr
18      && YY&       XX&&     ZZ&         2&\cr
19                                  \noalign{\hrule}
20      height 3pt depth 3pt&&
21        \omit&      \omit&& \omit&      \omit&\cr
22                                  \noalign{\hrule}
23      && AA&       BB&&     CCC&        3&\cr
24      && TT&       AA&&     TTTT&       4&\cr
25      && BIGG&     BIGGER&&VERY BIG&    300&\cr
26  }
27  \hrule
28  }$$
```

Sample Table 39.32 (Output).

XX__	. . . . . YY	Z	$\mathcal{A}_1$
YY__	 XX	ZZ	$\mathcal{A}_2$
AA__	. . . . . BB	CCC	$\mathcal{A}_3$
TT__	. . . . . AA	TTTT	$\mathcal{A}_4$
BIGG	BIGGER	VERY BIG	$\mathcal{A}_{300}$

39.6.4 Vertical Rules and \multispan

\multispan works with vertical rules, but you must be careful when counting columns now. Vertical rules are typically placed in their own columns and those columns have to be included when you compute the number of columns a \multicolumn spans. Also, be careful not to accidentally destroy the strut. Observe that \multispan should *not* span any columns that contain only vertical rules, or precede or follow the heading, because the templates of these vertical rule columns will be destroyed by \multispan. Here is an example:

Sample Table **39.33** (Source code).

```
 1  \InputD{setstrut.tip}                % 7.4.3.1, p. I-239.
 2  $$\vbox{
 3  \ComputeStrut
 4  \offinterlineskip
 5  \hrule
 6                                       \tabskip = 0pt
 7  \halign{
 8      \MyStrut#&
 9      \vrule#\relax            \tabskip = 15pt&
10      #\hrulefill&
11      \dotfill#&
12      \vrule#&
13      \bf#\hfil&
14      ${\cal A}_{#}$\hfil&
15      \vrule#\relax            \tabskip = 0pt
16  \cr
17      &&\multispan{5}\hfil\bf Some heading goes here\hfil&
18                               \cr    \noalign{\hrule}
19      &\multispan{6}\hfil\bf Wrong heading\hfil&
20                               \cr    \noalign{\hrule}
21      &\multispan{7}\hfil\bf Also a wrong heading\hfil
22                               \cr    \noalign{\hrule}
23      &&  XX&     YY&&    Z&         1&\cr    \noalign{\hrule}
24      &&  YY&     XX&&    ZZ&        2&\cr    \noalign{\hrule}
25          height 9pt&&
26          \omit&  \omit&& \omit&         \omit&\cr\noalign{\hrule}
27      &&  AA&     BE&&    CCC&       3&\cr    \noalign{\hrule}
28      &&  TT&     AA&&    TTTT&      4&\cr    \noalign{\hrule}
29      &&  BIGG&   BIGGER&&VERY BIG&  300&\cr
30  }
31  \hrule
32  }$$
```

Sample Table **39.33** (Output).

Some heading goes here		
Wrong heading		
Also a wrong heading		
XX___ ..... YY	**Z**	$\mathcal{A}_1$
YY___ XX	**ZZ**	$\mathcal{A}_2$
AA___ ..... BB	**CCC**	$\mathcal{A}_3$
TT___ ..... AA	**TTTT**	$\mathcal{A}_4$
BIGG BIGGER	**VERY BIG**	$\mathcal{A}_{300}$

39.6.5   First and Last Rows in a Table with Horizontal Rules

If you look at the previous tables, you will discover that the very first and the very last horizontal rules tend to be a little too close to the first and last rows of the table. This is corrected by increasing the height of the strut in the first row and the depth of the strut in the last row. The macros \HigherStrut and \DeeperStrut (defined in 7.4.3, p. I-239) allow you to do this very conveniently.

Sample Table **39.34** (Source code).

```
 1   \InputD{setstrut.tip}                    % 7.4.3.1, p. I-239.
 2   $$\vbox{
 3   \ComputeStrut
 4   \offinterlineskip
 5   \hrule
 6                                            \tabskip = 0pt
 7   \halign{
 8       \MyStrut#&
 9       \vrule#\relax                        \tabskip = 15pt&
10       #\hrulefill&
11       \dotfill##&
12       \vrule#&
13       \bf#\hfil&
14       ${\cal A}_{#}$\hfil&
15       \vrule#\relax                        \tabskip = 0pt
16   \cr
17       \HigherStrut{4.0pt}&
18       &   XX&       YY&&      Z&           1&\cr
19       &&  YY&       XX&&      ZZ&          2&\cr
20       height 4pt depth 4pt&&
21         \omit&      \omit&& \omit&         \omit&\cr
22       &&  AA&       BB&&     CCC&          3&\cr
23       &&  TT&       AA&&     TTTT&         4&\cr
24       \DeeperStrut{4.0pt}&
25       &   BIGG&    BIGGER&&VERY BIG&       300&\cr
```

```
26   }
27   \hrule
28   }$$
```

Sample Table 39.34 (Output).

XX__	..... YY	Z	$\mathcal{A}_1$
YY__	 XX	ZZ	$\mathcal{A}_2$
AA__	..... BB	CCC	$\mathcal{A}_3$
TT__	..... AA	TTTT	$\mathcal{A}_4$
BIGG	BIGGER	VERY BIG	$\mathcal{A}_{300}$

39.6.6   Horizontal Rules Spanning Selected Columns Only

One can also print horizontal rules that span only selected columns. These selective horizontal rules will occupy rows of their own. The height and depth of the rows is the same as the height and depth of the rules in them. The strut must be eliminated for those rows using \omit.

In the following, I discuss selective horizontal rules in connection with vertical rules, because this is a frequently-used application of selective horizontal rules. Here is how you should proceed:

1.   Use \omit to eliminate the strut.
2.   To generate a selective horizontal rule, determine the column numbers i and j between which the rules extend. Those column indices must be used for the column indices of the vertical rule columns that the selective horizontal rules will extend in.
3.   Insert a \multispan{$j - i + 1$}\hrulefill into the table to generate the horizontal rule.

Here is an example:

Sample Table 39.35 (Source code).

```
1    \InputD{setstrut.tip}              % 7.4.3.1, p. I-239.
2    $$\vbox{
3    \ComputeStrut
4    \offinterlineskip
5    \hrule
6                                       \tabskip = 0pt
7    \halign{
8       \MyStrut#&                                      % 1
9       \vrule#\relax          \tabskip = 15pt&         % 2 (1st vrule)
10      #\hrulefill&                                    % 3
11      \dotfill#&                                      % 4
```

```
12      \vrule#&                                                  % 5 (2nd vrule)
13      \bf#\hfil&                                                % 6
14      ${\cal A}_{#}$\hfil&                                      % 7
15      \vrule#\relax               \tabskip = 0pt      % 8 (3rd vrule)
16  \cr
17      \HigherStrut{4.0pt}&
18      &   XX&      YY&&     Z&           1&\cr
19      \omit&\multispan{4}\hrulefill&&\omit&\cr
20      &&  YY&      XX&&     ZZ&          2&\cr
21      \omit&&\omit&\omit&\multispan{4}\hrulefill\cr
22      &&  AA&      BB&&     CCC&         3&\cr
23      &&  TT&      AA&&     TTTT&        4&\cr
24      \DeeperStrut{4.0pt}&
25      &   BIGG&    BIGGER&&VERY BIG&     300&\cr
26  }
27      \hrule
28  }$$
```

Sample Table 39.35 (Output).

XX__	..... YY	Z	$\mathcal{A}_1$
YY__	 XX	ZZ	$\mathcal{A}_2$
AA__	..... BB	CCC	$\mathcal{A}_3$
TT__	..... AA	TTTT	$\mathcal{A}_4$
BIGG	BIGGER	VERY BIG	$\mathcal{A}_{300}$

Here is a more extensive example of a table with selective horizontal rules:

Sample Table 39.36 (Source code).

```
1   \InputD{setstrut.tip}                    % 7.4.3.1, p. I-239.
2   $$
3   \vbox{
4   \ComputeStrut
5   \offinterlineskip
6                           \tabskip = 0pt
7   \halign{
8       \MyStrut#&                           % 1
9       \vrule#\relax       \tabskip = 6pt&  % 2: 1st \vrule
10      \hfil#\hfil&                         % 3: 1st text column
11      \vrule#&                             % 4: 2nd \vrule
12      \hfil#\hfil&                         % 5: 2nd text column
13      \vrule#&                             % 6: 3rd \vrule
14      \hfil#\hfil&                         % 7: 3rd text column
15      \vrule#&                             % 8: 4th \vrule
16      \hfil#\hfil&                         % 9: 4th text column
17      \vrule#&                             % 10: 5th \vrule
18      \hfil#\hfil&                         % 11: 5th text column
19      \vrule#&                             % 12: 6th \vrule
20      \hfil#\hfil&                         % 13: 6th text column
21      \vrule#\relax       \tabskip = 0pt   % 14: 7th \vrule
```

```
22   \cr
23       \omit&\multispan{13}{\hrulefill}\cr
24       &&  A&\omit&      B&\omit&      C&\omit&      D&\omit&      E&&
25                       F&\cr
26       \omit&\multispan{3}{\hrulefill}&&\omit&&\omit&&\omit&&\omit&&\cr
27       &&  0&&          1&\omit&      2&\omit&      3&\omit&      4&&
28                       5&\cr
29       \omit&\omit&&\multispan{5}{\hrulefill}&&\omit&&\omit\cr
30       &&  S&\omit&      T&\omit&      U&&          V&\omit&      W&&
31                       X&\cr
32       \omit&\multispan{5}{\hrulefill}&&\multispan{7}{\hrulefill}\cr
33       &&  +&\omit&      -&&          *&\omit&      /&&          ?&\omit&
34                       !&\cr
35       &&  9&\omit&      8&&          7&\omit&      6&&          5&\omit&
36                       4&\cr
37       \omit& \multispan{13}{\hrulefill}\cr
38   }}$$
```

Sample Table 39.36 (Output).

A	B	C	D	E	F
0	1	2	3	4	5
S	T	U	V	W	X
+	-	*	/	?	!
9	8	7	6	5	4

39.7   The Proper Way to Input Tables

To input tables properly, it is essential that your input is easy to read so when
the table needs to be corrected or changed, you can understand your own input.
Compare, for instance, the following input

```
1                                       \tabskip = 15pt
2   \halign{
3       #\rightarrowfill           \tabskip = 45pt&        % 1
4       \dotfill#\relax            \tabskip = 0pt&         % 2
5       \bf#\hfil                  \tabskip = 15pt&        % 3
6       ${\cal A}_{#}$\hfil                                % 4
7   \cr
8       XX&      YY&      Z&        1&       XX& -1\cr
9       YY&      XX&      ZZ&       2&       YY& -2\cr
10      AA&      \omit LEFT\hfil&
11                       CCC&       \omit$\infty$\hfil&
12                                           LLL LLL\cr
13      TT&      AA&      TTTT&     4&       LL& -4\cr
14      BIGG&    BIGGER&  VERY BIG& 300&     L&  -300\cr
15  }
```

with the input:

```
1    \tabskip = 15pt
2    \halign{#\rightarrowfill\tabskip = 45pt&\dotfill#\tabskip = 0pt&&
3    \bf#\hfil\tabskip = 15pt&${\cal A}_{#}$\hfil\cr XX&
4    YY&Z&1&XX&-1\cr YY&XX&ZZ&2&YY&-2\cr
5    AA&\omit LEFT\hfil&CCC&\omit$\infty$\hfil&LLL
6    LLL\cr TT&AA&TTTT&4&
7    LL&-4\cr BIGG&BIGGER&VERY BIG&300&L&-300\cr}
```

Both inputs produce the same table. Clearly the first input is much easier to read because it more closely *resembles the typeset version of the table*. That last remark is actually the key to finding the proper way of inputting tables: make the table source resemble the output as closely as possible.

Here are some additional suggestions on how to input tables in T_EX:

1. Clearly mark the *end of the preamble*. For instance, put the \cr that ends the preamble on a line by itself.
2. In the preamble write *one template per line*. The \tabskip *specifications* of each template should be moved to the right and should all *align vertically*.
3. One line in the input should contain one row of the table.
4. Remember that *spaces* after & are ignored. Therefore, a table entry can be moved to the right as far as necessary, for a clean, structured input.
5. Align all \noalign instructions vertically.

39.8   Summary

In this chapter we learned:

- The vertical spacing in tables can be controlled by changing \baselineskip or by inserting struts.
- \noalign inserts unaligned material into a table. This can be used to change the spacing between two columns or to insert a horizontal rule within a table.
- Two techniques dealing with entries spanning multiple columns were discussed. One uses \hidewidth and the other uses \multispan.
- We also discussed how a left or right justified column can as a whole be centered under a heading which is wider than the widest entry in the column.
- Vertical rules require the use of struts to control the spacing in a table. It is advisable to use separate columns for vertical rules. No height or depth specification of these rules is usually necessary, because this information is taken from the strut being used. \offinterlineskip is used to disable the interline glue which is usually inserted by T_EX.
- Selective horizontal rules can be inserted using \multispan, where this \multispan occupies an almost zero thickness row by itself.

- The structure of input to a table should resemble the output. Neatness is important when coding TEX tables.

40
Even More Tables

This chapter will continue the discussion of tables. We will look at interesting cases such as tables which contain paragraphs or other tables as entries. We will also discuss macros for templates of tables. As discussed in 38.1, p. 199, the chapters on tables may show some rather poor page breaks. We apologize.

40.1 Centering a Table on the Page

There are three different ways to center a table:

1. Enclose the table inside a \vbox and then center this \vbox using display math mode; see 40.1.1 on this page.
2. Use \tabskip; see 40.1.2 on the next page.
3. Use \centerline; see 40.1.4, p. 297.

40.1.1 Centering a Table Using Display Mathmode

The standard way to center a table is to enclose the table inside a \vbox, and to center this \vbox using display math mode. This approach to centering tables was used for most of the tables presented so far. Note:

1. The table cannot be broken across pages, because it is contained inside a vbox. A paragraph inside a vbox also cannot be broken across pages; see 10.3.1, p. II-9.
2. The table is centered because the table is contained in a vbox, and display mathmode centers its material, in this case a vbox.
3. Vertical glue is inserted above and below the table, which again is due to display math mode; see 14.9.1, p. II-216.

For macros to center tables using the method discussed here see 40.5.8, p. 333.

40.1.2   Centering a Table Using \tabskip

A table can be centered by allowing tab_0 and tab_n to stretch (Opt plus 1fil) and by generating a table that is as wide as the page (\halign to \hsize). This approach has the disadvantage that as far as TeX is concerned, the width of the table, is now \hsize, and horizontal rules extend to the full width of the page.

Here is the example of a table which is centered with this method. The table is still enclosed inside a \vbox to prevent a page break in the middle of the table.

Sample Table **40.1** (Source code).

```
1    \vbox{
2                                   \tabskip = Opt plus 1fil
3    \halign to \hsize{
4        #\rightarrowfill          \tabskip = 10pt&
5        \dotfill#&
6        \bf #\hfil&
7        ${\cal A}_{#}$\hfil       \tabskip = Opt plus 1fil
8    \cr
9        XX&      YY&      Z&          1\cr
10       YY&      XX&      ZZ&         2\cr
11                                 \noalign{%
12                                     \vskip 2pt
13                                     \hrule
14                                     \vskip 2pt
15                             }
16       AA&      BB&      CCC&        3\cr
17       TT&      AA&      TTTT&       4\cr
18       BIGG&    BIGGGGGER&   VERY BIG&    300\cr
19   }}
```

Sample Table **40.1** (Output).

XX⟶	...........YY	Z	$\mathcal{A}_1$
YY⟶	XX	ZZ	$\mathcal{A}_2$
AA⟶	..........BB	**CCC**	$\mathcal{A}_3$
TT⟶	...........AA	**TTTT**	$\mathcal{A}_4$
BIGG→	BIGGGGGER	**VERY BIG**	$\mathcal{A}_{300}$

40.1.3   One Column Text, Widest Entry Centered

Let me briefly discuss a different solution to a problem which was already discussed in 6.10.4, p. I-203. Now, using \halign, a one column "table" (if you can call it that) is built, where that one column is right justified. This "table" is centered using tab_0 and tab_n. Here is the example:

Sample Table **40.2** (Source code).

The change of tab_0 is enclosed in a group so that tab_0 of later-following tables is not affected by this setting.

```
1    {
2                                    \tabskip = 0pt plus 1fil
3        \halign to \hsize{%
```

The one column of this table is right-justified.

```
4            \hfil#%
5        \cr
6            This is a centered line, like a quote or so.\cr
7                            The author's name.\cr
8        }
9    }
```

Sample Table **40.2** (Output).

This is a centered line, like a quote or so.
The author's name.

40.1.4   Centering a Table Using \centerline

Another way to center a table is to use \centerline. Enclose the table inside a \vbox and center this \vbox using \centerline. Here is an example of such a table. \medskip was inserted before and after the table to set off the table from the preceding and the following text; inserting vertical glue is unnecessary when display math mode is used, because the display mathmcde inserts such glue automatically (\abovedisplayskip, etc.).

Sample Table **40.3** (Source code).

```
1    \medskip
2    \centerline{%
3        \vbox{
4                                    \tabskip = 0pt
5            \halign{
6                #\rightarrowfill        \tabskip = 10pt&
7                \dotfill##&
8                \bf #\hfil&
9                ${\cal A}_{#}$\hfil        \tabskip = 0pt
10           \cr
```

```
11            XX&      YY&        Z&           1\cr
12            YY&      XX&        ZZ&          2\cr
13            AA&      BB&        CCC&         3\cr
14            TT&      AA&        TTTT&        4\cr
15            BIGG&    BIGGGGER&  VERY BIG&    300\cr
16        }
17      }
18  }
19  \medskip
```

Sample Table **40.3** (Output).

XX$\longrightarrow$	..........YY	Z	$\mathcal{A}_1$
YY$\longrightarrow$	XX	**ZZ**	$\mathcal{A}_2$
AA$\longrightarrow$	...........BB	**CCC**	$\mathcal{A}_3$
TT$\longrightarrow$	............AA	**TTTT**	$\mathcal{A}_4$
BIGG$\rightarrow$	BIGGGGGER	**VERY BIG**	$\mathcal{A}_{300}$

40.2   Sparse Tables

Sparse tables are tables with with relatively few entries. The term was defined with the term "sparse matrices" in mind which is used by mathematicians to describe matrices with relatively non-zero entries. Sparse tables are defined here as tables with relatively few (non-empty) entries.

40.2.1   An Example

The following table contains an entry in the first column, but there will be only one entry in one of the remaining columns (for each row). We assume here that the number of columns of the table is unknown and therefore the template of the second column is repeated for all subsequent columns (how this can be done is discussed in 38.3.16, p. 227).

First let me show the way the table is generated without using any fancy macros.

Sample Table **40.4** (Source code).

```
1   $$
2   \vbox{
3                          \tabskip = 0pt
4   \halign{
5       \hfil#\relax       \tabskip = 30pt&&
6       #\relax            \tabskip = 5pt
7   \cr
8       1.&x\cr
9       2.&&x\cr
10      3.&&&x\cr
11      4.&&&&x\cr
12      5.&&&&&x\cr
13      6.&&&&&&x\cr
14      7.&&&&&&&x\cr
15      8.&&&&&&&&x\cr
16      9.&&&&&&&&&x\cr
17      10.&&&&&&&&&x\cr
18  }}
19  $$
```

Sample Table **40.4** (Output).

```
 1.        x
 2.          x
 3.            x
 4.              x
 5.                x
 6.                  x
 7.                    x
 8.                      x
 9.                        x
10.                      x
```

Now I present the same table using a macro \AdvanceByTabStops. This macro takes one parameter, #1, the number of tabs to advance, as its argument. It then generates the appropriate number of "&"s.

$\mathcal{P}'$ • advtabst.tip •

```
15  \InputD{doloop.tip}                    % 27.1.8, p. III-412.
16  \catcode'\@ = 11
17  \newcount\AdvanceByTabStopsCount
18  \def\AdvanceByTabStops #1{%
19      \def\@MakeTabChars{}%
```

Generate the appropriate number of &s using \DoLoop.

```
20      \DoLoop{\AdvanceByTabStopsCount}{2}{1}{#1}%
21          {\edef\@MakeTabChars{\@MakeTabChars&}}%
22      \@MakeTabChars
23  }
```

```
24   \catcode'\@ = 12
```

• End of `advtabst.tip` •

Sample Table **40.5** (Source code).

```
1    \InputD{advtabst.tip}                    % 40.2.1, p. 299.
```

The preceding table is now repeated using macro \AdvanceByTabStops:

```
2    $$
3    \vbox{
4                                    \tabskip = 0pt
5    \halign{
6        \hfil#\relax               \tabskip = 30pt&&
7        #\relax                    \tabskip = 5pt
8    \cr
9        1.&\AdvanceByTabStops{1}x\cr
10       2.&\AdvanceByTabStops{2}x\cr
11       3.&\AdvanceByTabStops{3}x\cr
12       4.&\AdvanceByTabStops{4}x\cr
13       5.&\AdvanceByTabStops{5}x\cr
14       6.&\AdvanceByTabStops{6}x\cr
15       7.&\AdvanceByTabStops{7}x\cr
16       8.&\AdvanceByTabStops{8}x\cr
17       9.&\AdvanceByTabStops{9}x\cr
18       10.&\AdvanceByTabStops{10}x\cr
19   }}$$
```

Sample Table **40.5** (Output).

```
 1.     x
 2.       x
 3.         x
 4.           x
 5.             x
 6.               x
 7.                 x
 8.                   x
 9.                     x
10.                       x
```

Look at the definition of \AdvanceByTabChars closely; here is how this macro works:

1. The appropriate number of "&"s is collected in a macro (\@MakeTabChars).
2. The loop is terminated.
3. The collected "&" are inserted, all at once.

The following definition of \AdvanceByTabChars would *not* work:

```
1    \def\AdvanceByTabStops #1{%
2        \DoLoop{\AdvanceByTabStopsCount{1}{1}{#1}{&}}
```

```
3  }
```

The preceding definition would *not* work, because it is impossible to have a dangling `\fi` (which is generated through the `\DoLoop` macro) across columns of a table. Therefore, as done originally, all the "&"s must first be collected in a macro and then inserted as a whole into the table.

40.2.2 Naming Columns

In a sparse table it may be convenient to assign *names* to columns. This way it is no longer necessary to "count &s". Simply call a macro (the macro's name identifies which column the entry is inserted in) and provide the entry as argument to the macro: for instance, to put entry "ABC" into column "Prod" enter `\Prod{ABC}`. Let me show how this can be done in the following example. Also note that columns can be entered in an arbitrary order.

Before specifying the table let me provide you with an overview of the later used columns:

Column	Name	Description
1.		Strut
2.		`\vrule`
3.	Pro	Product
4.	PercA	Percentage of element A
5.	PercB	Percentage of element B
6.	PercC	Percentage of element C
7.		`\vrule`
8.	DensX	Density in X-dimen
9.	DensY	Density in Y-dimen
10.	DensZ	Density in Z-dimen
11.		`\vrule`

Sample Table **40.6** (Source code).

Here are the macros to define the entry for a column. To inform TEX about an entry for column Abc call `\Abc` with one argument, the entry. The entry is saved under the name `\XAbc` so it can be later expanded, when the entry is supposed to be printed. All definitions below are done on a global basis because calls to this macro happen "inside the table" and therefore the implicit grouping inside the table must be bypassed. Here are the macros for our example table:

```
1  \def\Pro   #1{\gdef\XPro{#1}\ignorespaces}
2  \def\PercA #1{\gdef\XPerA{#1}\ignorespaces}
3  \def\PercB #1{\gdef\XPerB{#1}\ignorespaces}
4  \def\PercC #1{\gdef\XPerC{#1}\ignorespaces}
5  \def\DenX  #1{\gdef\XDenX{#1}\ignorespaces}
6  \def\DenY  #1{\gdef\XDenY{#1}\ignorespaces}
7  \def\DenZ  #1{\gdef\XDenZ{#1}\ignorespaces}
```

The **\NewTemplateXs** resets all the saved column entries to empty. This macro is called after a row of the table was printed.

```
 8    \def\NewTemplateXs{%
 9        \Pro{}%
10        \PercA{}%
11        \PercB{}%
12        \PercC{}%
13        \DenX{}%
14        \DenY{}%
15        \DenZ{}%
16    }
```

The macro **\PrintOneRow** generates one table row. This macro maps the names of columns to their respective positions. This macro also "clears out" all entries recorded using the previous macro.

```
17    \def\PrintOneRow{%
18        &                  % 1
19        &                  % 2
20        \XPro&             % 3
21        \XPerA&            % 4
22        \XPerB&            % 5
23        \XPerC&            % 6
24        &                  % 7
25        \XDenX&            % 8
26        \XDenY&            % 9
27        \XDenZ&            % 10
28    \cr
29        \NewTemplateXs
30    }
```

The table starts here.

```
31    $$
32    \vbox{
```

Reset all column entries.

```
33        \NewTemplateXs
34        \offinterlineskip
35                                    \tabskip = 0pt
36        \halign{
```

The preamble starts here.

```
37            \MyStrut#&                              % 1
38            \vrule#\relax       \tabskip = 10pt&    % 2
39            \bf#\hfil&                              % 3
40            \hfil#&                                 % 4
41            \hfil#&                                 % 5
42            \hfil#&                                 % 6
43            \vrule#&                                % 7
44            \tt#\hfil&                              % 8
45            \tt#\hfil&                              % 9
46            \tt#\hfil&                              % 10
```

```
47              \vrule#\relax              \tabskip = 0pt        % 1:
48         \cr
```

The table body starts here.

```
49              \Pro{$<$none$>$}                      \PrintOneRow
50              \Pro{Ariolisis-I}   \PercA{0.4\%} \DenX{V type}
51                                                   \PrintOneRow
52              \Pro{Ariolisis-II} \PercB{0.8\%} \DenY{T type}
53                                                   \PrintOneRow
```

Demonstrate that the table is totally independent of the order it was entered in.

```
54              \DenX{L type}    \Pro{Lurilisis-I}  \PercC{0.3\%}
55                                                   \PrintOneRow
56              \Pro{Lurilisis-II} \PercC{0.4\%} \DenX{V type}
57                                                   \PrintOneRow
58              \Pro{Lurilisis-III}\PercB{0.8\%} \DenY{T type}
59                                                   \PrintOneRow
60              \Pro{Luriosis-III} \PercC{0.4\%} \DenZ{V type}
61                                                   \PrintOneRow
62              \Pro{Luriosis-I}    \PercC{0.3\%} \DenZ{L type}
63                                                   \PrintOneRow
64              \Pro{Luriosis-Ia}  \PercC{0.4\%} \DenZ{V type}
65                                                   \PrintOneRow
66              \Pro{Luriosis-Ib}  \PercB{0.8\%} \DenY{T type}
67                                                   \PrintOneRow
68              \Pro{Luriosis-II}               \DenZ{V type}
69                                                   \PrintOneRow
70         \cr
71         }
72    }
73    $$
```

Sample Table 40.6 (Output).

<none>						
Ariolisis-I	0.4%			V type		
Ariolisis-II		0.8%			T type	
Lurilisis-I			0.3%	L type		
Lurilisis-II			0.4%	V type		
Lurilisis-III		0.8%			T type	
Luriosis-III			0.4%			V type
Luriosis-I			0.3%			L type
Luriosis-Ia			0.4%			V type
Luriosis-Ib		0.8%			T type	
Luriosis-II						V type

40.3   Vboxes as Table Entries

I will now present tables with vboxes as table entries. This offers (among others) the following possibilities:

1. Tables with *paragraphs* and *displayed equations* as table entries; see 40 3.1 on this page.
2. Tables within tables; see 40.3.4, p. 309.

40.3.1   A Paragraph As a Table Entry

By enclosing a paragraph inside a vbox this paragraph can be made part of a table. Observe though that you must specify the width of the paragraph by setting \hsize inside the vbox in which the paragraph is enclosed. The width of this vbox entry, as far as the width computation for this entry's table computation is concerned, will be this locally set value of \hsize.

In the rest of this chapter the term *text column* is used to describe a column which has a \vtop-based template to include some paragraph or paragraphs of text.

Assume also that the table includes vertical rules, and therefore \offinter-lineskip is called to disable regular line spacing (based on \baselineskip). Therefore \normalbaselines must be called inside the \vbox containing the paragraph to undo the effect of \offinterlineskip. Here is the first example of such a table:

Sample Table **40.7** (Source code).

```
1   $$
2   \vbox{
3       \offinterlineskip
4       \hrule height 2pt
5                                           \tabskip = 0pt
6       \halign{
7           \vrule width 2pt#\relax      \tabskip = 10pt&      % 1
8           \hfil#\hfil&                                       % 2
9           \vrule width 3pt#&                                 % 3
10          \hfil#\hfil&                                       % 4
11          \vrule width 2pt#\relax      \tabskip = 0pt        % 5
12      \cr
13          &\multispan{3}{\hfil A Table Title\hfil}&\cr
14          &\vbox{%
15              \hsize = 1.4in
16              \normalbaselines
```

Columns are narrow, therefore a ragged-right layout is used.

```
17              \raggedright
18              This is a nice paragraph which
```

```
19          appears in the first column and the first row
20          of such table. We just plan to show the idea of
21          a little paragraph as part of a table.
22       }&&
23       Some entry&
24    \cr
25       \noalign{\hrule height 2pt}
26       &\vbox{%
27          \hsize = 2.0in
28          \normalbaselines
29          \raggedright
```

A "stronger version" of \raggedright is required here: increase \rightskip.

```
30          \rightskip = 0pt plus 30pt
31          This is a nice paragraph which
32          appears in the first column and the second row
33          of such table. This paragraph is a little wider.
34          Otherwise the same idea applies to the entries here.
35       }&&
36       \vbox{%
37          \hsize = 2.1in
38          $$\sum_{i=1}^n x_i + y_i$$}&
39    \cr
40    }
41    \hrule height 2pt
42 }$$
```

Sample Table **40.7** (Output).

<table>
<tr><td colspan="2" align="center">A Table Title</td></tr>
<tr>
<td>This is a nice paragraph which appears in the first column and the first row of such table. We just plan to show the idea of a little paragraph as part of a table.</td>
<td align="center">Some entry</td>
</tr>
<tr>
<td>This is a nice paragraph which appears in the first column and the second row of such table. This paragraph is a little wider. Otherwise the same idea applies to the entries here.</td>
<td align="center">$\sum_{i=1}^{n} x_i + y_i$</td>
</tr>
</table>

40.3.2   Improved Version of a Paragraph within a Table

Several corrections of the previous table seem to be appropriate.

1. Paragraphs in the same table row should line up with their *first* and not their last lines. Therefore \vtops instead of \vboxs must be used to enclose the paragraph.
2. As a matter of convenience, the \vtops and all other related instructions are made part of the template so that these instructions do not need to be repeated for each paragraph.
3. The vertical spacing around the first and last line of the various paragraphs is inappropriate. This is corrected by inserting a strut-like vertical rule into the first and last line of each paragraph to make these two lines artificially deeper and higher. In the example below these strut-like constructs were made visible by specifying a width difference from zero (the width would normally be set to zero).

Here is the corrected table:

Sample Table 40.8 (Source code).

```
 1   $$\vbox{
 2       \offinterlineskip
 3       \hrule height 2pt
 4                                            \tabskip = 0pt
 5       \halign{%
 6           \vrule width 2pt #\relax         \tabskip = 10pt&    % 1
 7           \vtop{%
 8               \hsize = 1.5in                                   % 2
 9               \normalbaselines
10               \raggedright
11               \vrule height 10pt depth 0pt width 2pt
12               #%
13               \vrule height  0pt depth 6pt width 2pt
14           }&
15           \vrule width 3pt#&                                   % 3
16           \vtop{%
17               \hsize = 2.0in                                   % 4
18               \normalbaselines
19               \raggedright
20               \vrule height 10pt depth 0pt width 2pt
21               #%
22               \vrule height  0pt depth 6pt width 2pt
23           }&
24           \vrule width 2pt#\relax          \tabskip = 0pt     % 5
25       \cr
```

Next specify a header in this table. Insert a "visible strut" to increase the spacing for the header.

```
26           & \multispan{3}%
```

```
27          \hfil
28          \vrule height 10pt depth 4pt width 2pt
29              \bf This Is the Table Title%
30          \hfil
31      &\cr
32      & This is just some text to show how you can
33        typeset a paragraph inside a table. Observe that we use
34        a ragged-right type of text. Let us see how it comes out.&&
35        Now this is text in the right column, which is a little
36        wider than the first left column. Observe that the first
37        line of each paragraph is on the same horizontal level.&\cr
38                          \noalign{\hrule}
39      & This is just some text to show how you can
40        typeset a paragraph inside a table. Observe that we use
41        a ragged-right type of text. Let us see how it comes out.
42      && Now this is text in the right column, which is a little
43        wider than the first left column. Observe that the first
44        line of each paragraph is on the same horizontal level.&\cr
45      }   % halign
46      \hrule height 2pt
47  }$$
```

Sample Table **40.8** (Output).

<table>
<tr><td colspan="2" align="center">**This Is the Table Title**</td></tr>
<tr>
<td>This is just some text to show how you can typeset a paragraph inside a table. Observe that we use a ragged-right type of text. Let us see how it comes out.</td>
<td>Now this is text in the right column, which is a little wider than the first left column. Observe that the first line of each paragraph is on the same horizontal level.</td>
</tr>
<tr>
<td>This is just some text to show how you can typeset a paragraph inside a table. Observe that we use a ragged-right type of text. Let us see how it comes out.</td>
<td>Now this is text in the right column, which is a little wider than the first left column. Observe that the first line of each paragraph is on the same horizontal level.</td>
</tr>
</table>

40.3.3   Another Case of a Paragraph in a Table

Following is another case of a paragraph as a table entry of some table. Assume
this paragraph is contained in the *second* column of a table with three columns
and, let's say, this paragraph when printed has *four lines*. The first and the last
column of the table contain centered entries, consisting of four rows each that
should match up with the four line paragraph. This case can be handled by using
a \vtop to generate the paragraph text. The *depth* of this \vtop-generated box
is set to zero, so that TeX is tricked into thinking that this paragraph has only
one line instead of four.

Instead of lengthy explanations let me show you the example now, first
without modifying the depth of the middle column box, leading to an *incorrect*
solution.

Sample Table 40.9 (Source code).

```
1    $$\vbox{
2                                    \tabskip = 0pt
3       \halign{%
4           \hfil\it#\hfil        \tabskip = 15pt&
5           \vtop{%
6               \hsize = 2.5in
7               \raggedright
8               \noindent
9               #
10          }&
11          \hfil\it#\hfil        \tabskip = 0pt
12      \cr
13          ABC& This is some text which will be too long to fit
14              on one line. Actually it will fit on four lines,
15              I chose a text which would do exactly that.& XXX\cr
16          DDD&     \omit& FFF\cr
17          ABCDEF& \omit& XYZXYZXYZ\cr
18          TT&      \omit& ABCDEF\cr
19      }
20   }
21   $$
```

Sample Table 40.9 (Output).

ABC	This is some text which will be too long to fit on one line. Actually it will fit on four lines, I chose a text which would do exactly that.	*XXX*
DDD		*FFF*
ABCDEF		*XYZXYZXYZ*
TT		*ABCDEF*

Here is the corrected solution, where the depth of the paragraph in the
middle column is set to zero.

Sample Table 40.10 (Source code).

```
1   $$\vbox{
2                              \tabskip = 0pt
3       \halign{%
4           \hfil\it#\hfil       \tabskip = 15pt&
```

Save the paragraph (column entry) in a box register.

```
5           \setbox 0 = \vtop{%
6               \hsize = 2.5in
7               \raggedright
8               \noindent
9               #
10          }%
```

Set the depth to zero, so the vertical spacing is not destroyed, then print the entry.

```
11              \dp0 = 0pt
12              \box0 &
13              \hfil\it#\hfil        \tabskip = 0pt
14      \cr
15          ABC& This is some text which will be too long to fit
16              on one line. Actually it will fit on four lines,
17              so I chose a text which would do exactly that.& XXX\cr
18          DDD&     \omit& FFF\cr
19          ABCDEF& \omit& XYZXYZXYZ\cr
20          TT&      \omit& ABCDEF\cr
21      }
22  }
23  $$
```

Sample Table 40.10 (Output).

ABC	This is some text which will be too long	*XXX*
DDD	to fit on one line. Actually it will fit on	*FFF*
ABCDEF	four lines, so I chose a text which would	*XYZXYZXYZ*
TT	do exactly that.	*ABCDEF*

40.3.4   Tables Inside Tables

The table of the preceding subsections used \vtop to include a whole paragraph inside a table. Again we will use a vertical box, but this time to enclose another table (leading to tables inside tables) instead of enclosing a paragraph.

Sample Table **40.11** (Source code).

Everything is enclosed in the following vbox.

```
1   $$\vbox{
```

First define a macro \SmallTable, which typesets a small table with two columns and two rows. All elements are set in math mode. This macro has three parameters:

- #1. The element in the first row and the first column.
- #2. The element in the first row and second column, and at the same time the element in the second row and the first column.
- #3. The element in the second row, second column.

```
2       \def\SmallTable #1#2#3{%
3           \vbox{
4               \baselineskip = 20pt
5               \ComputeStrut
6               \offinterlineskip
```

In order to generate a # token inside a macro definition you must use ##. Otherwise TeX expects the # followed by some number specifying a parameter of the macro being defined. Here is the preamble of the table generated by \Small-Table.

```
7                                       \tabskip = 0pt
8           \halign{
9               \MyStrut##&                              % 1
10              \hfil##\hfil          \tabskip = 10pt&%   2
11              \vrule##&                                % 3
12              \hfil##\hfil          \tabskip = 0pt    % 4
13          \cr
14              &   $#1$&    &   $#2$\cr
15                  \noalign{\hrule}
16              &   $#2$&    &   $#3$\cr
17          }
18      }
19  }
```

Now build a larger table that includes this smaller table defined above. Turn off automatic line spacing (the table being generated contains vertical rules).

```
20          \offinterlineskip
```

The outermost table that is currently being built needs no strut: spacing is controlled by the little tables. This table has five columns, and column 1, 3 and 5 contain vertical rules. The other two columns will be filled with "SmallTable."

```
21          \hrule height 2pt
22                                      \tabskip = 0pt
23          \halign{
24              \vrule width 2pt#\relax \tabskip = 10pt&    % 1
```

```
25          \hfil#\hfil&                              % 2
26          \vrule width 3pt#&                        % 3
27          \hfil#\hfil&                              % 4
28          \vrule width 2pt#\relax \tabskip = 0pt    % 5
29      \cr
30          &\SmallTable{A}{B}{C}&&
31           \SmallTable{B}{C}{X}&\cr
32                                  \noalign{\hrule height 2pt}
33          &\SmallTable{A^+}{B^+}{C^+}&&
34           \SmallTable{B^+}{C^+}{X^+}&\cr
35                                  \noalign{\hrule height 2pt}
36          &\SmallTable{A^-}{B^-}{001}&&
37           \SmallTable{00}{100}{001}&\cr
38                                  \noalign{\hrule height 2pt}
39          &\SmallTable{A}{B}{C}&&
40           \SmallTable{B}{C}{X}&\cr
41      }
42      \hrule height 2pt
43  }$$
```

Sample Table 40.11 (Output).

A	B	B	C
B	C	C	X
A^+	B^+	B^+	C^+
B^+	C^+	C^+	X^+
A^-	B^-	00	100
B^-	001	100	001
A	B	B	C
B	C	C	X

40.4   Tables and Paragraphs

Next we will discuss the combination of tables and paragraphs. This time the paragraphs in question will precede a table or be inserted into a table using \noalign; these paragraphs are not columns within the tables.

40.4.1   A Paragraph Followed by a Table

Consider the following problem: assume you would like to place a paragraph on top of a table. This paragraph should have the same width as the table.

The problem here is that you cannot typeset the paragraph *before* you typeset the table, because the table width is unknown at this point in time. Therefore typesetting table and paragraph must be reversed. Here is how we will proceed:

1. Typeset the table *without* printing it. This is done by enclosing the table inside a \vbox and assigning this box to a box register. There are two consequences to this action:

    (a)   The typeset table is available in a box register for later printing.
    (b)   The table's width can be accessed through \wd of the box register in which the table is stored.

2. Print the paragraph. The width as just explained is known.
3. Output the table from the box register.

Here is an example.

Sample Table **40.12** (Source code).

As stated, typeset the table, but do not print it: this is done by storing the table in a box and assigning this box to a box register.

```
 1   \setbox 0 = \vbox{
 2                                          \tabskip = 0pt
 3       \halign{
 4           \hfil#\hfil             \tabskip = 15pt&
 5           \hfil#&
 6           #\hfil&
 7           -#-\hfil                \tabskip = 0pt
 8       \cr
 9           XX&       YY&       Z&            1\cr
10           YY&       XX&       ZZ&           2\cr
11           AA&       BB&       CCC&          3\cr
12           TT&       AA&       TTTT&         4\cr
13           BIGG&     BIGGGGGER&   VERY BIG&     300\cr
14       }
15   }
```

Now start the \vbox that will contain the paragraph followed by the table.

```
16   $$
17   \vbox{
```

Set the width of the paragraph to be printed to the width of the table (that is stored in box register 0):

```
18       \hsize = \wd0
```

Print the paragraph now.

```
19      \noindent
20      Here is a paragraph which goes on top of the table. This
21      could be in real life a table heading or some other
22      information. It does not really matter. It is important
23      though to see that the text is as wide as the table itself.
```

The paragraph was printed: throw in some vertical space and a horizontal line.

```
24      \vskip 5pt
25      \hrule
26      \vskip 5pt
```

Print the table now (simply extract it from box register 0).

```
27      \box0
28   }$$
```

Sample Table 40.12 (Output).

> Here is a paragraph which goes on top of the table. This could be in real life a table heading or some other information. It does not really matter. It is important though to see that the text is as wide as the table itself.

XX	YY	Z	-1-
YY	XX	ZZ	-2-
AA	BB	CCC	-3-
TT	AA	TTTT	-4-
BIGG	BIGGGGGER	VERY BIG	-300-

A paragraph could be added at the end of the preceding table very easily if desired: simply enclose it into the preceding vbox.

40.4.2   Footnotes in Tables

The previous example showed how it may be useful to typeset a table and store it in a box register to have access to the width of the table. This idea is used to typeset footnotes in tables: the text of all table footnotes is first collected using \edef (see 22.1.8, p. III-209, for a more detailed explanation). The idea of the previous Subsection is also applied here: first the table is typeset to establish the width of the table, then the table is printed, and finally the footnotes are printed (at that point in time the required width is known).

The table footnote markers are numbers which will be printed right-justified for proper alignment, if these numbers should increase above 9.

Sample Table **40.13** (Source code).

Load the macro to collect information and initialize this macro.

```
1    \InputD{collect.tip}                    % 22.1.8, p. III-210.
2    \InitialCollectInfo
```

Define macro \TableFootNote with two parameters:

- #1. Footnote number.
- #2. Text of the footnote.

```
3    \def\TableFootNote #1#2{%
```

Print the footnote marker in the table.

```
4        \hbox{$^{#1}$}%
```

Save the footnote text. Make a complete paragraph out of it so it can be printed later.

```
5        \AddInfo{%
6            \leavevmode
7            \hbox to 0pt {%
8                \hss
9                $^{#1}$%
10               \space
11           }%
12           #2%
```

Each footnote is a paragraph.

```
13           \par
14       }%
15   }
```

First step: build the table, collect table footnotes in \Collect.

```
16   \setbox 0 = \vbox{
17                                           \tabskip = 0pt
18       \halign{
19           #\hfil                          \tabskip = 30pt&
20           #\hfil&
21           \hfil :#:\hfil&
22           $a\over #$\relax              \tabskip = 0pt
23   \cr
24       ABC&           BCA&            *&   x^2\cr
25       ABC-1&         BCA&            *&   x^2\cr
26       ABCD\TableFootNote{1}{{\bf D} stands for dubidulo}&
27           XX-X\TableFootNote{2}{{\bf X} stands for eXceptional}&
28                                       132/7\cr
29       ABC-2&         BCA&            *&   x^2\cr
30       ABC-20&        BCA&            *&   x^2\cr
31       ABCD\TableFootNote{3}{{\bf D} stands for dodola}&
32           XX-X\TableFootNote{4}{{\bf X} stands for eXtraordinary}&
```

```
33                          132/7\cr
34       XXX\TableFootNote{5}{Such a table footnote can be pretty long
35           actually, and then you see that the text wraps around. But
36           what else did you expect?}\cr
37       }
38    }
```

Print the table. Set the width of the \vbox enclosing the table to the width as already established, so that the footnote paragraphs have the same width as the table itself.

```
39    $$
40    \vbox{
41        \hsize = \wd0
```

Print table.

```
42        \box0
```

Now print the footnotes. First some space between table and footnotes.

```
43        \vskip 5pt
```

Set the layout for printing the footnotes.

```
44        \parindent = 0pt
45        \parskip = 5pt
46        \leftskip = 10pt
```

Print the footnote text.

```
47        \Collect
48    }
49    $$
```

Sample Table 40.13 (Output).

ABC	BCA	:*:	$\frac{a}{x^2}$
ABC-1	BCA	:*:	$\frac{a}{x^2}$
ABCD[1]	XX-X[2]	:132/7:	
ABC-2	BCA	:*:	$\frac{a}{x^2}$
ABC-20	BCA	:*:	$\frac{a}{x^2}$
ABCD[3]	XX-X[4]	:132/7:	
XXX[5]			

[1] **D** stands for dubidulo

[2] **X** stands for eXceptional

[3] **D** stands for dodola

[4] **X** stands for eXtraordinary

[5] Such a table footnote can be pretty long actually, and then you see that the text wraps around. But what else did you expect?

To illustrate that the footnotes were only *collected* when the table was typeset and that they were actually typeset later, I will print the contents of \Collect using \meaning. The input

```
1   \tt\raggedright
2   \meaning\Collect
```

generates the following text:

```
macro:->\unhbox \voidb@x \hbox to 0pt {\hss $^{1}$ }{\fam = \bffam
\Xbf D} stands for dubidulo\par \unhbox \voidb@x \hbox to 0pt {\hss
$^{2}$ }{\fam = \bffam \Xbf X} stands for eXceptional\par \unhbox
\voidb@x \hbox to 0pt {\hss $^{3}$ }{\fam = \bffam \Xbf D} stands
for dodola\par \unhbox \voidb@x \hbox to 0pt {\hss $^{4}$ }{\fam =
\bffam \Xbf X} stands for eXtraordinary\par \unhbox \voidb@x \hbox to
0pt {\hss $^{5}$ }Such a table footnote can be pretty long actually,
and then you see that the text wraps around. But what else did you
expect?\par
```

It obviously would be rather simple to number the table footnotes automatically by the introduction of a counter register.

40.4.3   Paragraphs Between Table Rows

We will continue using the idea of the preceding subsections. A table with paragraphs *between* the rows of the table will be presented now. The paragraphs are supposed to be as wide as the table itself. Again, it is necessary to first determine the width of the table (without the paragraphs between the table rows) before the table is typeset with the paragraphs.

Sample Table 40.14 (Source code).

This time it is not possible to simply typeset the table and to extract the table from a box register. This time the table must be set *twice*. Therefore, first a macro, containing the complete table including the \noaligns containing the paragraphs, is defined.

```
1   \def\MyTable{
2                                           \tabskip = 0pt
3       \halign{
4           \hfil##\hfil            \tabskip = 15pt&
5           \hfil##&
6           ##\hfil&
7           -##-\hfil               \tabskip = 0pt
8       \cr
```

The first paragraph, preceding the table itself, is built now (all remaining paragraphs are built the same way). The settings of \hsize inside the \noaligns will be explained momentarily.

```
9                       \noalign{
```

```
10                          \vskip 5pt  \hrule  \vskip 5pt
11                          \vbox{
12                              \hsize = \wd0
13                              \noindent
14                              This is a little paragraph, which goes on
15                              top of the table. Short and easy. Just
16                              so we will have two lines or so.
17                          }
18                          \vskip 5pt  \hrule  \vskip 5pt
19                      }
```

Here are the first three rows of the table.

```
20          XX&     YY&     Z&          1\cr
21          YY&     XX&     ZZ&         2\cr
22          AA&     BB&     CCC&        3\cr
```

Second paragraph.

```
23                  \noalign{
24                      \vskip 5pt  \hrule  \vskip 5pt
25                      \vbox{
26                          \hsize = \wd0
27                          \noindent
28                          This is a little paragraph, which goes in
29                          the middle of the table. Short and easy. Just
30                          so we will have two lines or so.
31                      }
32                      \vskip 5pt  \hrule  \vskip 5pt
33                  }
```

Here are the next two rows.

```
34          TT&     AA&         TTTT&       4\cr
35          BIGG&   BIGGGGGER&  VERY BIG&   300\cr
```

Here is the third and last paragraph.

```
36                  \noalign{
37                      \vskip 5pt  \hrule  \vskip 5pt
38                      \vbox{
39                          \hsize = \wd0
40                          \noindent
41                          This is a little paragraph, which goes at
42                          the end of the table. Short and easy. Just
43                          so we have two or three lines. By the way the
44                          table is \the\wd0 \space wide.
45                      }
46                      \vskip 5pt  \hrule  \vskip 5pt
47                  }
48          }
49      }
```

Now typeset the table that all \noaligns disabled. Originally \noalign is a primitive with one argument. Now \noalign is being redefined as a macro with one argument, where the macro simply discards the argument (and also otherwise does not generate any further tokens; this is called *gobbling* in T_EX; see 21.8.6, p. III-186). This redefinition is local, because it occurs inside a vbox.

```
50   \setbox 0 = \vbox{
51       \def\noalign #1{}
52       \MyTable
53   }
```

Finally print the table including the paragraphs. All \noaligns act normally now. Note that now the \wd0s in the \noaligns of the preceding table expand to the width of the table, because the table was once typeset before, without the intervening paragraphs, and the table was assigned to box register 0.

```
54   $$
55       \vbox{\MyTable}
56   $$
```

Sample Table **40.14** (Output).

This is a little paragraph, which goes on top of the table. Short and easy. Just so we will have two lines or so.			
XX	YY	Z	-1-
YY	XX	ZZ	-2-
AA	BB	CCC	-3-
This is a little paragraph, which goes in the middle of the table. Short and easy. Just so we will have two lines or so.			
TT	AA	TTTT	-4-
BIGG	BIGGGGGER	VERY BIG	-300-
This is a little paragraph, which goes at the end of the table. Short and easy. Just so we have two or three lines. By the way the table is 207.36134pt wide.			

The above problem could also be solved by first typesetting the whole table (without paragraphs), and then using \vsplit to break the table into partial tables which then can be combined with the paragraphs.

40.4.4   Computing the Width of a Paragraph Table Column

In the preceding example where text in a paragraph fashion as a column entry
was included, we had to explicitly specify the width of the paragraph and assign
it to `\hsize`. Let me present the a problem where this width is computed automatically. Assume that with the exception of the last column all other columns
are regular table columns, i.e., columns which do *not* contain paragraph structured texts. For these columns the concept of the widest entry determining the
width of a column is therefore still valid. We would like to typeset the table in
such a way that the last column, the only column with text in a paragraph-like
fashion, has such width that the width of the whole table is `\hsize`.

Again an idea we used before will be applied now: the table is typeset twice,
the first time only to determine the width of the table *without* the last column.
Then the width of the last column is computed and the table is typeset again,
this time obviously including the last column.

Here is the example:

Sample Table 40.15 (Source code).

First define a macro `\MyTable`, which expands to the table, because the table
must be processed twice.

```
 1   \def\MyTable{%
 2                                    \tabskip = 0pt
 3       \halign{
 4           \vrule width 2pt##\relax     \tabskip = 15pt&     % 1
 5           ##\hfil&                                          % 2
 6           \hfil##\hfil&                                     % 3
 7           \vrule width 3pt##&                               % 4
```

Observe that for the last text column a macro is defined so that the template
of this column can be modified easily (the term "last") refers to what appears
to be the last column when the table is printed (column five), rather than what
TeX thinks is the last column, which is column six with the vertical rule in it.

```
 8           \TextColumn{##}&                                  % 5
 9           \vrule width 2pt##\relax     \tabskip = 0pt       % 6
10       \cr
11           & XXX & YYYYYYYY &&
12           This is just some text to show how you can
13           typeset a paragraph inside a table. Observe that we use
14           a ragged-right type of text. Let us see how it comes out. &
15       \cr                                          \ncalign{\hrule}
16           & XXXXX & YYYY &&
17           This is just some text to show how you can
18           typeset a paragraph inside a table. Observe that we use
19           a ragged-right type of text. Let us see how it comes out. &
20       \cr
21       }
22   }
```

Typeset the table without the last column. The template of the last column is defined to throw away its entry. Using the gobbling macros of 21.8.6, p. III-186, \TextColumn could also have been defined as \let\TextColumn = \GobbleOne.

```
23    \def\TextColumn #1{}
24    \setbox 0 = \vbox{
25        \MyTable
26    }
```

Now \wd0 contains the width of the table without the last column. Compute the remaining width (\hsize − width of table) and store this value in \ColumnWidth.

```
27    \newdimen\ColumnWidth
28    \ColumnWidth = \hsize
29    \advance\ColumnWidth by -\wd0
```

Define a new macro \TextColumn, the new template of the last column. This time the last column will be actually typeset the desired way.

```
30    \def\TextColumn #1{%
31        \vtop{%
32            \hsize = \ColumnWidth
33            \normalbaselines
34            \raggedright
35            \vrule height 10pt depth 0pt width 0pt
36            #1%
37            \vrule height  0pt depth 6pt width 0pt
38        }
39    }
```

Finally print the complete table.

```
40    \vbox{
41        \hrule
42        \MyTable
43        \hrule
44    }
```

Sample Table 40.15 (Output).

XXX YYYYYYYY	This is just some text to show how you can typeset a paragraph inside a table. Observe that we use a ragged-right type of text. Let us see how it comes out.
XXXXX YYYY	This is just some text to show how you can typeset a paragraph inside a table. Observe that we use a ragged-right type of text. Let us see how it comes out.

40.5   Tables Can Run Across Multiple Pages

If a table is not enclosed inside a \vbox, then the page-breaking algorithm will break the table because the table is nothing else than a sequence of \hboxes with interline glue inserted in between.

Here is the example of a table running across two pages. The change of \tabskip is enclosed in a group to localize the effect of the tab_0 setting.

Sample Table **40.16** (Source code).

```
 1  {
 2                                          \tabskip = 0pt plus 1fil
 3  \halign to \hsize{
 4       \tt#\hfil                          \tabskip = 15pt&
 5       \bf#\hfil&
 6       \it#\hfil                          \tabskip = 0pt plus 1fil
 7  \cr
 8       X&        Y&        Z\cr
 9       XX&       YY&       ZZ\cr
10       XXX&      YYY&      ZZZ\cr
11       XXXX&     YYYY&     ZZZZ\cr
12       XXXXX&    YYYYY&    ZZZZZ\cr
13       XXXX&     YYYY&     ZZZZ\cr
14       XXX&      YYY&      ZZZ\cr
15       XXXX&     YYYY&     ZZZZ\cr
16       XXXXX&    YYYYY&    ZZZZZ\cr
17       XXXX&     YYYY&     ZZZZ\cr
18       XXX&      YYY&      ZZZ\cr
19       XX&       YY&       ZZ\cr
20       X&        Y&        Z\cr
21       X&        Y&        Z\cr
22       XX&       YY&       ZZ\cr
23       XXX&      YYY&      ZZZ\cr
24       XXXX&     YYYY&     ZZZZ\cr
25       XXXXX&    YYYYY&    ZZZZZ\cr
26       XXXX&     YYYY&     ZZZZ\cr
27       XXX&      YYY&      ZZZ\cr
28       XX&       YY&       ZZ\cr
29       X&        Y&        Z\cr
30       X&        Y&        Z\cr
31       XX&       YY&       ZZ\cr
32       XXX&      YYY&      ZZZ\cr
33       XXXX&     YYYY&     ZZZZ\cr
34       XXXXX&    YYYYY&    ZZZZZ\cr
35       XXXX&     YYYY&     ZZZZ\cr
36       XXX&      YYY&      ZZZ\cr
37       XX&       YY&       ZZ\cr
38       X&        Y&        Z\cr
39       X&        Y&        Z\cr
40       XX&       YY&       ZZ\cr
41       XXX&      YYY&      ZZZ\cr
```

```
42    XXXX&    YYYY&    ZZZZ\cr
43    XXXXX&   YYYYY&   ZZZZZ\cr
44    XXXX&    YYYY&    ZZZZ\cr
45    XXX&     YYY&     ZZZ\cr
46    XX&      YY&      ZZ\cr
47    X&       Y&       Z\cr
48    X&       Y&       Z\cr
49    XX&      YY&      ZZ\cr
50    XXX&     YYY&     ZZZ\cr
51    XXXX&    YYYY&    ZZZZ\cr
52    XXXXX&   YYYYY&   ZZZZZ\cr
53    XXXX&    YYYY&    ZZZZ\cr
54    XXX&     YYY&     ZZZ\cr
55    XX&      YY&      ZZ\cr
56    X&       Y&       Z\cr
57    X&       Y&       Z\cr
58    XX&      YY&      ZZ\cr
59    XXX&     YYY&     ZZZ\cr
60    XXXX&    YYYY&    ZZZZ\cr
61    XXXXX&   YYYYY&   ZZZZZ\cr
62    XXXX&    YYYY&    ZZZZ\cr
63    XXX&     YYY&     ZZZ\cr
64    XX&      YY&      ZZ\cr
65    X&       Y&       Z\cr
66    X&       Y&       Z\cr
67    XX&      YY&      ZZ\cr
68    XXX&     YYY&     ZZZ\cr
69    XXXX&    YYYY&    ZZZZ\cr
70    XXXXX&   YYYYY&   ZZZZZ\cr
71    XXXX&    YYYY&    ZZZZ\cr
72    XXX&     YYY&     ZZZ\cr
73    XX&      YY&      ZZ\cr
74    X&       Y&       Z\cr
75    X&       Y&       Z\cr
76    XX&      YY&      ZZ\cr
77    XXX&     YYY&     ZZZ\cr
78    XXXX&    YYYY&    ZZZZ\cr
79    XXXXX&   YYYYY&   ZZZZZ\cr
80    XXXX&    YYYY&    ZZZZ\cr
81    XXX&     YYY&     ZZZ\cr
82    XX&      YY&      ZZ\cr
83    X&       Y&       Z\cr
84  }
85  }
```

Sample Table **40.16** (Output).

X	Y	Z
XX	YY	ZZ
XXX	YYY	ZZZ
XXXX	YYYY	ZZZZ
XXXXX	YYYYY	ZZZZZ

XXXX	YYYY	ZZZZ
XXX	YYY	ZZZ
XXXX	YYYY	ZZZZ
XXXXX	YYYYY	ZZZZZ
XXXX	YYYY	ZZZZ
XXX	YYY	ZZZ
XX	YY	ZZ
X	Y	Z
X	Y	Z
XX	YY	ZZ
XXX	YYY	ZZZ
XXXX	YYYY	ZZZZ
XXXXX	YYYYY	ZZZZZ
XXXX	YYYY	ZZZZ
XXX	YYY	ZZZ
XX	YY	ZZ
X	Y	Z
X	Y	Z
XX	YY	ZZ
XXX	YYY	ZZZ
XXXX	YYYY	ZZZZ
XXXXX	YYYYY	ZZZZZ
XXXX	YYYY	ZZZZ
XXX	YYY	ZZZ
XX	YY	ZZ
X	Y	Z
X	Y	Z
XX	YY	ZZ
XXX	YYY	ZZZ
XXXX	YYYY	ZZZZ
XXXXX	YYYYY	ZZZZZ
XXXX	YYYY	ZZZZ
XXX	YYY	ZZZ
XX	YY	ZZ
X	Y	Z
X	Y	Z
XX	YY	ZZ
XXX	YYY	ZZZ
XXXX	YYYY	ZZZZ
XXXXX	YYYYY	ZZZZZ
XXXX	YYYY	ZZZZ
XXX	YYY	ZZZ
XX	YY	ZZ
X	Y	Z
X	Y	Z
XX	YY	ZZ
XXX	YYY	ZZZ

XXXX	YYYY	ZZZZ
XXXXX	YYYYY	ZZZZZ
XXXX	YYYY	ZZZZ
XXX	YYY	ZZZ
XX	YY	ZZ
X	Y	Z
X	Y	Z
XX	YY	ZZ
XXX	YYY	ZZZ
XXXX	YYYY	ZZZZ
XXXXX	YYYYY	ZZZZZ
XXXX	YYYY	ZZZZ
XXX	YYY	ZZZ
XX	YY	ZZ
X	Y	Z
X	Y	Z
XX	YY	ZZ
XXX	YYY	ZZZ
XXXX	YYYY	ZZZZ
XXXXX	YYYYY	ZZZZZ
XXXX	YYYY	ZZZZ
XXX	YYY	ZZZ
XX	YY	ZZ
X	Y	Z

There is of course a limit with respect to the maximum length a table built with \halign can have: TeX has to read in the whole table to determine the widest entry before it can start writing the table out.

Normally there is no problem with this limitation: the version of TeX that I used to write this series, would report an error in the case of the above table only if this table had more than 680 lines (this amounts to a table running across more than 14 pages!).

40.6   Macros to Build Tables and Preambles

I will now discuss issues that come up when you try to define macros to simplify the making of tables. It is possible to separate the specification of a preamble, the table entries and the specification of the end of a table.

40.6.1   Beginning and Ending Tables using \bgroup and \egroup

The curly braces of an \halign enclosing a table can be replaced by \bgroup
and \egroup. Observe also that the curly braces of a \vbox can be replaced by
\bgroup and \egroup. Here is a simple example table to show exactly that:

Sample Table 40.17 (Source code).

```
 1  $$\vbox\bgroup                          \tabskip = 15pt
 2      \halign\bgroup
 3        \hfil#\hfil&
 4        \hfil#&
 5        #\hfil&
 6        -#-\hfil
 7      \cr
 8        XX&       YY&       Z&            1\cr
 9        YY&       XX&       ZZ&           2\cr
10        AA&       BB&       CCC&          3\cr
11        TT&       AA&       TTTT&         4\cr
12        BIGG&     BIGGGGGER&   VERY BIG&300\cr
13      \egroup                            % Used \egroup instead of '}'.
14  \egroup                                % Used \egroup instead of '}'.
15  $$
```

Sample Table 40.17 (Output).

XX		YY	Z	-1-
YY		XX	ZZ	-2-
AA		BB	CCC	-3-
TT		AA	TTTT	-4-
BIGG	BIGGGGGER		VERY BIG	-300-

40.6.2   Macros Defining a Preamble

The previous Subsection suggests that it must be possible to write macros generating a complete preamble. Here is an example.

Sample Table 40.18 (Source code).

First define macro \MyTablePreamble.

```
 1  \def\MyTablePreamble{%
 2                                          \tabskip = 0pt
 3      \halign\bgroup
```

To generate a # token in a macro definition you must write ##.

```
 4        \hfil##\hfil              \tabskip = 15pt&
 5        \hfil##&
 6        ##\hfil&
```

```
7              -##-\hfil                              \tabskip = 0pt
8          \cr
9      }
```

Now use this macro to generate a preamble and to print a table.

```
10   $$
11   \vbox{
12       \MyTablePreamble
13          XX&      YY&      Z&           1\cr
14          YY&      XX&      ZZ&          2\cr
15          AA&      BB&      CCC&         3\cr
16          TT&      AA&      TTTT&        4\cr
17          BIGG&    BIGGGGGER&   VERY BIG&300\cr
```

Terminate the \halign of \MyTablePreamble.

```
18       \egroup
19   }$$
```

Sample Table **40.18** (Output).

XX	YY	Z	-1-
YY	XX	ZZ	-2-
AA	BB	CCC	-3-
TT	AA	TTTT	-4-
BIGG	BIGGGGGER	VERY BIG	-300-

Here is the example of a preamble-generating macro \MyTablePreambleA that has two parameters.

- #1. This parameter specifies some text inserted before each entry of the first column.
- #2. This parameter specifies some text inserted after each entry of the first column.

Sample Table **40.19** (Source code).

Start with the macro definition.

```
1   \def\MyTablePreambleA #1#2{
2                                       \tabskip = 15pt
3       \halign\bgroup
```

Read the next line carefully: the first template written as a list of tokens reads as follows: \hfil • #1 • ## • #2 • \hfil.

```
4          \hfil#1###2\hfil&              % 1
5          \hfil##&                       % 2
6          ##\hfil&                       % 3
7          -##-\hfil                      % 4
8      \cr
9   }
```

Use this macro and print a table.

```
10    $$
11    \vbox{
12        \MyTablePreambleA{::}{++}
13            XX&      YY&      Z&          1\cr
14            YY&      XX&      ZZ&         2\cr
15            AA&      BB&      CCC&        3\cr
16            TT&      AA&      TTTT&       4\cr
17            BIGG&    BIGGGGGER&   VERY BIG&300\cr
```

Terminate the \halign of \MyTablePreambleA.

```
18        \egroup
19    }$$
```

Sample Table **40.19** (Output).

::XX++		YY	Z	-1-
::YY++		XX	ZZ	-2-
::AA++		BB	CCC	-3-
::TT++		AA	TTTT	-4-
::BIGG++	BIGGGGGER	VERY BIG		-300-

40.6.3   Macros for Templates

Now I will discuss using macros for *defining a template*. For instance, it would
be nice if you could write \TCen (template, centered) instead of \hfil#\hfil,
\TRight (template, right-justified) for \hfil# and \TLeft (template, left justified) for #\hfil.

When TeX reads in the preamble, it is read in *without* expansion, i.e., any
macros calls contained in the templates are not expanded at this point in time
(the same way no tokens in the replacement text of a macro are expanded if
this macro is being defined using \def). The only changes TeX will do to the
template is to save and then remove all \tabskip specifications, because these
specifications really have nothing to do with the templates itself. There is one
exception though, and that is that TeX will expand the token following a \span
in a preamble. This will be discussed shortly.

The question therefore is how we can define and then later make TeX recognize the desired macros \TCen, \TLeft and \TRight to build a preamble at
the time the preamble is read in, and not at the time when the templates are
expanded because table entries are substituted. In other words, after the following definitions (again the double ## stands for a simple # token inside a macro
definition):

```
1    \def\TCen{\hfil##\hfil}
2    \def\TLeft{##\hfil}
3    \def\TRight{\hfil##}
```

it is *not* possible to typeset a table as follows:

```
1   \halign{
2       \TCen&              % 1
3       \TLeft&             % 2
4       \TRight             % 3
5   \cr
6       ...
7   }
```

There are two solutions to make TeX recognize macros in a template when the table is read in:

1. Force the expansion using `\edef` as discussed in the next Subsection.
2. Force the expansion inside the template using `\span` (see 40.6.5, p. 330, for details).

40.6.4   Template Macros, Forced Expansion Using \edef

First define the desired macros `\TCenX`, `\TLeftX` and `\TRightX` as follows (the definitions are different from the preceding definitions of `\TCen`, etc.):

```
1   \def\TCenX{\hfil####\hfil}
2   \def\TLeftX{####\hfil}
3   \def\TRightX{\hfil####}
```

Now define `\TableBegin` as follows (because `\edef` is used, the expansion of `\TCen` etc., takes place now):

Sample Table **40.20** (Source code).

```
1   \edef\TableBegin{%
2                                   \tabskip = 10pt
3       \halign\bgroup
4           \TCenX&                 % 1
5           \TRightX&               % 2
6           \TLeftX&                % 3
7           \TLeftX                 % 4
8       \cr
9   }
```

Before we continue let us reprint the replacement text of `\TableBegin` (this output will not appear until we come to the output part of this table):

```
10  {
11      \raggedright
12      \rightskip = 0pt plus 60pt
13      Here is the token list of \verb+\TableBegin+:
14      \tt
15      \meaning\TableBegin
16      \par
17  }
```

Here is the example of a table using the previously defined macro. Observe that if the macro \TableBegin contained calls to other macros, that are supposed to be expanded when the templates are expanded, then these macro calls must be "protected" by \noexpand so that the macro call is not expanded at the time of the \edef-based definition of \TableBegin.

```
18   $$\vbox{
19                                         \tabskip = 10pt
20       \TableBegin
21          XX&      YY&      Z&           1\cr
22          YY&      XX&      ZZ&          2\cr
23          AA&      BB&      CCC&         3\cr
24          TT&      AA&      TTTT&        4\cr
25          BIGG&    BIGGGGGER&   VERY BIG&300\cr
```

Terminate the \halign\bgroup of \TableBegin.

```
26       \egroup
27   }$$
```

Sample Table **40.20** (Output).

Here is the token list of \TableBegin: macro:->\tabskip = 10pt \halign \bgroup \hfil ##\hfil & \hfil ##& ##\hfil & ##\hfil \cr

XX	YY	Z	1
YY	XX	ZZ	2
AA	BB	CCC	3
TT	AA	TTTT	4
BIGG	BIGGGGGER	VERY BIG	300

Assume now that *only* the preamble itself is contained in a macro, excluding the \halign\bgroup which begins the table. Here is how this case can be handled. The \expandafter forced the preamble macro to be expanded before \hb, ultimately \halign\bgroup is inserted into the main token list.

Sample Table **40.21** (Source code).

```
1    \edef\MyPreamble{%
2        \TCenX&
3        \TRightX&
4        \TLeftX&
5        \TLeftX
6        \cr
7    }
8    $$\vbox{
9        \def\hb{\halign\bgroup}
10                                       \tabskip = 10pt
11       \expandafter\hb\MyPreamble
12          XX&      YY&      Z&          1\cr
13          YY&      XX&      ZZ&         2\cr
14          AA&      BB&      CCC&        3\cr
15          TT&      AA&      TTTT&       4\cr
16          BIGG&    BIGGGGGER&   VERY BIG&300\cr
```

Closes `\halign\bgroup` of `\MyPreamble`.

```
17        \egroup
18    }$$
```

Sample Table 40.21 (Output).

XX		YY Z	1
YY		XX ZZ	2
AA		BB CCC	3
TT		AA TTTT	4
BIGG	BIGGGGGER	VERY BIG	300

40.6.5   Template Macros, Forced Expansion Using `\span`

There is a way in TEX to force the expansion of tokens inside templates at the time when the template is read in: a token in the template of a table is expanded if it is preceded by `\span` (this application of `\span`, by the way, has absolutely nothing to do with the application of the same instruction in 39.3.2, p. 264).

In other words, the token following `\span` (and only that token) in a template is expanded when the template is read in. Here is an example (based on the original `\TLeft`, `\TCen` and `\TRight`):

Sample Table 40.22 (Source code).

Define the three template macros first.

```
1    \def\TLeft{##\hfil}
2    \def\TCen{\hfil##\hfil}
3    \def\TRight{\hfil##}
```

Now typeset the table.

```
4    $$
5    \vbox{
6                        \tabskip = 20pt
7    \halign{
8        \tt\span\TCen&              % \tt NOT expanded right away!
9        \span\TLeft&
10       \span\TRight
11   \cr
12       ABC&        DEF&        KLM\cr
13       12345&      54321&      XX YY ZZ\cr
14       12345 678&  A1 B2 C3 D4&  XX YY ZZ KK LL\cr
15   }}$$
```

Sample Table **40.22** (Output).

ABC	DEF	KLM
12345	54321	XX YY ZZ
12345 678	A1 B2 C3 D4	XX YY ZZ KK LL

40.6.6 Changing Templates

Let me quickly point out that while it is not directly possible to *change* templates in the middle of a table (and in that sense the title of this Subsection is misleading). The desired effect can be achieved though by having the template invoke a macro, and then change the definition of that macro from inside the table.

I mention this here because it reveals an interesting timing issue. You should compare the following discussion with the discussion in 38.6.2, p 245.

Sample Table **40.23** (Source code).

```
 1  $$
 2  \vbox{
 3                                       \tabskip = 20pt
```

At first both columns of the following table should generate centered entries.

```
 4      \def\ColumnOne #1{\hfil#1\hfil}
 5      \def\ColumnTwo #1{\hfil#1\hfil}
 6      \halign{
 7          \ColumnOne{#}&
 8          \ColumnTwo{#}%
 9      \cr
10          XXX&                YY\cr
11          XXXXXX&             YYYYYYY\cr
12          X&                  YYYY\cr
13                                       \noalign{\bigskip}
```

Now redefine the templates, so that entries in the first column are right-justified, and those of the second column are left-justified. If you look at the output you will discover that the next entry coming (first entry in the first column) is still centered and not yet right-justified. Obviously somehow the new template macro was not used yet. This will be explained shortly.

```
14      \gdef\ColumnOne #1{\hfil #1}%       % 1st redefinition.
15      \gdef\ColumnTwo #1{#1\hfil}%
16          XXX&                YY\cr
17          XXXXXX&             YYYYYYY\cr
18          X&                  YYYY%
```

Now template changes are done correctly: the template macros are redefined before the next row that is supposed to refer to the redefined macros is started.

```
19      \gdef\ColumnOne #1{#1\hfil}%        % 2nd redefinition.
```

```
20    \gdef\ColumnTwo #1{\hfil#1}%
21                    \cr
22                                    \noalign{\bigskip}
23        XXX&            YY\cr
24        XXXXXX&         YYYYYYY\cr
25        X&             YYYY\cr
26    }}$$
```

Sample Table **40.23** (Output).

XXX	YY
XXXXXX	YYYYYYY
X	YYYY
XXX	YY
XXXXXX	YYYYYYY
X	YYYY
XXX	YY
XXXXXX	YYYYYYY
X	YYYY

What is wrong with the first redefinition? When TeX starts to parse a new table entry it does so expanding tokens to determine whether the first token is an \omit or not (space tokens during this process are skipped). If it is *not* an \omit (\gdef) the template as defined is inserted. This means in our case that the old definition of \ColumnOne is expanded, *then* redefined, but this redefinition is too late to have any effect on the current entry.

In the second approach the redefinition of the template macro was made part of the preceding row so that the redefined template macro would already be present when the entry in the first column of the next row is processed.

40.6.7   A Forgotten \cr Can Be Inserted Automatically, \crcr

We have seen that it is possible to define macros expanding into the preamble of a table. It is also possible to define macros "closing tables." Let me look at those macros now.

When you write a macro that is supposed to *end* a table, then you would like to be able to silently insert a \cr in case the user has forgotten the closing \cr of the last row. On the other hand, nothing should happen if the closing \cr was already entered by the user. Therefore it is wrong to simply begin a table ending macro with \cr, because if the user has already provided the \cr, then two \crs would result (normally causing an empty row to be added to the bottom of the table).

TEX has the `\crcr` primitive which solves the previous problem: in case `\crcr` follows a `\cr` or a `\noalign` it does nothing. Otherwise `\crcr` is equivalent to `\cr`. Almost all table-ending macros therefore begin with `\crcr`. Here is the example of such a macro (assume the table was started with `\halign\bgroup`):

```
1    \def\TableEnd{%
2         \crcr
3         \egroup
4    }
```

Naturally `\everycr` is evaluated in case `\crcr` had the effect of a `\cr` and was not ignored; see 41.8, p. 373.

There is one more additional rather peculiar detail to the application of `\crcr`. First note that `\crcr` can also be used to end a preamble. This is somehow strange in the sense that there is obviously nothing optional about the requirement of ending a preamble: a preamble must be ended explicitly, and the question is why one may want to use `\crcr` instead of `\cr`. The reason for this is very simple: in a few instances it might be that `\cr` has been redefined. And in that case, instead of preceding the redefinition of `\cr` by a `\let\@cr = \cr` (or something like that) and ending the preamble using `\@cr`, one simply uses `\crcr`. The point here is that in that case `\crcr` is not really used with its original intention in which which is "insert a `\cr` if there was not yet a `\cr` inserted preceding it." Instead `\crcr` is used in place of `\cr`, because `\cr` was redefined.

40.6.8   Macros to Begin and End Centered Tables

What we have learned so far allows us to write macros to begin and end centered tables. Assume you would like to define macros to begin and end centered tables, where these tables are centered by using display math mode. Here is how it can be done:

$$\mathcal{P}'\qquad\bullet\ \texttt{tabcent.tip}\ \bullet$$

```
15    \def\TableBeginCentered{%
16         $$
17             \vbox\bgroup
18                 \offinterlineskip
19                              \tabskip = 0pt
20             \halign\bgroup
21    }
```

And here is the definition of a macro ending such a table.

```
22    \def\TableEndCentered{%
```

The `\crcr` inserts a `\cr`, in case the user forgot one in the last row.

```
23                 \crcr
```

Terminate the `\halign\bgroup`.

```
24                 \egroup
```

Terminate the `\vbox\bgroup`.

```
25            \egroup
26        $$
27    }
```

• End of <code>tabcent.tip</code> •

The above macros simplify the following table source code from

```
1    $$
2    \vbox{
3        \halign{
4            #\hfil          \tabskip = 10pt&
5            #\hfil          \tabskip = 0pt
6        \cr
7            ABC&     DEF\cr
8            XX XX&   YY YY YY\cr
9        }
10    }
11    $$
```

to

```
1    \TableBeginCentered
2            #\hfil          \tabskip = 10pt&
3            #\hfil          \tabskip = 0pt
4        \cr
5            ABC&     DEF\cr
6            XX XX&   YY YY YY\cr
7    \TableEndCentered
```

Note that the `\cr` of line 6 is not required because of the `\crcr` of `\TableEnd-Centered`.

40.7   Summary

In this chapter we learned:

- Tables can be centered in various ways. The most frequently-used method is to enclose the table inside a `\vbox`, which is centered using display math mode.
- Techniques for the typesetting of sparse tables were demonstrated. It is possible to use a macro to advance by a certain number of columns and to name columns. It is also possible to use the idea of named columns and to therefore define column entries by names.
- Using vboxes as table entries allows the inclusion of complete paragraphs in tables, the inclusion of other tables as table entries, and the usage of display math mode equations as table entries.

- There are a variety of problems in typesetting tables that can be solved by typesetting the table enclosed in a vbox and assigning this vbox to a box register that allows the determination of the width of the table. Among those problems are printing a paragraph of the width of a table preceding the table, and printing the footnotes in tables.
- If a table is not enclosed inside a vbox and exceeds one page in length, it will break automatically, as does regular text.
- Instead of the curly braces of \halign, \bgroup and \egroup can be used because it is possible to include the preamble into a macro, which is invoked before the table body is specified. In such macros the # of templates must be specified as ##. \span can be used to force the expansion of the token following it when the preamble is parsed, rather than when the table entry is substituted.
- \crcr acts like \cr if there is an open row, that is, if there is a row which can be terminated. Otherwise it does not do anything.

41
Still Tables . . .

This chapter ends the discussion of tables. We will look at numerical computations in tables, splitting of tables with \vsplit, the vertical counter part of \halign called \valign (you probably had guessed that one right). We will discuss "double tables," expansion issues and \everycr.

As mentioned in 38.1, p. 199, you will find a couple of pages in this chapter with some rather poor page breaks. We apologize.

41.1   Numerical Computations in Tables

The following table has some numerical computations in it and shows how flexible TeX really is.

Teaching TeX for the TeX Users Group I had to submit expenses bills after each trip. I organized these bills as a table. After I had entered all expenses in such a table I then would take out my pocket calculator to determine the sum of all expenses.

After "being caught" (having made an error in my favor of course), I decided to let TeX do the work (it is therefore natural to dedicate this section to Ray Goucher, the former executive director of the TeX Users Group); it was rather inefficient anyway to type in the same numbers twice.

To do this type of arithmetic (TeX *cannot* handle floating point numbers) requires TeX to perform all arithmetic using integers, i.e., the computations are done in cents, rather than dollars. Before printing such an amount in cents it is "converted back into dollars" using the macro \PrintInDollar. The conversion from dollars to cents is done easily by telling TeX to ignore the decimal point (category code 9 is assigned to the period). This requires you to enter an amount like $5.00 as 5.00 and *not* 5; see 18.1.7, p. III-8, item 9.

Sample Table **41.1** (Source code).

```
 1   \InputD{pdollars.tip}                          % 25.1.13, p. III-331.
```

\TLineNumber is used to number the table lines, \ColumnSum to add the values in the last column, and \ThisAmount is used to store the current entry in the third column.

```
 2   \newcount\TLineNumber
 3   \TLineNumber = 1
 4   \newcount\ColumnSum
 5   \ColumnSum = 0
 6   \newcount\ThisAmount
```

Print the table as usual (centered, using display math mode, enclosed inside a \vbox):

```
 7   $$\vbox{
 8                                                  \tabskip = 0pt
 9   \halign{
```

The first column is a column with no table entries, but the template generates a number to number the lines of the table.

```
10       \hfil\the\TLineNumber.%
11           \global\advance\TLineNumber by 1 #%
12                                                  \tabskip = 5pt&
```

The second column contains the information about what the amount is all about.

```
13       #\hfil                                     \tabskip = 15pt&
```

The third column contains the dollar amount which must be added up. The period is made an ignored character (see above for an explanation). Then the amount is read in and \PrintInDollar is used to print the amount.

```
14       \catcode`\. = 9
15       \ThisAmount = #%
16       \global\advance\ColumnSum by \ThisAmount
17       \hfil\tt\PrintInDollar{\ThisAmount}%
18                                                  \tabskip = 0pt
19   \cr
```

The preamble is over, the table body begins now.

```
20       & Hotel                           & 245.98\cr
21       & Airline fare                    & 135.09\cr
22       & Long distance phone call        &   2.00\cr
23       & Local phone call                &   0.20\cr
24       & Rental car                      & 137.23\cr
25       & Credit travel advance           &-150.00\cr
26                                         \noalign{\medskip}
```

In the last table row print the sum of the last column.

```
27   \omit&\omit\hfil\bf SUM               &
28                   \omit\hfil\bf\$ \PrintInDollar{\ColumnSum}\cr
29   }}$$
```

Sample Table 41.1 (Output).

1.	Hotel	245.98
2.	Airline fare	135.09
3.	Long distance phone call	2.00
4.	Local phone call	0.20
5.	Rental car	137.23
6.	Credit travel advance	−150.00
	SUM	**$ 370.50**

41.2   Parentheses Around and Within Tables

Now let me show some tables with delimiters around the tables and parenthesis inside the table, spanning multiple columns. Whenever we use \left and \right below, other delimiters than the one shown can be used, of course.

41.2.1   Parenthesis Around a Table

To print a table surrounded by parenthesis, curly braces or other types of delimiters, use the appropriate delimiters in math mode, and center the table horizontally by enclosing it inside a \vcenter. In the following example, observe that tab_0 and tab_n are used to control the distance between the table and the braces surrounding the table. If the table contains horizontal and vertical rules, this may not be appropriate and you may have to use horizontal glue to increase the distance between the delimiters and the table itself.

Sample Table 41.2 (Source code).

```
1   $$
2   \left\lbrace
3       \vcenter{
4                                           \tabskip = 3pt
5           \halign{
6               \it#\rightarrowfill        \tabskip = 15pt&% 1
7               \hfil\tt#&                                 % 2
8               \bf #\hfil&                                % 3
9               ${\cal A}_{#}$\hfil        \tabskip = 3pt  % 4
10          \cr
11              XX&       YY-Y&      Z&         1\cr
12              YY&       XX/L&      ZZ&        2\cr
13              AA&       1020&      CCC&       3\cr
14              TT&       AATT&      TTTT&      4\cr
15              BIGG&     (80)&      VERY BIG&  300\cr
```

```
16            }
17        }
18   \right\rbrace
19   $$
```

Sample Table 41.2 (Output).

$$\left\{ \begin{array}{llll} XX\longrightarrow & \text{YY-Y} & \mathbf{Z} & \mathcal{A}_1 \\ YY\longrightarrow & \text{XX/L} & \mathbf{ZZ} & \mathcal{A}_2 \\ AA\longrightarrow & \text{1020} & \mathbf{CCC} & \mathcal{A}_3 \\ TT\longrightarrow & \text{AATT} & \mathbf{TTTT} & \mathcal{A}_4 \\ BIGG\rightarrow & \text{(80)} & \mathbf{VERY\ BIG} & \mathcal{A}_{300} \end{array} \right\}$$

It is important to use \vcenter, not \vtop or \vbox, because otherwise the reference point of the table box is not vertically centered. Here is the previous table with \vcenter replaced by \vbox.

Sample Table 41.3 (Source code).

```
1    $$
2    \left\lbrace
3        \vbox{
4                                    \tabskip = 3pt
5            \halign{
6                \it#\rightarrowfill    \tabskip = 15pt&% 1
7                \hfil\tt#&                          % 2
8                \bf #\hfil&                         % 3
9                ${\cal A}_{#}$\hfil    \tabskip = 3pt  % 4
10           \cr
11               XX&      YY-Y&     Z&          1\cr
12               YY&      XX/L&     ZZ&         2\cr
13               AA&      1020&     CCC&        3\cr
14               TT&      AATT&     TTTT&       4\cr
15               BIGG&    (80)&     VERY BIG&   300\cr
16           }
17       }
18   \right\rbrace
19   $$
```

Sample Table 41.3 (Output).

$$\left\{ \begin{array}{llll} XX\longrightarrow & \text{YY-Y} & \mathbf{Z} & \mathcal{A}_1 \\ YY\longrightarrow & \text{XX/L} & \mathbf{ZZ} & \mathcal{A}_2 \\ AA\longrightarrow & \text{1020} & \mathbf{CCC} & \mathcal{A}_3 \\ TT\longrightarrow & \text{AATT} & \mathbf{TTTT} & \mathcal{A}_4 \\ BIGG\rightarrow & \text{(80)} & \mathbf{VERY\ BIG} & \mathcal{A}_{300} \end{array} \right\}$$

41.2.2 Increasing the Height and Depth of Parentheses Around Tables

Assume that you want to make the parentheses around your table higher and lower than the table itself. The problem can be solved easily: typeset the table assigning it to a box register, and then manipulate the dimensions of this box register (we increase height and depth of the box register by 10 pt in the example below). Then the box register containing the table is printed. Here is the example:

Sample Table **41.4** (Source code).

```
\InputD{chboxd.tip}                          % 4.5.10, p. I-102.
$$ \setbox 0 = \hbox{$\vcenter{
                        \tabskip = 15pt
    \halign{
        \it#\rightarrowfill&                 % 1
        \hfil\tt#&                           % 2
        \bf #\hfil&                          % 3
        ${\cal A}_{#}$\hfil                  % 4
    \cr
        XX&        YY-Y&     Z&               1\cr
        YY&        XX/L&     ZZ&              2\cr
        AA&        1020&     CCC&             3\cr
        TT&        AATT&     TTTT&            4\cr
        BIGG&      (30)&     VERY BIG&        300\cr
    }
}$}
\AdvanceBoxDimension{\ht0}{10pt}%
\AdvanceBoxDimension{\dp0}{10pt}%
    \left(
        \box 0
    \right)
$$
```

Sample Table **41.4** (Output).

$$
\left(
\begin{array}{llll}
XX\longrightarrow & \text{YY-Y} & \text{Z} & \mathcal{A}_1 \\
YY\longrightarrow & \text{XX/L} & \text{ZZ} & \mathcal{A}_2 \\
AA\longrightarrow & 1020 & \textbf{CCC} & \mathcal{A}_3 \\
TT\longrightarrow & \textbf{AATT} & \textbf{TTTT} & \mathcal{A}_4 \\
BIGG\rightarrow & (80) & \textbf{VERY BIG} & \mathcal{A}_{300}
\end{array}
\right)
$$

Now solve the same problem using a different approach: insert a strut-like construct into the first row to make the first row artificially higher and insert a similar construct into the last row. In the following example, the rule was made visible by giving it a non-zero width; in a "real-world" table you would set the width of this rule to zero, of course.

Sample Table **41.5** (Source code).

```
1   \InputD{chboxd.tip}                          % 4.5.10, p. I-102.
2   $$
3   \left[
4      \vcenter{
5                              \tabskip = 15pt
6          \halign{
7              \it#\rightarrowfill&          % 1
8              \hfil\tt##&                    % 2
9              \bf #\hfil&                    % 3
10             ${\cal A}_{#}$\hfil           % 4
11         \cr
12             XX\vrule height 20pt width 1pt&
13                     YY-Y&    Z&               1\cr
14             YY&     XX/L&    ZZ&              2\cr
15             AA&     1020&    CCC&             3\cr
16             TT&     AATT&    TTTT&            4\cr
17             BIGG\vrule height 0pt depth 16pt width 1pt&
18                     (80)&    VERY BIG&        300\cr
19         }
20      }
21   \right]
22   $$
```

Sample Table **41.5** (Output).

$$
\left[
\begin{array}{llll}
XX\longrightarrow & \text{YY-Y} & \mathbf{Z} & \mathcal{A}_1 \\
YY\longrightarrow & \text{XX/L} & \mathbf{ZZ} & \mathcal{A}_2 \\
AA\longrightarrow & \text{1020} & \mathbf{CCC} & \mathcal{A}_3 \\
TT\longrightarrow & \text{AATT} & \mathbf{TTTT} & \mathcal{A}_4 \\
BIGG\rightarrow & \text{(80)} & \mathbf{VERY\ BIG} & \mathcal{A}_{300}
\end{array}
\right]
$$

41.2.3   Parentheses Spanning Multiple Rows Within a Table

Now let me show how you can typeset a table with parentheses spanning selected rows. You must do some arithmetic yourself in this case.

First here is the definition of the macro \MultiRowDel, which has the following parameters:

- #1. A delimiter, any delimiter legal in the context of \left and \right.
- #2. Height (= depth) desired for this symbol.

The important point is that while a symbol of the requested dimensions is being generated, the height and the depth of this symbol is set to zero. Therefore the line spacing within a table is not destroyed.

$$\mathcal{P}' \quad \bullet \, \texttt{mrdel.tip} \, \bullet$$

```
15   \def\MultiRowDel #1#2{%
```

Start a group, and set horizontal glue surrounding inline math mode equations to zero.

```
16       {%
17           \mathsurround = 0pt
```

Build the "math equation" inside an hbox.

```
18           \setbox 0 = \hbox{%
19             $%
20                 \vcenter{%
21                     \hbox{%
22                         $
23                             \left#1%
24                             \vrule height #2 depth #2 width 0pt
25                             \right.
26                         $%
27                     }%
28                 }%
29             $%
30           }%
```

Here the height and the depth of the symbol just generated are set to zero.

```
31           \ht0 = 0pt
32           \dp0 = 0pt
33           \box 0
34       }%
35   }
```

An alternate definition.

```
36   \def\MultiRowDel #1#2{%
```

Start a group, and set horizontal glue surrounding inline math mode equations to zero.

```
37       {%
38           \mathsurround = 0pt
```

Build the "math equation" inside an hbox.

```
39           \setbox 0 = \hbox{%
40                 $
41                     \left#1%
42                     \vrule height #2 depth #2 width 0pt
43                     \right.
44                 $%
45             }%
```

Here the height and the depth of the symbol just generated are set to zero.

```
46              \ht0 = 0pt
47              \dp0 = 0pt
48              \box 0
49         }%
50    }
```

• End of <code>mrdel.tip</code> •

And here is the example of a table using the previous macro:

Sample Table 41.6 (Source code).

```
1    \InputD{mrdel.tip}                        % 41.2.3, p. 343.
2    $$
3    \vbox{
4                        \tabskip = 10pt
5    \halign{
6        #\hfil&                          % 1. Text column
7        #\hfil&                          % 2. Delimiter column
8        #\hfil&                          % 3. Text column
9        #\hfil&                          % 4. Delimiter column
10       #\hfil                           % 5. Text column
11   \cr
12       ABC DEF&&            XXX&&        NONE OF\cr
13             &&            X&&           MORE\cr
14       ABC&\MultiRowDel{\lbrace}{15pt}&
15                           XX&\MultiRowDel{)}{27pt}&
16                                        ZZ TT UU\cr
17               &&          X&&          AB CD AB CD AB CD OF\cr
18       LL TT UU&&          ***&&        NONE NONE\cr
19   }}$$
```

Sample Table 41.6 (Output).

$$
\begin{array}{l l l}
\text{ABC DEF} & \text{XXX} & \text{NONE OF} \\
 & \text{X} & \text{MORE} \\
\text{ABC} \left\{ & \text{XX} \right) & \text{ZZ TT UU} \\
 & \text{X} & \text{AB CD AB CD AB CD OF} \\
\text{LL TT UU} & \text{***} & \text{NONE NONE}
\end{array}
$$

41.3 Vertical Splitting Tables by Hand

It is possible to allow TeX to break tables across pages. This was already shown (see 40.5, p. 321), and by using tab_0 and tab_n such tables can even be centered.

What we try to do here is a little more complicated. We store a table inside a \vbox and then use \vsplit (see 8.2, p. I-267) to break the table apart. This allows for the automatic duplication of table headers. Using the macro \ComputeFreeSpaceOnPage of 32.2.4, p. 6, you can completely automize the length computations in such a way that all the part tables fill up the current page automatically.

41.3.1 An Easy Example

Define three box registers, which we will use frequently.

Sample Table 41.7 (Source code).

```
1   \newbox\TableBox
2   \newbox\TableHeaderBox
3   \newbox\TablePartBox
```

Define \AStrut to control the strut's height and depth precisely. This is needed to be able to compute the dimensions for \vsplit properly.

```
4   \def\AStrut{\vrule height 8pt depth 4pt width 0pt}
```

Store a table with two header lines in the box register \TableBox.

```
5   \setbox\TableBox = \vbox{
6       \offinterlineskip
7                                   \tabskip = 0pt
8       \halign {
9           \AStrut#\relax          \tabskip = 10pt&
10          #\hfil&
11          \hfil #\hfil            \tabskip = 0pt
12      \cr
13          & \it Header I& \it Header II\cr
14          & \it Subheader I& \it Subheader II\cr
15          & L-1& Line 1 of table\cr
16          & L-2& Line 2 \dots\cr
17          & L-3& Some more stuff in Line 3\cr
18          & L-4& This is Line 4\cr
19          & L-5& More stuff\cr
20          & L-6& Even more stuff\cr
21          & L-7& The last line\cr
22      }
23  }
```

First print the table.

```
24  $$
25      \copy\TableBox
26  $$
```

Sample Table **41.7** (Output).

Header I	Header II
Subheader I	Subheader II
L-1	Line 1 of table
L-2	Line 2 ...
L-3	Some more stuff in Line 3
L-4	This is Line 4
L-5	More stuff
L-6	Even more stuff
L-7	The last line

41.3.2   Splitting the Table Twice

The table is split as follows:

1. The header (two lines long) is split from the table and stored in its own box register (`\HeaderBox`). The necessary dimension for `\vsplit` to split the header is computed as follows: height (8 pt) plus depth (4 pt) of the first header line plus height (8 pt) of the second header line. Together this adds up to 20 pt.
2. Then the table is split after the third row (this count excludes the header lines).

41.3.2.1   Splitting Off the Header Lines

Here is the code that splits up the header:

Sample Table **41.8** (Source code).

```
1   \InputD{box-mac.tip}                    % 9.3.14, p. I-343.
2   \splittopskip = 8pt
3   \setbox\TableHeaderBox = \vsplit\TableBox to 20pt
```

Now print the table header.

```
4   \centerline{\bf 1. Table header}
5   $$
6       \CopyR\TableHeaderBox
7   $$
```

Then print the rest of the table, the table body.

```
8   \centerline{\bf 2. Rest of table.}
9   $$
10      \CopyR\TableBox
11  $$
```

Finally print the header and table body together.

```
12    \centerline{\bf 3. Table header and table body reunited.}
13    $$
14        \vbox{
15            \cffinterlineskip
16            \copy\TableHeaderBox
17            \copy\TableBox
18        }
19    $$
```

Sample Table **41.8** (Output).

1. Table header

Header I	Header II
Subheader I	Subheader II

2. Rest of table.

L-1	Line 1 of table
L-2	Line 2 ...
L-3	Some more stuff in Line 3
L-4	This is Line 4
L-5	More stuff
L-6	Even more stuff
L-7	The last line

3. Table header and table body reunited.

Header I	Header II
Subheader I	Subheader II
L-1	Line 1 of table
L-2	Line 2 ...
L-3	Some more stuff in Line 3
L-4	This is Line 4
L-5	More stuff
L-6	Even more stuff
L-7	The last line

41.3.2.2   Splitting the Table Body Itself

Now split the table body into two smaller tables, one containing the first three
rows and the other containing the remaining four rows. The \vsplit dimension
is computed as follows: height (8 pt each) plus depth (4 pt each) of the first two
rows, and height of the third row (8 pt). The sum of all these values is 32 pt.

Sample Table **41.9** (Source code).

```
 1   \splittopskip = 8pt
 2   \setbox\TablePartBox = \vsplit\TableBox to 32pt
```

Print the first three lines of the table body.

```
 3   \centerline{\bf 4. The first three lines of the table body.}
 4   $$
 5       \copy\TablePartBox
 6   $$
```

Print the four remaining lines of the table body.

```
 7   \centerline{\bf 5. The remaining four lines of the table body.}
 8   $$
 9       \copy\TableBox
10   $$
```

Because the header was split and stored in a separate box, the header and
the first part of the table body can be combined into a complete table, and also
the header and the second part of the table can be combined into a new table.
Here is how this is done.

```
11   \centerline{\bf 6. Combine the header and the first three lines.}
12   $$
13       \vbox{
14           \copy\TableHeaderBox
15           \copy\TablePartBox
16       }
17   $$
18
19   \centerline{\bf 7. Combine the header and the last four lines.}
20   $$
21       \vbox{
22           \copy\TableHeaderBox
```

Print whatever is left over.

```
23           \copy\TableBox
24       }
25   $$
```

Sample Table **41.9** (Output).

4. The first three lines of the table body.

L-1	Line 1 of table
L-2	Line 2 ...
L-3	Some more stuff in Line 3

5. The remaining four lines of the table body.

L-4	This is Line 4
L-5	More stuff
L-6	Even more stuff
L-7	The last line

6. Combine the header and the first three lines.

Header I	*Header II*
Subheader I	*Subheader II*
L-1	Line 1 of table
L-2	Line 2 ...
L-3	Some more stuff in Line 3

7. Combine the header and the last four lines.

Header I	*Header II*
Subheader I	*Subheader II*
L-4	This is Line 4
L-5	More stuff
L-6	Even more stuff
L-7	The last line

41.3.3   Penalty Controlled Splits

Computing the size of the header to use \vsplit can be avoided (compare this with 41.3.1, p. 345). I will apply what was discussed in 8.2.8, p. 285.

After the table header is generated, a penalty of -10000 is inserted, which forces the \vsplit command to break off the header from the table body at this point, assuming that the \vsplit dimension is large enough. Here is the example:

Sample Table **41.10** (Source code).

```
1    \setbox\TableBox = \vbox{
2       \def\AStrut{\vrule height 8pt depth 4pt width 0pt}
3       \offinterlineskip
4                                              \tabskip = 0pt
5       \halign {
6          \AStrut#&
7           #\hfil                             \tabskip = 10pt&
8          \hfil #\hfil                        \tabskip = 0pt
9       \cr
10           & \it Header I& \it Header II\cr
11           & \it Subheader I& \it Subheader II\cr
```

This penalty separates header and body.

```
12                       \noalign{\penalty -10000}
```

Here is the body.

```
13           & L-1& Line 1 of table\cr
14           & L-2& Line 2 \dots\cr
15           & L-3& Some more stuff in Line 3\cr
16           & L-4& This is Line 4\cr
```

```
17          & L-5& More stuff\cr
18          & L-6& Even more stuff\cr
19          & L-7& The last line\cr
20      }
21  }
```

Now split off the header controlled by a built-in break point. The chosen dimension of \vsplit is much higher than necessary, but the forced break point will break off the header. The resulting underfull \vbox must then be adjusted (all this is explained in much more detail at the given reference).

```
22  \splittopskip = 8pt
23  \setbox\TableHeaderBox = \vsplit \TableBox to 1000pt
24  \setbox\TableHeaderBox = \vbox{\unvbox \TableHeaderBox}
```

Now print the table header, followed by the rest of the table, followed by the table header and the reunited table body.

```
25  \centerline{\bf 8. Table header.}
26  $$
27      \copy\TableHeaderBox
28  $$
29  \centerline{\bf 9. Rest of table.}
30  $$
31      \copy\TableBox
32  $$
33
34  \centerline{\bf 10. Table header and body reunited.}
35  \nobreak
36  $$
37      \vbox{
38          \offinterlineskip
39          \copy\TableHeaderBox
40          \copy\TableBox
41      }
42  $$
```

Sample Table **41.10** (Output).

8. Table header.

Header I	*Header II*
Subheader I	*Subheader II*

9. Rest of table.

L-1	Line 1 of table
L-2	Line 2 ...
L-3	Some more stuff in Line 3
L-4	This is Line 4
L-5	More stuff
L-6	Even more stuff
L-7	The last line

10. Table header and body reunited.

Header I	*Header II*
Subheader I	*Subheader II*
L-1	Line 1 of table
L-2	Line 2 ...
L-3	Some more stuff in Line 3
L-4	This is Line 4
L-5	More stuff
L-6	Even more stuff
L-7	The last line

41.3.4   Page Breaks in Tables with Horizontal Rules

Assume that a table with horizontal rules between its rows should be broken-up.
If this breaking occurs at a horizontal rule; then this horizontal rule should be
duplicated so it prints on the bottom of the split table, as well as on top of the
remainder of the table. This necessary duplication of the horizontal rule can be
automized: all rules in the table are simply printed twice (see 39.2.4, p. 260).
The splitting occurs at the glue between the two overprinting horizontal rules;
the first rule will be printed at the bottom of the split part of the table and the
second rule will be printed at the top of the remaining part of the table, exactly
as required.

41.4   Table-Related Macros in the Plain Format

I briefly discuss some \halign-based macros in the plain format. As usual I
took the liberty of reformatting the code for improved readability. We already
discussed \hidewidth (see 39.3.1, p. 263) and \ialign (see 38.4.6.1, p. 238)
which are important for the understanding of the following macros. You may
also want to read about using \crcr (in particular the aspect of its use to end a
preamble) again in 40.6.7, p. 332.

1. \oalign. This macro is used to place characters on top of each other. Note
    that because \baselineskip is set to zero and \lineskip is set to 0.25 ex
    the distance between the stacked characters will be 0.25 ex. Also, because the
    template of the table being built by this macro contains no glue, the entries
    will be left-justified. Here is an example application: \oalign{X\cr*} prints
    X. An identical output is produced by \oalign{X\cr*\cr}; the trailing \cr
    *
    missing in the first call is inserted by the \oalign macro's \crcr (line 10).
        Here is the definition of macro \oalign.

• <code>tab-pln.tip</code> •

1   `\def\oalign #1{%`

Make sure that you are in horizontal mode (this allows an `\oalign` call to occur at the beginning of a paragraph).

2   `\leavevmode`

A `\vtop` is used that makes the first of the two lines generated by a call to this macro line up with the rest of the text (the second line therefore appears below the first line). This means, assuming reasonably sized characters, etc., that the second character is placed under the first character, rather than the first character on top of the second character (which would be the case if `\vbox` had been used in place of `\vtop`).

3   `\vtop{%`

The following two lines in the source code mean that the two table lines are regarded as "too close." Therefore `\lineskip` glue, vertical space of 0.25 ex is inserted between these two lines. Because the changes to these two parameters occur inside a group (the implicit group established by `\vtop`), there is no problem with those changes influencing line spacing otherwise.

4   `\baselineskip = 0pt`
5   `\lineskip = .25ex`

Start an initialized `\halign`.

6   `\ialign{%`

No alignment glue is specified: everything is left-justified, unless glue is provided as part of the table entries.

7   `##%`

End the preamble (a one column table is built).

8   `\crcr`

Print the two lines. #1 might look something like "`X\crcr*`."

9   `#1%`

The user does not need to (but can) end the second line with a `\cr`.

10   `\crcr`

End the table started by `\ialign`.

11   `}%`

End the `\vtop`.

12   `}%`

End the macro.

13   `}`

If you look at the preceding example closely then you realize that the asterisk
is not centered below the X, because \oalign by default treats everything
left-flush. If, on the other hand we write

\oalign{\hfil X\hfil\crcr\hfil*\hfil\cr}

the output is X and indeed the asterisk is centered below the X. Note, because
in this case we know that the X is wider than the asterisk, we could have
omitted the \hfils around the X.

2. \ooalign calls \oalign, the macro just explained, after it has changed the
    value of \lineskiplimit to \maxdimen. This is also what macro \Always-
    Baselineskip does (see 7.3.13, p. I-232). In effect no lines are too close now,
    and \baselineskip is used for interline glue. \baselineskip is set to zero
    by \oalign, that is the two symbols print on top of each other.

    \ooalign changes \lineskiplimit and this change is not by default
    enclosed inside a group. Therefore, usually when you call \ooalign you
    need to enclose the call inside a group.

    See the later-following definition of the \copyright macro for an appli-
    cation of \ooalign.

```
14   \def\ooalign{%
15       \lineskiplimit = -\maxdimen
16       \oalign
17   }
```

3. The \copyright macro prints the copyright symbol (©). It uses \ooalign
    and is defined as follows:

```
18   \def\copyright{%
```

Enclose the following \ooalign because of the changes to the various line
spacing parameters (such as \baselineskip).

```
19       {%
20           \ooalign{%
```

A centered lower case "c" positioned appropriately.

```
21               \hfil
22               \raise .07ex \hbox{c}%
23               \hfil
```

End the first line containing a centered "c."

```
24           \crcr
```

Select character "0D from font family 2, which is the circle we need.

```
25               \mathhexbox 20D
```

End the \ooalign.

```
26           }%
```

End the group of this macro.

```
27       }%
```

End the macro.

```
28    }
```

4. A dot-under accent is generated using the \d macro. This macro has one
    parameter, #1, the character under which the period is placed. Here is the
    definition of this macro (the dot is centered with respect to the character;
    \d{x} prints x̣):

```
29    \def\d #1{%
30        \oalign{%
31            #1\crcr
```

Print a centered period, ignoring its width. As a practical matter usually
the symbol under which the period is printed, is wider than a period and
instead of \hidewidth we could have used \hfil.

```
32            \hidewidth.\hidewidth
33        }%
34    }
```

5. The bar-under accent macro \b is defined as follows (\b{x} prints x̲).

```
35    \def\b #1{%
36        \oalign{%
37            #1\crcr
```

The following \hidewidth together with the one further below means print
the underbar (a moved down overbar) centered, but ignore the width if this
underbar should be wider than the character under which it occurs.

```
38            \hidewidth
```

The following vbox will be 0.2 ex high, and have a zero depth.

```
39            \vbox to .2ex{%
```

The following character code is the character code of an overbar.

```
40                \hbox{\char '22}%
41                \vss
42            }%
43            \hidewidth
44        }%
45    }
```

6. The macro \c produces the cedilla accent, is defined next. For instance,
    \c{c} prints ç.

```
46    \def\c #1{%
47        \setbox 0 = \hbox{#1}%
48        \ifdim\ht0 = 1ex
49            \accent '30 #1%
50        \else
51            {%
52                \ooalign{%
53                    \hidewidth
54                    \char '30
55                    \hidewidth
56                    \crcr
```

```
57                    \unhbox 0
58                }%
59            }%
60        \fi
61    }
```

• End of `tab-pln.tip` •

41.5   Vertical Alignment, \valign

\valign is, simply spoken, the vertical counterpart of \halign. Here is a brief description:

1. There is one template per *row*.
2. The table is entered on a *column-by-column* basis.
3. & is used to separate one row's entry from the next row's entry.
4. \cr is used to end the preamble and to end a column.
5. \crcr, as before, works as a non-redundant \cr.
6. The following standard templates are used in \valign:

    (a) Top-justified entries are generated by #\vfil; compare this with left-justified entries (#\hfil) in \halign-based tables.
    (b) Vertically-centered entries are generated by \vfil#\vfil; compare this with horizontally centered entries in \halign-based tables which are generated by \hfil#\hfil.
    (c) Bottom-justified entries are generated by \vfil#. Compare this with right-justified entries in \halign -based tables using the \hfil# template.

7. determines the distance between *rows*.
8. allows the insertion of material between *columns*.

Here is an example of a \valign table.

Sample Table **41.11** (Source code).

```
1  $$
2  \vbox{
```

Define three macros to create three pieces of text of various length.

```
3        \def\TextA{This is a little short paragraph.}
4        \def\TextB{This is a medium sized paragraph. We make it a little
5            longer, so we get at least three lines.}
6        \def\TextC{This is the third and longest paragraph, it will have
7            five or six lines. We need this so we can see how
8            {\tt\string\valign} works and aligns text.}
```

Define the macro `\OneEntry` to generate short paragraph-like table entries. This macro has one parameter, #1, the text to be used in these short paragraphs.

```
 9        \def\OneEntry #1{%
10            \vtop{%
11                \hsize = 1.4in
12                \raggedright
13                \parindent = 0pt
14                \vrule height 10pt depth 0pt width 0pt
15                #1%
16                \vrule height 0pt  depth 5pt width 0pt
17            }
18        }
```

Now define macro `\VfilLeader` to print a vertical leader instead of `\vfil`.

```
19    \def\VfilLeader{\leaders\hrule width 3pt\vfil}
```

Now the table begins.

```
20                                                \tabskip = 0pt
21        \valign{
22            \OneEntry{#}\VfilLeader
23                                                \tabskip = 10pt&% 1st row
24            \VfilLeader\OneEntry{#}\VfilLeader&           % 2nd row
25            \VfilLeader\OneEntry{#}\relax    \tabskip = 0pt  % 3rd row
26        \cr
27            1: \TextA& 2: \TextA& 3: \TextA\cr
28                     \noalign{\hskip 5pt \vrule \hskip 5pt}
29            4: \TextB& 5: \TextB& 6: \TextB\cr
30                     \noalign{\hskip 5pt \vrule \hskip 5pt}
31            7: \TextC& 8: \TextC& 9: \TextC\cr
32        }
33    }
34    $$
```

Sample Table **41.11** (Output).

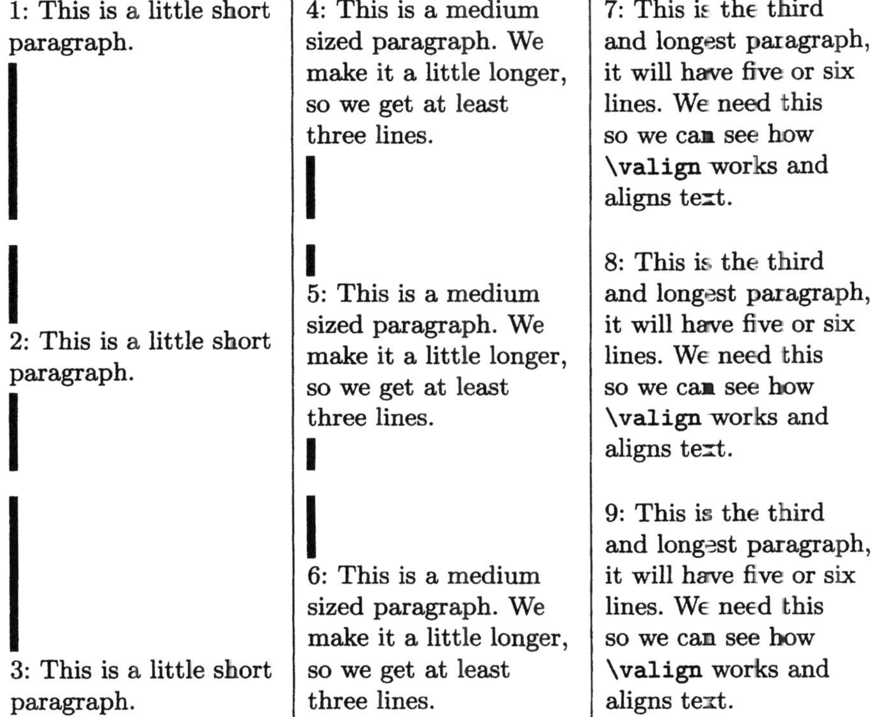

Observe that the input is now on a column-by-column basis, and no longer on a row-by-row basis! now controls the distance between rows, rather than columns.

41.6   Double Tables

The term "double table" describes the case where two tables are printed side-by-side and where these two tables are really *one* table (*one* \halign construct) as far as T_EX is concerned.

41.6.1   Some Double Table Examples

One type of table uses selective horizontal rules which make the tables appear to be two tables side-by-side.

Sample Table **41.12** (Source code).

```
1   \InputD{setstrut.tip}                           % 7.4.3.1, p. I-239.
2   $$
3   \vbox{
4       \ComputeStrut
5       \offinterlineskip
6                           \tabskip = 0pt
7   \halign{
```

The left table starts here.

```
8       \MyStrut#&                                   % 1: strut
9       \vrule#\relax       \tabskip = 10pt&         % 2: vrule
10      #\hfil&                                       % 3: left-justified col.
11      \vrule#&                                      % 4: vrule
12      #\hfil&                                       % 5: left-justified col.
13      \vrule#\relax       \tabskip = 1.0in&         % 6: vrule
```

The right table starts here.

```
14      \vrule#\relax       \tabskip = 10pt&         % 7: vrule
15      \hfil#\hfil&                                  % 8: centered col.
16      \vrule#&                                      % 9: vrule
17      \hfil#\hfil&                                  % 10: centered col.
18      \vrule#\relax       \tabskip = 0pt           % 11: vrule
19  \cr
20      \omit&\multispan{5}{\hrulefill}&\multispan{5}{\hrulefill}\cr
21          && ABC&&       DEF&&&            XXX&&    YYY&\cr
22          && CBA-CBA&& FED&&&              TT&&     YY&\cr
23          && ABC-0&&    DEF&&&            X&&      Y&\cr
24      \omit&\multispan{5}{\hrulefill}&\multispan{5}{\hrulefill}\cr
25          && ABC&&      DEF&&&           XXX-2&& YYY-2&\cr
26          && CBA-CBA&& FED&&&             TT&&     YY&\cr
27          && ABC-0&&    DEF&&&            X&&      Y&\cr
28      \omit&\multispan{5}{\hrulefill}&\multispan{5}{\hrulefill}\cr
29      }
30  }
31  $$
```

Sample Table **41.12** (Output).

ABC	DEF
CBA-CBA	FED
ABC-0	DEF
ABC	DEF
CBA-CBA	FED
ABC-0	DEF

XXX	YYY
TT	YY
X	Y
XXX-2	YYY-2
TT	YY
X	Y

The preceding example may not have convinced you of the necessity of treating what appears to be two tables as one table in TeX. The following table should! In this table the height of the rows varies and therefore the heights of the rows of each table must be synchronized. By treating the two tables as one table, this problem can be solved easily.

Sample Table **41.13** (Source code).

```
 1  \InputD{setstrut.tip}                          % 7.4.3.1, p. I-239.
 2  $$
 3  \vbox{
 4      \ComputeStrut
 5      \offinterlineskip
 6                              \tabskip = 0pt
 7  \halign{
```

The left table starts here.

```
 8      \MyStrut#&                              % 1: strut
 9      \vrule#\relax        \tabskip = 10pt&   % 2: vrule
10      #\hfil&                                 % 3: left-justified col.
11      \vrule#&                                % 4: vrule
12      \vtop{                                  % 5: text col.
13          \hsize = 1.8in
14          \raggedright
15          \parindent = 0pt
16          \normalbaselines
17          \MyStrut#\MyStrut
18      }&
19      \vrule#\relax        \tabskip = 25pt&   % 6: vrule
```

The right table starts here.

```
20      \vrule#\relax        \tabskip = 10pt&   % 7: vrule
21      \hfil#\hfil&                            % 8: centered col.
22      \vrule#&                                % 9: vrule
23      \hfil#\hfil&                            % 10: centered col.
24      \vrule#\relax        \tabskip = 0pt     % 11: vrule
25  \cr
26      \omit&\multispan{5}{\hrulefill}&\multispan{5}{\hrulefill}\cr
27      && ABC&&      DEF&&&         XXX&&    YYY&\cr
28      && CBA-CBA&& DEF and quite some additional
29                      stuff so that the text actually wraps around.&&&
30                                      TT&&    YY&\cr
31      && ABC-O&&    DEF&&&         X&&     Y&\cr
32      \omit&\multispan{5}{\hrulefill}&\multispan{5}{\hrulefill}\cr
33      && ABC&&      DEF&&&         XXX-2&& YYY-2&\cr
34      && CBA-CBA&& FED&&&          TT&&    YY&\cr
35      && ABC-O&&    Oh yes, and let's have another one of those
36          real long entries here.&&&
37                              X&&     Y&\cr
38      \omit&\multispan{5}{\hrulefill}&\multispan{5}{\hrulefill}\cr
39      }
40  }
```

41   **$$**

Sample Table **41.13** (Output).

ABC	DEF
CBA-CBA	DEF and quite some additional stuff so that the text actually wraps around.
ABC-0	DEF
ABC	DEF
CBA-CBA	FED
ABC-0	Oh yes, and let's have another one of those real long entries here.

XXX	YYY
TT	YY
X	Y
XXX-2	YYY-2
TT	YY
X	Y

41.6.2   Organizing a Double Table

Let us rewrite the previous table as follows (our goal is to be able to typeset the left and right tables independently of each other but still treat both tables as a unit; this is something we want to do if the two tables are later to be placed on two facing pages in a book, for instance. In that case the two tables belong together but are printed on two different pages). Here is what we will do:

1. Insert the complete table into the macro `\TableBody` so that the table can be easily typeset more than once.

2. Use various dimension registers to determine the `\tabskip` values as follows (the first and last `\tabskip` are set to zero):

    (a) `\TabskipLeftTable` is the value for `\tabskip` to be used between all columns of the left table.

    (b) `\TabskipBetweenTables` determines the distance between the two tables.

    (c) `\TabskipRightTable` is the value for `\tabskip` to be used for all columns of the right table.

3. The templates themselves are left almost unchanged, except for the macros `\TempLeft` and `\TempRight` that are called for the templates of the left and right columns.

    Here is the new table:

Sample Table **41.14** (Source code).

```
1    \InputD{setstrut.tip}                    % 7.4.3.1, p. I-239.
```

Get three dimension registers to hold the various values for the \tabskip macro used in this table. The dimensions were found by trial and error, just so that the two tables would fit side-by-side on a page.

```
2    \newskip\TabskipLeftTable          \TabskipLeftTable = 10pt
3    \newskip\TabskipBetweenTables      \TabskipBetweenTables = 25pt
4    \newskip\TabskipRightTable         \TabskipRightTable = 10pt
```

Now define the macros \TempLeft and \TempRight in such a way that these macros echo their arguments.

```
5    \def\TempLeft  #1{#1}
6    \def\TempRight #1{#1}
```

Define \DoubleTable next.

```
7    \def\DoubleTable{
8                                   \tabskip = 0pt
9        \halign{
```

First column: strut.

```
10              \MyStrut##&
```

Second column: \vrule.

```
11              \TempLeft{\vrule##}%
12                                  \tabskip = \TabskipLeftTable&
```

Third column: first text column of left table.

```
13              \TempLeft{##\hfil}&
```

Fourth column: \vrule.

```
14              \TempLeft{\vrule##}&
```

Fifth column: second text column of the left table.

```
15              \TempLeft{%
16                 \vtop{
17                     \hsize = 1.8in
18                     \raggedright
19                     \parindent = 0pt
20                     \normalbaselines
21                     \MyStrut##\MyStrut
22                 }}&
```

Sixth column: \vrule terminating left table.

```
23              \TempLeft{\vrule##}%
24                                  \tabskip = \TabskipBetweenTables&
```

Seventh column: \vrule starting right table.

```
25              \TempRight{\vrule##}%
26                                  \tabskip = \TabskipRightTable&
```

8th column: first text column, right table.

```
27            \TempRight{\hfil##\hfil}&
```

9th column: \vrule.

```
28            \TempRight{\vrule##}&
```

10th column: second text column, right table.

```
29            \TempRight{\hfil##\hfil}&
```

11th column: \vrule.

```
30            \TempRight{\vrule##}%
31                        \tabskip = 0pt
32     \cr
33         \omit&\multispan{5}{\hrulefill}&\multispan{5}{\hrulefill}\cr
34            && ABC&&        DEF&&&             XXX&&    YYY&\cr
35            && CBA-CBA&& DEF and quite some additional
36                        stuff so that the text actually
37                        wraps around.&&&
38                                        TT&&      YY&\cr
39            && ABC-O&&    DEF&&&            X&&       Y&\cr
40         \omit&\multispan{5}{\hrulefill}&\multispan{5}{\hrulefill}\cr
41            && ABC&&        DEF&&&            XXX-2&& YYY-2&\cr
42            && CBA-CBA&& FED&&&              TT&&      YY&\cr
43            && ABC-O&&    Oh yes, and let's have another one of those
44               real long entries here.&&&
45                                   X&&       Y&\cr
46         \omit&\multispan{5}{\hrulefill}&\multispan{5}{\hrulefill}\cr
47      }
48  }
```

Now define the macro \PrintDoubleTable to print the whole table.

```
49  \def\PrintDoubleTable{%
50      $$
51         \vbox{
52            \ComputeStrut
53            \offinterlineskip
54            \DoubleTable
55         }
56      $$
57  }
```

Next print the whole table.

```
58  \PrintDoubleTable
```

Sample Table **41.14** (Output).

ABC	DEF		XXX	YYY
CBA-CBA	DEF and quite some additional stuff so that the text actually wraps around.		TT	YY
ABC-0	DEF		X	Y
ABC	DEF		XXX-2	YYY-2
CBA-CBA	FED		TT	YY
ABC-0	Oh yes, and let's have another one of those real long entries here.		X	Y

41.6.3   Printing the Left and Right Table Separately

Notice that with the approach of the previous Subsection it is possible to print the left and the right table separately. This is necessary, for instance, if the left and right tables are supposed to be placed on the left-hand and right-hand page of two facing pages in a book, and the "synchronization" between the left and right tables must be preserved (which was the original reason why the left and right tables were not treated as two independent tables, but as the left and right tables of a double table).

Together with the preparation of the preceding explanations, printing the left and right tables separately can be solved as follows. I will now describe how the left table is printed (and the right table is eliminated).

Note that the height and depth of entries in the right table must be preserved even when only the left table is printed. Here is how we proceed

1. The \tabskip glues between the two tables and in the right table itself are set to zero by setting \TabskipBetweenTables and \TabskipRightTable to zero.
2. The text in the right table is typeset (to get its height and depth) but it is not printed. In essence the right table is eliminated, which is achieved as follows:

   (a) The text of those columns is still typeset.
   (b) The text is not printed because it is assigned to a box register (box register 1 below), but the box register is never printed.
   (c) Note though that the height and depth of this box register is used to generate a strut of the appropriate height and depth which causes the height and depth information from this column not to be lost (see the following definition for \ReduceToStrut for more information).

In other words, the right table is reduced to zero width columns of the height and depth of each row as if the text in those columns were typeset.

The left table is printed by itself now.

Sample Table 41.15 (Source code).

```
1   \InputD{redtost.tip}                    % 7.4.4, p. I-241.
2   \InputD{setstrut.tip}                   % 7.4.3.1, p. I-239.
```

First a header above the table is printed.

```
3   \par\centerline{\bf The left table.}
```

Now the left table is printed and centered using display math mode. The right table is eliminated. \ReduceToStrut is used to prevent any loss of height information resulting from typesetting the right table.

```
4   {
5       \let\TempRight = \ReduceToStrut
```

Note that the changes to \TabskipBetweenTables, \TabskipRightTable and \TempRight are executed inside a group.

```
6       \TabskipBetweenTables = 0pt
7       \TabskipRightTable = 0pt
8       \PrintDoubleTable
9   }
```

Sample Table 41.15 (Output).

The left table.

ABC CBA-CBA ABC-0	DEF DEF and quite some additional stuff so that the text actually wraps around. DEF
ABC CBA-CBA ABC-0	DEF FED Oh yes, and let's have another one of those real long entries here.

Sample Table **41.16** (Source code).

```
1   \InputD{redtost.tip}                    % 7.4.4, p. I-241.
2   \InputD{setstrut.tip}                   % 7.4.3.1, p. I-239.
```

Repeat everything for the right table.

```
3   \par\centerline{\bf The right table.}
4   {
5       \let\TempLeft = \ReduceToStrut
6       \TabskipLeftTable = 0pt
7       \TabskipBetweenTables = 0pt
8       \PrintDoubleTable
9   }
```

Sample Table **41.16** (Output).

The right table.

XXX TT X	YYY YY Y
XXX-2 TT X	YYY-2 YY Y

41.6.4   Double Tables with Semiautomatic Determination of the Column Widths of Text Columns

Let me now discuss another variation of the double table problem. Assume the left and right table of the double table are to be printed on facing pages in a double page document, and both tables extend over the whole width of each page. In addition, assume that a semiautomatic way of determining the width of each paragraph text column is devised (remember a paragraph text column is a column built by using \vtop in which \hsize is set to the desired width of the text, and the text in this column is automatically broken along lines).

This problem can be solved as follows:

1. The same approach as before will be used as far as the settings of \tabskip are concerned. All distances between the columns in the left table will be fixed, and so will the distances between the columns in the right table. This restriction could be easily removed.

2. The macros to be described below are supposed to compute the widths of each paragraph text column in a semiautomatic way along the following lines: the user specifies, for the left and the right table on an independent basis, the relative width available for each paragraph text column. For instance, if there is only one paragraph text column available in the left table, its relative width would be 1. If there were two such paragraph text columns available, the user might choose a relative width of 0.5 for both, or 0.3 for one and 0.7 for the other.

3. Note that "regular" columns with left- or right-justified entries, or centered entries do not participate in any of this "relative weight" business. The widths are determined by the width of the widest entry in each column.

Before the example, we need to define some macros.

• <code>dtabmac.tip</code> •

```
 1    \InputD{redtost.tip}                        % 7.4.4, p. I-241.
```

The macro `\DoubleTableVtop` defines a standard `\vtop` as it is used in a double table. The macro has two parameters.

- #1. The `\hsize` to be used in this `\vtop`.
- #2. The text to be typeset.

The macro is defined as follows:

```
 2    \def\DoubleTableVtop #1#2{%
 3        \vtop{%
 4            \hsize = #1
 5            \raggedright
 6            \parindent = 0pt
 7            \normalbaselines
 8            \MyStrut#2\MyStrut
 9        }%
10    }
```

The following macro has the same parameters as the preceding macro, but generates a strut only which corresponds to the height and depth of the material being typeset.

```
11    \def\DoubleTableVtopStrut #1#2{%
12        \ReduceToStrut{%
13            \DoubleTableVtop{#1}{#2}}%
14    }
```

• End of <code>dtabmac.tip</code> •

An example of how this problem can be solved in TeX follows. Here is a new table.

Sample Table **41.17** (Source code).

```
1    \InputD{setstrut.tip}                              % 7.4.3.1, p. I-239.
2    \InputD{gobble.tip}                                % 21.8.6, p. III-186.
3    \InputD{dtabmac.tip}                               % 41.6.4, p. 363.
```

Declare \NewDoubleTable now.

```
4    \def\NewDoubleTable{%
5                          \tabskip = 0pt
6        \halign{
```

The left table starts here.
    First column: strut.

```
7                \MyStrut##&
```

Second column: \vrule.

```
8                \TempLeft{\vrule##}%
9                          \tabskip = \TabskipLeftTable&
```

Third column: first text column, left table.

```
10               \TempLeftVtop{0.6\WidthLeftTable}{##}&
```

Fourth column: \vrule.

```
11               \TempLeft{\vrule##}&
```

Fifth column: second text column, left table.

```
12               \TempLeftVtop{0.4\WidthLeftTable}{##}&
```

Sixth column: \vrule terminating left table.

```
13               \TempLeft{\vrule##}%
14                          \tabskip = \TabskipBetweenTables&
```

The right table starts here. Seventh column: \vrule starting right table.

```
15               \TempRight{\vrule##}%
16                          \tabskip = \TabskipRightTable&
```

Eighth column: first text column, right table.

```
17               \TempRight{\hfil##\hfil}&
```

Ninth column: \vrule.

```
18               \TempRight{\vrule##}&
```

Tenth column: second text column, right table.

```
19               \TempRightVtop{1.0\WidthRightTable}{##}&
```

Eleventh column: \vrule.

```
20               \TempRight{\vrule##}%
21                          \tabskip = 0pt
22       \cr
```

The table body starts here.

```
23          \omit&\multispan{5}{\hrulefill}&\multispan{5}{\hrulefill}\cr
24             && ABC and such is just some nice text for this wonderful
25                column. Note that there is a lot of text in this
26                column, or at least we will enter quite a bit here.
27                Let that be it.&&
28              Then there is more again, and we have to keep writing
29                to fill this column also. &&&          XXX&&
30            And here is now some text which is supposed to
31                wrap around in the right table. The right table
32                only contains one paragraph text column so we
33                must make it long enough so we can actually see the
34                text wrap around.&\cr
35          \omit&\multispan{5}{\hrulefill}&\multispan{5}{\hrulefill}\cr
36             && CBA-CBA. Why don't we make this one short.&&
37                DEF and quite some additional
38                stuff so that the text actually
39                wraps around.&&&
40                                      TT&&      YY&\cr
41          && ABC-0&&    DEF&&&      X&&      Y&\cr
42          \omit&\multispan{5}{\hrulefill}&\multispan{5}{\hrulefill}\cr
43          && ABC&&      DEF&&&         XXX-2&& YYY-2&\cr
44          && CBA-CBA&& FED&&&        TT&&      YY&\cr
45          && ABC-0&&    Oh yes, and let's have another one of those
46                real long entries here.&&&
47                           X&&      Y&\cr
48          \omit&\multispan{5}{\hrulefill}&\multispan{5}{\hrulefill}\cr
49       }%
50    }
```

Now define two more new macros \TempLeftVtop and \TempRightVtop, which
have two arguments. The first one is the \hsize value to be used in this template,
the second one is the table entry.

```
51    \let\TempLeftVtop = \DoubleTableVtop
52    \let\TempRightVtop= \DoubleTableVtop
```

Define two dimension registers, which are loaded with the widths of the right
and left tables, ignoring the text columns.

```
53    \newdimen\WidthLeftTable
54    \newdimen\WidthRightTable
```

Now define the macro \PrintNewDoubleTable to print the whole table. Display
math mode is not used because, by definition, each of the two tables will be as
wide as the whole page.

```
55    \def\PrintNewDoubleTable{%
56       \vbox{%
57          \ComputeStrut
58          \offinterlineskip
59          \NewDoubleTable
60       }%
61    }
```

Now execute the following steps (these steps are discussed for the left table).

1. Compute the width of the left table, ignoring all its paragraph text columns.
2. Compute the remaining widths for all its paragraph text columns.
3. Compute the width of the right table ignoring all its paragraph text columns.
4. Compute the remaining widths for all its paragraph text columns.
5. Print the left table by itself.
6. Print the right table by itself.

Step 1 is executed now.

```
62   \setbox 0 = \hbox{%
63       \let\TempRight = \GobbleOne
64       \let\TempRightVtop = \GobbleTwo
65       \let\TempLeftVtop  = \GobbleTwo
66       \TabskipBetweenTables = 0pt
67       \TabskipRightTable = 0pt
68       \PrintNewDoubleTable
69   }
```

Now compute how much space is left for all the text columns of the left table. This is step 2 of the previous list.

```
70   \WidthLeftTable = \hsize
71   \advance\WidthLeftTable by -\wd0
```

Now repeat the same steps for the right table which is step 3 of the previous list.

```
72   \setbox 0 = \hbox{%
73       \let\TempLeft = \GobbleOne
74       \let\TempLeftVtop = \GobbleTwo
75       \let\TempRightVtop  = \GobbleTwo
76       \TabskipBetweenTables = 0pt
77       \TabskipLeftTable = 0pt
78       \PrintNewDoubleTable
79   }
```

Now compute how much space is left for all the text columns of the right table (step 4 of the previous list).

```
80   \WidthRightTable = \hsize
81   \advance\WidthRightTable by -\wd0
```

Now print the left table, discarding the right table, with the correct widths of the text columns of this table (step 5 of the previous list).

```
82   {
83       \par
84       \let\TempRight = \ReduceToStrut
85       \let\TempRightVtop = \DoubleTableVtopStrut
86       \TabskipBetweenTables = 0pt
87       \TabskipRightTable = 0pt
88
89       \centerline{\bf The left table.}
90       \medskip
91       \PrintNewDoubleTable
92   }
```

Now print the right table, discarding the left table, with the correct widths of
the text columns of this table (step 5 of the previous list).

```
93   \bigskip
94   {
95       \par
96       \let\TempLeft = \ReduceToStrut
97       \let\TempLeftVtop = \DoubleTableVtopStrut
98       \TabskipBetweenTables = 0pt
99       \TabskipLeftTable = 0pt
100
101      \centerline{\bf The right table.}
102      \medskip
103      \PrintNewDoubleTable
104  }
```

Sample Table 41.17 (Output).

The left table.

ABC and such is just some nice text for this wonderful column. Note that there is a lot of text in this column, or at least we will enter quite a bit here. Let that be it.	Then there is more again, and we have to keep writing to fill this column also.
CBA-CBA. Why don't we make this one short. ABC-0	DEF and quite some additional stuff so that the text actually wraps around. DEF
ABC CBA-CBA ABC-0	DEF FED Oh yes, and let's have another one of those real long entries here.

The right table.

XXX	And here is now some text which is supposed to wrap around in the right table. The right table only contains one paragraph text column so we must make it long enough so we can actually see the text wrap around.
TT X	YY Y
XXX-2 TT X	YYY-2 YY Y

41.6.5   Stretching Tables Without Paragraph Text Columns

Assume that the right table of a double table does *not* contain any paragraph text columns. Still you would like to typeset this table in such a way that its width stretches over the whole width of the page. The solution is very straightforward.

1.   Set `\TabskipRightTable` to a glue value with unlimited stretchability such as `0pt plus 1fil`.
2.   Execute steps 1 and 2 of p. 369.
3.   Execute step 5 of the same list to print the left table (note that steps 3 and 4 were not executed because they have no practical value in this case).
4.   To print the right table you must print the table using `\halign to \hsize` instead of simply `\halign`. This is most easily achieved by adding a parameter, `#1`, to the definition of `\PrintNewDoubleTable` and a parameter to `\NewDoubleTable`, which allows the insertion of to `\hsize` after the `\halign` of `\NewDoubleTable`.

41.7   The Placement of Tables

The reader may have the impression that tables are always enclosed inside a `\vbox`, and this `\vbox` is centered with the help of display math mode. While this is frequently true, it is not the only way to print a table.

41.7.1   Tables within Paragraphs

In the example where two small tables are part of a paragraph. the tables are centered with respect to the text by using `\vcenter`. `\vcenter` can be used only in math mode (this is the only reason why math mode is used here). The example is:

Sample Table 41.18 (Source code).

```
1   \medskip
2   This is text $\vcenter{\tabskip = 5pt
3                \halign{\hfil#\hfil\cr
4                abc\cr def\cr xyz\cr}}$ and some more
5            $\vcenter{\tabskip = 5pt
6                \halign{#\hfil& \hfil#\cr
7                X& Y\cr XX&YY\cr XXX&YYY\cr XXXX&YYYY\cr}}$ and
8   that's an example, where we have two tables as part of a paragraph.
9   Naturally the line spacing is now messed-up.
10  \medskip
```

Sample Table 41.18 (Output).

This is text

abc / def / xyz — and some more — X, XX, XXX, XXXX / Y, YY, YYY, YYYY — and that's an example, where we have two tables as part of a paragraph. Naturally the line spacing is now messed-up.

41.7.2   Two Tables Side-By-Side

Here is an example where two tables appear side-by-side with equal space before the first and after the second table. The distance between the two tables has been fixed to 1.0 in.

Sample Table 41.19 (Source code).

```
1    \hbox to \hsize{%
2        \hfil
3        $\vcenter{
4            \ComputeStrut
5            \offinterlineskip
6                                              \tabskip = 0pt
7            \halign{
8                \MyStrut#&
9                \vrule#\relax               \tabskip = 10pt&
10               \tt#\hfil&
11               \it\hfil#\hfil&
12               \vrule#\relax               \tabskip = 0pt
13           \cr
14               &&XX&          YY&\cr
15               &&XXXX&        Y&\cr
16               &&X X X X&   YYYYYYYY&\cr
17           }
18       }$%
19       \hskip 1.0in
20       $
21       \vcenter{
22           \ComputeStrut
23           \offinterlineskip
24                                              \tabskip = 0pt
25           \halign{
26               \MyStrut#&
27               \vrule#\relax               \tabskip = 10pt&
28               \tt\hfil#&
29               \it\hfil#&
30               \vrule#\relax               \tabskip = 0pt
31           \cr
32               &&X&          YYYYY&\cr
```

```
33              &&XX&           YYYY&\cr
34              &&XXX&          YYY&\cr
35              &&XXXX&         YY&\cr
36              &&XXXXX&        Y&\cr
37              &&X X X X&      YYYYYYYY&\cr
38          }
39        }
40        $%
41        \hfil
42      }
```

Sample Table 41.19 (Output).

XX	YY
XXXX	Y
X X X X	YYYYYYYY

X	YYYYY
XX	YYYY
XXX	YYY
XXXX	YY
XXXXX	Y
X X X X	YYYYYYYY

41.8   The Token Register \everycr

The token register \everycr is evaluated after each \cr (and after a nonredundant \crcr), as the name suggests.

Because it is evaluated immediately after a \cr, it can be used for two purposes:

1. Insert vertical material between the rows of a table with the help of a .
2. Insert material in the first column. This is discouraged, because if there is any material to be generated for the first column, it should be done so by inserting the appropriate instructions into the first template. This is where such instructions belong.

Now note that in the following application of \everycr,

```
1   \everycr = {\noalign {\vskip 2pt}},
```

the same effect could have been achieved by defining a macro \Cr as follows:

```
1   \def\Cr{%
2       \cr
3       \noalign {\vskip 2pt}%
4   }
```

and then, instead of ending a row using \cr, one would have to end such a row using \Cr (if you are hooked on using \cr then read 21.2.4, p. III-154, which explains how \cr could be redefined).

There is an advantage of using `\everycr` rather than the macro `\Cr`: the following definition of `\CrCr` does not relate to `\crcr` in the same way as does `\Cr` to `\cr`:

```
1   \def\CrCr{%
2       \crcr
3       \noalign{\vskip 2pt}%
4   }
```

There is no way in TeX to say "do this and that" if the last `\crcr` was "really" a `\cr`, and do nothing if the last `\crcr` did not do anything. By storing `\noalign{\vskip 2pt}` in token register `\everycr`, TeX causes this `\noalign` instruction also to be executed if `\crcr` acted like `\cr`, and only then.

See also 38.4.6.2, p. 239, for a macro which resets `\everycr` and `\tabskip`.

41.8.1   `\everycr` and `\cr` of the Preamble

Observe that `\everycr` is also executed when the `\cr` closing the preamble is expanded. Here is a proof for this:

Sample Table 41.20 (Source code).

Count how often `\everycr` was executed.

```
1   \newcount\EveryCrCount
2   \EveryCrCount = 0
3   $$
4   \vbox{
5       \everycr = {\global\advance \EveryCrCount by 1 }
6                                           \tabskip = 10pt
7       \halign{
8           #\hfil&
9           #\hfil
```

The following `\cr` causes the first execution of `\everycr`.

```
10          \cr
11              ABC& DEF\cr
```

The following `\cr` causes the second execution of `\everycr`.

```
12              KLM& XX%
13                  \global\everycr = {}%
```

When this `\cr` is executed, `\everycr` is already cleared out.

```
14              \cr
15          }
16  }
17  $$
18  {\tt\string\everycr} was called \the\EveryCrCount\space times.
```

Sample Table **41.20** (Output).

$$ABC \quad DEF$$
$$KLM \quad XX$$

\everycr was called 2 times.

You may wonder why \everycr was reset before the second and last row
was terminated. The reason is that when TeX expands \everycr, the first token
it will see is \global (of \global\advance), and TeX will assume that this
material belongs to the first column, because the material is neither an \omit
nor a \noalign, so another row would be started in that case.

41.8.2   An Example

In the following example \everycr is used to increase the distance between every
fifth row and the row following it. For that purpose, I use a counter register
(counting rows). This counter register is reset to zero every fifth row and then
generates a \noalign{\vskip 2pt}.

Sample Table **41.21** (Source code).

```
 1    \InputD{modonead.tip}                   % 3.3.9, p. I-52.
 2    \newcount\RowCount
 3    $$
 4    \vbox{
 5        \RowCount = -1
 6        \everycr = {%
 7            \noalign{%
 8                \ifnum\ModuloOneAdvanceNumCond{\RowCount}{5} = 0
 9                    \medskip
10                \fi
11            }%
12        }%
13                            \tabskip = 10pt
14        \halign{
15            #\hfil&
16            #\hfil
17        \cr
18            XXX&            YY-YY-YY\cr
19            XXX-XXX&        YY-YY\cr
20            XXX&            YY-YY-YY\cr
21            XXX-XXX&        YY-YY\cr
22            XXX&            YY-YY-YY\cr
23            XXX-XXX&        YY-YY\cr
24            XXX&            YY-YY-YY\cr
25            XXX-XXX&        YY-YY\cr
26            XXX&            YY-YY-YY\cr
27            XXX-XXX&        YY-YY\cr
28            XXX&            YY-YY-YY\cr
```

```
29          XXX-XXX&      YY-YY\cr
30       }
31    }
32    $$
```

Sample Table **41.21** (Output).

XXX	YY-YY-YY
XXX-XXX	YY-YY
XXX	YY-YY-YY
XXX-XXX	YY-YY
XXX	YY-YY-YY
XXX-XXX	YY-YY
XXX	YY-YY-YY
XXX-XXX	YY-YY
XXX	YY-YY-YY
XXX-XXX	YY-YY
XXX	YY-YY-YY
XXX-XXX	YY-YY

It is interesting to note that the following definition of \everycr in the previous example does not work.

```
1    \everycr = {%
2        \ifnum\ModuloOneAdvanceNumCond{\RowCount}{5} = 0
3            \noalign{\medskip}%
4        \fi
5    }%
```

The error message will read "misplaced \noalign". What is the reason for this? When TeX expands \everycr, it will expand \ModuloOneAdvance to find out whether there is a \noalign or \omit following. Therefore TeX expands the first template assuming that \ModuloOneAdvance belongs to the first column.

41.9   Expansion Issues in Tables

The following table shows some interesting expansion issues.

Sample Table **41.22** (Source code).

```
1    \InputD{doloop.tip}                        % 27.1.8, p. III-412.
```

Print a discount table. The first column contains a base price, the second column contains the price assuming a discount of 40% is applied, and the third column assumes a discount of 55%.

```
2    \newcount\BaseAmount
3    \newcount\FourtyDiscount
4    \newcount\FiftyFiveDiscount
```

Store the loop body in a separate macro because there will be more than one loop below!

```
5    \def\LoopBodyAction{%
6        {%
7            \let\THE = \relax
```

The table body which is recursively built in a loop is the old table body followed by certain discounted amounts. Note that while the \DoLoop below executes, \thes are expanded, but \THE is *not* expanded. Because \THE is made equivalent to \relax, \THE stays in the token list of \TableBody as \THE. Note: when \edefs (or \xdefs) are executed, no *computations* take place, and therefore the computation of the two discounted prices cannot be done inside the loop (at least the way the loop is written below) but must be done when the table is actually typeset.

```
8            \xdef\TableBody{%
9                \TableBody
10               \the\BaseAmount&
11               \FourtyDiscount = \the\BaseAmount
```

A 40% discount means that the product costs 60% of the sticker price.

```
12               \multiply\FourtyDiscount by 60
13               \divide\FourtyDiscount by 100
14               \THE\FourtyDiscount&
15               \FiftyFiveDiscount = \the\BaseAmount
```

A 55% discount means that the product costs 45% of the sticker price.

```
16               \multiply\FiftyFiveDiscount by 45
17               \divide\FiftyFiveDiscount by 100
18               \THE\FiftyFiveDiscount
19               \cr
20           }%
21       }%
22   }
```

Next is a loop in which we generate the rows.

```
23   \def\TableBody{}
24   \DoLoop{\BaseAmount}{500}{500}{10000}%
25       {\LoopBodyAction}%
```

Now \THE acts as \the.

```
26    \let\THE = \the
```

Build a table with the base amount in the first column and two discounted amounts in the second and third column.

```
27    $$
28        \vbox{
29                              \tabskip = 15pt
30          \halign{%
31              \hfil #.00&        % 1. Baseamount
32              \hfil #.00&        % 2. 40% discount
33              \hfil #.00%        % 3. 55% discount
34          \cr
```

Print a header.

```
35              \omit\hfil\bf Baseprice\hfil&
36              \omit\hfil\bf 40\% discount\hfil&
37              \omit\hfil\bf 55\% discount\hfil
38          \cr\noalign{\smallskip}
39              \TableBody
40          }
41        }
42    $$
```

Sample Table **41.22** (Output).

Baseprice	40% discount	55% discount
500.00	300.00	225.00
1000.00	600.00	450.00
1500.00	900.00	675.00
2000.00	1200.00	900.00
2500.00	1500.00	1125.00
3000.00	1800.00	1350.00
3500.00	2100.00	1575.00
4000.00	2400.00	1800.00
4500.00	2700.00	2025.00
5000.00	3000.00	2250.00
5500.00	3300.00	2475.00
6000.00	3600.00	2700.00
6500.00	3900.00	2925.00
7000.00	4200.00	3150.00
7500.00	4500.00	3375.00
8000.00	4800.00	3600.00
8500.00	5100.00	3825.00
9000.00	5400.00	4050.00
9500.00	5700.00	4275.00
10000.00	6000.00	4500.00

This Section is dedicated to a company which shall remain anonymous. This company's standard discount towards a certain organization (which shall also remain anonymous) is ordinarily 40%, but which in one particular instance offered a 55% discount. I would not mention that if I had not profited from this one time special discount....

41.10 Summary

In this chapter we learned:

- It is possible to do numerical computations in a table such as adding all entries in a column.
- Instead of the curly braces of \halign, \bgroup and \egroup can be used because it is possible to include the preamble into a macro, which is invoked before the table body is specified. In such macros the # of templates must be specified as ##. \span can be used to force the expansion of the token following it when the preamble is parsed, rather than when the table entry is substituted.
- Tables can be vertically split using \vsplit. This way it is possible to also replicate headers or horizontal rules automatically in split tables. Splitting can be controlled by penalties inserted using \noalign.
- The vertical alignment instruction \valign was also introduced.
- \crcr acts like \cr if there is an open row, that is, if there is a row which can be terminated. Otherwise it does not do anything.
- \everycr is a token register evaluated at every \cr and non-redundant \crcr.
- The concept of a double table was introduced.
- The timing of the expansion of table entries was discussed.
- Using display math mode to surround tables by parenthesis or other delimiters is fairly straightforward. Delimiters, within the table spanning multiple columns, can be generated easily too.

Bibliography

Let me begin this bibliography with some general remarks about books relating to T_EX and typesetting.

For books which may improve your knowledge with respect to typesetting and design, I would give the following recommendations: Tufte (1983) is a very interesting book concerning the design of tables. In Chicago (1982) you find a very valuable source concerning styles and any other information pertaining to writing books and many other types of documents. This book provides information about how to do indices, whether or not to capitalize a figure caption, what words to capitalize in a chapter or other title, and so forth. I have tried to follow the standards of this book.

In Lee (1979) you find an excellent book about bookmaking Strunk (1979) is an old time classic. Concerning typesetting terminology, I find Craig (1978) extremely useful. The explanations in this book are very straightforward and to the point. This book's dedication reads, "Dedicated to the designer who uses phototypesetting without really understanding it." I found this to be true.

Note that Knuth (1986a) is frequently referred to as the T_EXbook in this series.

If a reference below contains "WEB program," then this means that the documentation is part of a WEB program (which by definition contains its own documentation). All of the WEB programs cited below belong to a standard T_EX distribution. There is now a very nice book you can read about WEB which I can highly recommend, because it very eloquently makes the case for "literate programming." See Knuth (1992).

In case you are looking for one of the file format definitions (such as the pk file type) be sure to use the references listed below, because some of the file types (for instance, the dvi file type) were redefined during the history of T_EX.

As far as POSTSCRIPT is concerned note that there are three books listed below. I added the specification "the red book," the "blue book" or the "green book" to it, because all three books carry distinctive colors and in particular in the early days when there were no alternative POSTSCRIPT books available one usually referred to these books by their color. The former red POSTSCRIPT book is now red and white (I personally don't like the new design that much) and that took away some from the colored references to these books.

Anyway, it is time for the bibliography itself now.

Bechtolsheim S (1988) Using the Emacs Editor to Safely Edit TeX Sources, Conference Proceedings, TeX Users Group, Ninth Annual Meeting, Montreal, August 24–24, 1988, TeXniques, Publications for the TeX Community, Number 7.

Bechtolsheim S (1989) A dvi File Processing Program, TUGBOAT, 10:3, pp. 329–332.

Bechtolsheim S (1990a) A TeX Font Catalogue, Integrated Computer Software, Inc., West Lafayette, IN.

Bechtolsheim S (1990b) TeXPS, A TeX PostScript Software Package, Integrated Computer Software, Inc., West Lafayette, IN.

Bechtolsheim S (1990c) A Universal TeX Preprocessor and make Related Utility, Integrated Computer Software, Inc., West Lafayette.

Chicago Manual of Style (1982), Thirteenth Edition, The University of Chicago Press, Chicago

Craig J (1978) Phototypesetting: A Design Manual, Watson-Guptill Publications, New York.

Victor Eijkhout (1992) TeX by Topic, a TeXnician's Reference, Addison-Wesley Publishing Company, Inc., Reading, MA.

Fuchs DR (1981) The Format of PXL Files, TUGBOAT 2:3, pp. 8–12.

Fuchs DR (1982) The Format of TeX's dvi Files, TUGBOAT 3:2, pp. 14–19.

Guidelines (1990) Guidelines for the Preparation of a Typeset Camera-Ready Manuscript, Springer-Verlag, New York.

Herwijnen Eric van (1990) Practical SGML, Kluwer Academic Publishers, Dordrecht, Netherlands.

Kabelschacht A (1987) \expandafter vs. \let and \def in Conditionals and a Generalization of PLAIN's \loop, TUGBOAT 8:2, pp. 184–185.

Knuth DE (1983) The PROFILE Processor, Version 1.1, November 1983, WEB program.

Knuth DE (1986a) The TeX Book, Addison-Wesley Publishing Company, Inc., Reading, MA.

Knuth DE (1986b) TeX: The Program, Addison-Wesley Publishing Company, Inc., Reading, MA.

Knuth DE (1986c) The METAFONT Book, Addison-Wesley Publishing Company, Inc., Reading, MA.

Knuth DE (1986d) The Metafont Program, Addison-Wesley Publishing Company, Inc., Reading, MA.

Knuth DE (1986e) Computer Modern Fonts, Addison-Wesley Publishing Company, Inc., Reading, MA.

Knuth DE (1988) The WEB System of Structured Documentation, WEB program (tangle, weave) (documentation is part of the software).

Knuth DE (1989) The GFtoDVI Processor, Version 3.0, a WEB program.

Knuth DE (1990) The New Versions of TeX and METAFONT, TUGBOAT, 10:3, pp. 325–328.

Lamport L (1985) The LaTeX Manual, Addison-Wesley Publishing Company, Inc., Reading, MA.

Lamport L (1987) Makeindex: an Index Processor for LaTeX. Source code of document comes with the makeindex software.

Lee M (1979) Bookmaking: the illustrated guide to the design/production/editing, R. R. Bowker, New York.

Patashnik O (1985), BibTeXing, Version of May 23, 1985 (this documentation comes with the BibTeX software).

Platt C (1985) Macros for Two-Column Format, TUGBOAT 6:1, pp. 29–30.

PostScript (1985a) PostScript Language, Tutorial and Cookbook, Adobe Systems, Inc., Addison-Wesley Publishing Company, Inc., Reading, MA. The *blue* book.

PostScript (1985b) PostScript Language, Reference Manual, 2nd Edition, Adobe Systems, Inc., Addison-Wesley Publishing Company Inc., Reading, MA. The *red* book.

PostScript (1988) PostScript Language, Language Program Design, Adobe Systems, Inc., Addison-Wesley Publishing Company, Inc., Reading, MA. The *green* book.

Rokicki T (1985) Packed (PK) Font File Format, TUGBOAT 6:1, pp. 115–120.

Rokicki T (1988) The GFread Processor, Version 1.1, July 11, 1988, WEB program.

Sewell W (1989) Weaving a Program, Literate Programming in WEB, Van Nostrand Reinhold, New York.

Shakespeare W (1605) King Lear, Edited by Alfred Harbage, Penguin Books, New York, revised edition 1970.

Solomon D (1990a) Output Routines: Examples and Techniques. Part I: Introduction and Examples, TUGBOAT 11:1, pp. 69–84.

Solomon D (1990b) Output Routines: Examples and Techniques. Part II: OTR Techniques, TUGBOAT 11:2, pp. 212–249.

Spivak M (1986) The Joy of TeX, American Mathematical Society, Providence, RI.

Spivak M (1989) LAMS-TeX The Synthesis, TeXplorators Corporation, Houston, TX.

Stallmann R (1986) GNU Emacs Manual, Free Software Foundation, Cambridge, MA.

Strunk W, Jr. and White EB (1979) The Elements of Style, Macmillan Publishing Co., Inc., New York, NY.

Tufte ER (1983) The Visual Display of Quantitative Information, Graphics Press, Cheshire, CT.

Webster (1985) Webster's Standard American Style Manual, Merriam-Webster, Inc., Springfield, MA.

Index

Each volume contains a comprehensive index. The index is followed by a separate index of all published source code files. Primitives in the index are marked by an asterisk, e.g., *\par.

Source Code
File Index

This index is an index of all the *published* source code files in this series. Most macro source code files belong to the **texip** format (defined in 3_.3, p. III-612). If you use this format no special action is necessary to use most of the published macro source files of this series. Those files which do *not* belong to the **texip** format are marked by an asterisk (*).